WHAT PEOPLE ARE SAYING ABOUT VALUE ADDED AUDITING!

Value Added Auditing is the first textbook that I have read cover-to-cover. This approach to auditing represents a total paradigm shift. Instead of viewing auditors as policemen and auditing as punitive and after the fact, this new approach stresses collaborative team effort, participatory problem solving, and proactive improvement rather then reactive corrections. I was always taught that 'you support what you are involved in making happen' and I fully support the philosophy of Value Added Auditing.
David Wallace, Former Director of Audit of Compliance, Office of the Governor of Texas

I have reviewed all the materials and I think that they are really on-target. I believe that this is the type of material that 'modern' auditors must get a handle on, specifically learn new ways of doing things to add value.
Pat Keith, CQA, MBA, CIO – Texas State Auditor's Office

The world of auditing is growing by leaps and bounds. Many 'quality' auditors are going to be left behind. To avoid becoming obsolete, reading this book is essential. **Value Added Auditing** contains the latest information in an easy to understand style. Must reading!
Don Dewar, President QCI & Quality Digest, author, consultant

Value Added Auditing is auditing for increased profitability and improved customer satisfaction. You can't afford not to do it.
Jim Lamprecht, ISO 9001 'best selling' author, consultant, educator

Value Added Auditing is auditing on steroids. Managing risk determines whether the business lives another day.
Phil Schwaab, QMS-A, ISO 9001 auditor

Regain public trust. To CEO's of Enron, WorldCom, and Qwest: You should have read **Valued Added Auditing**. It brings trust back into business. It's a powerful manual for today's business environment.
Karl Schultz, President – Schultz Associates

Value Added Auditing is the right way to uncover risks in the today's organization. **VAA** delivers the in-depth diagnoses needed for proactive risk analysis and improvement.
Edward J. Metzler, President - aCCredo Corporation

VALUE ADDED AUDITING

Greg Hutchins PE CERM
Quality Plus Engineering
Portland, OR 97232
800.COMPETE
503.233.1012
www.CERMAcademy.com

VALUE ADDED AUDITING

HOW TO ORDER:

Cost is $89.00 per copy plus $6.00 S/H. Quantity discounts are available from the publisher

Quality Plus Engineering	503.233.1012
4052 NE Couch	800.COMPETE
Portland, OR 97232	800.266.7383
USA	gregh@QualityPlusEngineering.com

On your letterhead, please include information concerning the intended use of the books and the number of books you wish to purchase.

First Edition: 2006 Quality Plus Engineering –Value Added Auditing, Certified Enterprise Risk Manager, CERM, etc. are trademarks of Quality Plus Engineering.

VALUE ADDED AUDITING

TABLE OF CONTENTS

PREFACE

This book has its roots in the quality revolution. Its original title was **Standard Manual of Quality Auditing** (Prentice Hall 1992). Things happened. The market changed. Enron tanked. Financial restatements abound. Shareholders and senior management now want more detailed operational information. Risk management and risk assessments are required more often. ISO 9001 – 2000 expanded its focus to process and effectiveness based assessments. ISO 9001: 2015 incorporates risk as well as other ISO families of standards. The Institute of Internal Auditors (IIA) changed the rules of the game by developing new internal auditing standards. Publicly listed companies must now have internal audit departments that conduct value added audits.

HOW TO USE THIS BOOK

You pays your money and takes your choice.
Punch

What are value added audits? They can be used to satisfy the intent of internal, compliance, ISO 9001, and management audits. Value added auditing is sometimes called management audits, operational audits, and performance audits.

VALUE ADDED AUDITING – THE BOOK'S VALUE PROPOSITION

Value Added Auditing explains and shows you how value added auditing is conducted; in other words 'how to do it.' There is amazingly little information on the subject. Much of what exists is technical and often lists definitions and procedures from standards that are difficult to interpret and apply. So to remedy this problem, this guide to the actual practice of how value added auditing was developed.

Value Added Auditing incorporates the Generally Accepted Government Auditing Standards (2011) Revision (GAGAS). We believe GAGAS standards offer a solid template and standard for value added auditing. The standard is a crucial and fundamental value-add of this book. This book incorporates the Government Auditing Standards in sidebars at the top of appropriate pages. Then, the text in that section addresses the context, value, and implementation of the standard.

Value Added Auditing has been used to conduct 'Homeland Security' risk assessments of operations and infrastructure projects. We believe the risk – control approach will be the standard by which all 'Homeland Security' assessments will be conducted.

PREFACE

> **PURPOSE AND APPLICABILITY OF GAGAS**
>
> **GAGAS 1.05** Audits performed in accordance with GAGAS provide information used for oversight, accountability, transparency, and improvements of government programs and operations. GAGAS contains requirements and guidance to assist auditors in objectively acquiring and evaluating sufficient, appropriate evidence and reporting the results. When auditors perform their work in this manner and comply with GAGAS in reporting the results, their work can lead to improved government management, better decision making and oversight, effective and efficient operations, and accountability and transparency for resources and results.

Value Added Auditing meets the internal auditing requirements of the publicly listed companies as well as governmental agencies. All publicly traded companies must develop and deploy some form of value added auditing as part of their new corporate governance and assurance programs. The US Congress, Securities and Exchange Commission (SEC) and even the New York Stock Exchange (NYSE) have developed requirements for companies to manage risk, have internal controls systems, and have mechanisms to determine operational effectiveness. These are all fundamental elements of value added auditing. As well, the Government Accountability Office (GAO) has adopted the IIA standards as part of its Yellow Book. This means that every governmental agency will be conducting value added audits. The purpose and applicability of GAGAS is described in the above sidebar.

Value Added Auditing introduces new corporate governance requirements, auditing standards, and auditing methodologies to operational and quality auditors who will be impacted by the rules and regulations coming out of Congress, the SEC, and other regulatory authorities.

Value Added Auditing has a process emphasis. Most auditing books or manuals focus on evaluations from a vertical or specialized perspective. They only address quality, operational, or internal auditing. We've strived to take a process orientation, cutting across appraisal and auditing functions. Why? More businesses are run along core processes. Core process thinking permeates 'value added auditing'.

Value Added Auditing is the first book that addresses the convergence of operational, quality, customer-supplier, internal, compliance and other types of audits. The future of audits will be to add overall stakeholder value.

Value Added Auditing brings to you the latest value added auditing technologies. As such, this manual describes a process for conducting practical, cost effective, risk based audits in many different functional areas.

Value Added Auditing offers an Objective – Risk – Control – Analysis/Assurance approach to auditing.[1] Our approach is harmonized with the COSO Integrated Framework of Internal Control and encompasses a menu of processes, technologies, tools and techniques. Not all of these practices, tools, and methodologies have to be used in the audit. You, the value added auditor, must make an educated or cost/benefit judgment on what to do to satisfy audit objectives.

Value Added Auditing presents a framework that can be used by those who conduct value added audits. Our approach encompasses a menu of processes, technologies, tools and techniques. Not all of these practices, tools, and methodologies have to be used in the audit. You, the value added auditor, must make an educated or cost/benefit judgment on what to do to satisfy audit objectives.

Value Added Auditing can be used in several ways. You can read the book from beginning to end to develop an overall understanding of value added auditing. Or, you can refer to any chapter or section for specific information and answers to your questions. You can also use this book to develop your own specialized value added audit questionnaires. Whether you are about to conduct your first value added audit or are experienced and just want to expand your knowledge, you will find this manual useful.

STATE OF AUDITING – IT'S AT A CROSSROADS

Fundamental progress has to do with the reinterpretation of basic ideas.
Alfred North Whitehead

The critical question for all internal assessments and evaluations is: 'Do they add value?' More often, critical organizational stakeholders want more value from audits than simply compliance.

WHAT ADDS VALUE?

This is probably the most important question management is asking these days. Functions and activities that don't add value are outsourced. Quality, operational, internal, and financial audits are outsourced if they don't add real operational and managerial value. If auditing doesn't add real and immediate value, then audits are outsourced to be more flexible, objective, and cost effective.

Value means different things to different people. What is important to one organization may be different to another. One company's best practice may be another company's standard operating procedure.

Today, audit value comes from managing risks, measuring operational effectiveness, and assuring stakeholders business requirements are satisfied. Also, periodic assessments provide trend lines to determine improvement, performance gaps, and performance improvement. New issues arise. Processes, systems, and products are better understood. Risk correction, prevention, preemption, and prediction are emphasized. Mistakes are detected and controlled further up the value chain. Business objectives are operationalized. Risks are managed. Opportunities are found. Priorities are refocused.

PREFACE

AUDITING CROSSROADS

Internal and quality auditing have matured and are at a crossroads. In the early days, internal auditing was successful by adding credibility to financial statements. The concept of internal financial control became institutionalized in many organizations and became fundamental to most management decision-making.

But things change. Critical stakeholders now aren't happy with the value they are receiving from financial statements. Why? These statements often reflect a situation that is three months or a year old. This seems like an eternity in today's rapidly changing market. Operational and management data are poorly represented in the statements. Financial analysts and investors want more reliable and accurate financial and most importantly operational data. How do they get this? The term that we are hearing a lot today is 'value added audits.'

Why did we choose the term 'value added auditing' for this book? Traditional internal auditing, quality auditing, environmental auditing, ISO 9001 systems auditing and other types of auditing are being integrated into a unified body of assurance, assessment, and consulting services.

INTEGRATION OF INTERNAL AUDITING AND QUALITY AUDITING PRACTICES

All organizations seem to have gone through at least one reengineering or reinvention cycle. What adds and detracts value is being looked at ever more closely. Value detracting activities are eliminated. So, we're beginning to see the integration of quality auditing and internal auditing into a generic value added auditing model of partnering, process assurance, and risk management services. As you read through this book, you'll come across sidebars that present Generally Accepted Government Auditing Standards (GAGAS) that reflect the theme of performance auditing and value added auditing.

Internal auditing, quality auditing, and other internal assurance/control functions are being reinvented. We're seeing more cross functionality, where many assurance functions including security, safety, environment, quality, customer-supplier evaluations, and internal auditing are being subsumed under an Internal Auditing or Risk Management organization. This group consists of a multidisciplinary group that provides best practices, internal auditing, consulting, and regulatory compliance auditing.

The term quality auditor is even disappearing in many organizations. Quality auditors, environmental auditors, operational auditors, customer-supplier auditors, or systems auditors, and internal auditors will have simply one title – auditor.

IT'S ALL IN A NAME

Labels are important. They denote status, respect, level, authenticity, confidence and even fear. The word 'audit' implies an overview and to some even a policing of activities. There is much discussion on what to call value-based audits: management audits, performance assessments, risk audits, or process audits. We chose 'value added audits' as our preferred term largely as a result of the Institute of Internal Auditor's adopting the following definition of internal auditing:

> "Internal auditing is an independent, objective assurance and consulting activity designed to **add value** (our emphasis) and improve an organization's operations."[2]

Also, Enron, WorldCom and even 9/11 changed the emphasis of auditing from an 'after the fact' review to one that focuses on prevention, managing risk, reducing waste, and in other words 'adding value.' Adding value also became the mission and purpose of all auditors, regardless of their background, discipline, and focus.

When we refer to the term 'audit' in this book, we are referring to 'value added audits' If we refer to a specific type of assessment, we'll refer to it as a 'quality audit, 'ISO 9001 audit' or 'internal audit.'

SPECIFIC BENEFITS TO YOU

Success, as I see it, is a result, not a goal.
Gustave Flaubert, Writer

The **Value Added Auditing** offers the following benefits to you, specifically explaining:

- How to plan, conduct and report value added audits so that internal customers are delighted.
- How to comply with SOX reporting requirements for operations, IT, supply chain, and other operational risks.
- How to develop the appropriate level of assurance for your company, program, or project based on risk appetite and risk tolerance.
- How to manage, plan, conduct, and report value added audits.
- How to clarify and understand the audit customer's requirements.
- What are nonthreatening methods to notify the auditee of the impending audit.
- How to evolve from audit policing to value added problem solving.
- How to identify and manage process risks.
- How and when value added auditors should be internal consultants.
- Who are the key audit players and what are their responsibilities and authorities.
- What are the characteristics that make a successful auditor.
- How to get value-added auditing started.
- What are the six steps to managing and planning value added audits.
- How to develop a realistic audit schedule.
- What is process management and why it is critical to value added auditors.
- How to identify, minimize, and control audit risk.
- How to use questionnaires to evaluate internal process controls.
- How to develop a tailored value added audit questionnaire.

- What are the eight methods of evaluating service and administrative internal process controls.
- What is a step-by-step process for establishing internal process management, assurance and control.
- What are the steps to gaining an understanding of the audit client.
- How-to tips for conducting an interview.
- How to develop doable and accurate value added audit objectives.
- How to prepare the value added audit plan.
- How to go beyond compliance to business and process improvement.
- What is the most critical red flag in value added auditing.
- What are six techniques for gathering evidence.
- What are six effective steps for testing quality systems and processes.
- What are the six basic questions of process management, assurance, and control.
- How to evaluate process management, assurance, and control.
- What are eight examples of value added audit reports.
- What are the essential elements of an action-oriented value added audit report.
- How to use assessment-rating plans to create objective value added audit reports.
- How to report positive and negative findings.
- How to conduct the exit interview of a value added audit.
- How to package and report value added audit results.
- How to report corrective action and preventive action properly.
- How to audit using specific, detailed questionnaires.
- What is the future of value added auditing.

VALUE ADDED AUDITING READERS

Knowledge is power.
Proverb

As the value added auditing concept expands to incorporate issues of corporate social responsibility, safety, livability, the environment, competitiveness, quality of work life, health, and others, more groups are using value added audits, including the following:

- Governments.
- Security agencies.
- Regulatory agencies.
- State governments.
- Military.
- Commercial manufacturers.
- Consumer product manufacturers.
- Service providers.
- Suppliers.

PREFACE

WHO ARE THE READERS AND USERS OF THIS BOOK?

The reader of this book is anyone who conducts assessments to improve processes, manage risks, provide assurance services, implement internal controls, assure effectiveness or as the title of this book states 'Adding Value'. The following readers can benefit from this book:

- Supply management.
- Operational management.
- Senior management.
- Process owners.
- Security managers.
- Quality directors and managers.
- Quality auditors.
- Customer – supplier auditors.
- Quality consultants.
- Internal auditing.
- Risk managers.
- Financial auditors.
- Internal auditors.
- Management accountants.
- Internal auditing directors.
- Performance improvement consultants.
- ISO 9001 consultants.
- ISO 9001 registrars.
- Quality Management System (ISO 9001) auditors.
- ISO consultants.

USE COMMON SENSE

This book consists of a set of guidelines, not prescribed rules. The methods and ideas in this book must be understood and accepted by organizations in which they will be used. Then, the value added auditor can then tailor them to the organization.

Use common sense. This book covers much value added auditing material and particularly process auditing. Do what makes sense and adds value. In general, there are some universal truths to auditing, management, and good decision-making. Here are some 'rule of thumb' auditing suggestions:

- Assurance level is based on stakeholder requirements, risks, and controls.
- High-risk areas, activities, systems, and projects require more attention.
- Higher risk areas create higher probability of an unwanted occurrence or event.
- Higher risk areas create more variation around requirements.
- Higher risk requires more process management and control.
- Manage and audit using facts and evidence.
- Value added audit benefits must substantially exceed costs.

- Assurance level is based on stakeholder requirements, risks, and controls based on risk appetite and risk tolerance.

VALUE ADDED AUDITING CONTROL CHECKLISTS

Internal control surveys are also helpful in determining the size and direction of the gap between present management controls and what is acceptable risk to the organization

We have developed a number of detailed self-assessment surveys to facilitate control/risk assessments, which can be found in **Value Added Auditing Surveys**. Any organization that uses self-assessments to assess and evaluate control procedures should have a formal, documented process. The documented process should:

1. Identify enterprise/process/project/product business objectives.
2. Identify risk and exposures.
3. Assess control processes that manage risk.
4. Develop action plans to reduce risk to acceptable levels.
5. Assess the probability of achieving business objectives.

THANK YOU

I want to thank SME, APICS, IEEE, ASCE, ASQ and ISM for sponsoring and endorsing many of our auditing products. Finally, I want to keep this book current. You are my customer. I want your suggestions on how to improve this product. So please call me with your suggestions.

We interviewed many people for this book and reviewed many articles. We've done our best to ensure that we have acknowledged and sourced everyone's contributions. If we have missed someone, please give us a call and we will remedy this in our reprinting or revision.

Thank you for buying **Value Added Auditing**. We look forward to hearing from you. Best,

Greg Hutchins PE CERM
800.COMPETE or 503.233.1012

PART I: VALUE ADDED AUDITING FUNDAMENTALS

CHAPTER 1
TODAY'S COMPETITIVE MARKETPLACE

Value adding auditing only adds organizational value if it is closely aligned with an organization's strategy, customer requirements, and senior management concerns. If not, critical stakeholders will ask: 'Why are we doing this?' or 'Where's the beef?' Value added auditors must understand their customer's current business model as well as the company's competitive marketplace.

Changing marketplace, escalating competition, and changing business conditions are forcing organizations to reinvent their business model. In much the same way, audit organizations must reassess how they conduct audits.

MANAGE CORE PROCESSES

Doing little things well is the way towards doing big things better.
Anonymous

Along with the new competitive focus, managers in all organizations realize that a new way to manage must be developed to deliver quality products and services. This is generally called quality management, six sigma, business process improvement, or simply work improvement. To satisfy the final customer requires the total involvement of all stakeholders in the organization in the pursuit of continual work improvement, risk management, and customer satisfaction.

Traditional management focused on maintaining the status quo. Organizations were loaded with managers. Management style was authoritarian and top-down. Compliance auditing was the favored type of assessment to determine whether the status quo was properly maintained and controlled.

This is no longer good enough for companies competing in a global economy. The survival issues confronting many organizations deal with how to increase profitability, improve market share, enhance profit margins, and shorten product life cycles. What ties all these elements together? Process management. Risk management.

NEW WAY TO MANAGE

The new way to manage is based on process management, specifically on designing and producing low-cost products and delivering courteous service. The process focus is the common theme to ISO 9001: 2015 revision, Quality Management, Lean Enterprise, Six Sigma, and Theory of Constraints. The important point is that quality is no longer the domain of a single department called Quality Control, Industrial Engineering, or Operational Excellence. Value adding work is the responsibility of everyone in the organization.

Value added auditing is based on developing a new organizational culture, changing attitudes, and implementing new systems at all levels and in all areas including the supplier base. The new organizational culture is based on enhancing competitiveness through achieving total customer satisfaction through continual process improvement and enterprise risk management. The changing attitude is that everyone is accountable for satisfying internal and external customers. More importantly, workers are given the guidance, skills, and authority to satisfy the customer. New processes and systems are implemented throughout the supply chain, so that customer satisfaction objectives can be reached. Finally, risks are identified and controls are evaluated for effectiveness.

CORE PROCESSES AND CORE STRENGTHS

To be successful in business, be daring, be first, be different.
William Merchant

Core process competencies are a company's fundamental strengths. These fundamental assets and abilities separate, distinguish, and position a company from its competition.

Each company should have a set of core processes and capabilities, in other words something that the company is exceptionally good at. The company can use its core capabilities to distinguish itself to customers or to differentiate itself from competitors. For example, Chrysler quickly develops innovative autos; Wal-Mart has efficient distribution, planning, and logistics; and Nordstrom offers exceptional customer service in the apparel trades.

HOW CORE PROCESSES ADD VALUE

Core processes are the organization's core strengths. Core processes are essential to the operation and success of any business. Core processes are often horizontal and multifunctional spanning different functions, plants, and departments through the organization. For example, Purchasing may deal with Engineering to develop prints and bills of material, to Production that may have special handling, delivery, or packaging requirements to Quality that conducts special testing and conducts supplier audits to Accounting that deals with accounts payable and accounts receivable.

Often a company may only have one strategic area of excellence, internal capability, or core competency which may involve quality systems, stable/capable processes, low costs, state-of-the-art research and development, critical management abilities, special equipment, quality culture, team effectiveness, technology/know-how, or distribution strengths.[1]

ACQUIRING AND DEVELOPING CORE STRENGTHS

Theoretically if a company builds on its core strengths where it knows it can add market value, it will constantly achieve superior performance or competitive returns from its business. Customers will flock to it because of its value distinguishing products and services that result from its core processes. A cohesive mission, strategies, and objectives allow business units, plants, teams, and processes to be linked and synergized. The challenge is to develop, nurture, evoke, and reinforce these unique value-adding processes.

MARKET/CUSTOMER ORIENTATION ARE KEY TO ADDING VALUE

Core competencies often emphasize marketing and distribution processes. U.S. growth is slowing in many mature industries. Overseas growth opportunities are often higher than domestic opportunities. Companies are placing production and design facilities in markets that provide the most potential for profits and growth. Often, global businesses are fully integrated and operate as a matrixed organization or through functional, marketing, or production facilities in different countries.

CORE VALUES

Core values, such as ethics, may be as important as an organization's core strengths. Core values include beliefs, attitudes, principles ethics, and culture. Each reinforces the other. Core values help shape organizational principles and attitudes. Core values may mean the difference between success and failure to the value added initiatives.

In light of the recent accounting scandals, more companies consider values, culture, principles and ethics to be core ingredients of their success. Core values include respecting individual dignity, individual rights, integrity, and honesty. Core values become ingrained in the way employees act and think. External stakeholders and customers over time adopt the same attitudes and perceptions. Values and principles guide employees in the right course of actions, establish directions, define challenges, and suggest preferred responses.

CORE ELEMENTS OF ORGANIZATIONAL SUCCESS

An organization can't impose its values on employees. A person's beliefs, principles or convictions are a personal option. So, an organization adopts and endorses societal values and standards of behavior.

Shared commercial values, such as a customer focus, bind organizational stakeholders and direct behavior along a commonly accepted path. Values also provide a framework for making decisions, encourage involvement, define personal growth, and business growth.

Who develops the core values of an organization? Sometimes, a charismatic CEO may prod an organization to adopt value added auditing practices. A mature company may have values that were developed by the initial founder and became institutionalized over time. Values may also be developed through consensus building that distills the views of most organizational stakeholders.

Core values are an important element of soft controls. Soft controls are those that may not be written down but are still powerful in terms of how an organization adds value and controls its work. Ethics is an example of a soft control. If an organization has a code of ethics that senior management actively supports then there is a higher level of confidence that fraud and other abuses will less likely occur.

VALUE ADDED AUDITING DRIVERS

There is no such thing as a free lunch.
Anonymous

Understanding the following marketplace drivers will assist the value added auditor in conducting an audit:

- Competition is global and fierce.
- Competitively priced quality products and services sell.
- Sales produce profits and jobs.
- Total customer satisfaction produces sales.
- Total customer satisfaction is achieved through adding value.
- Total customer satisfaction is the goal of all stakeholders.
- 'Voice of the customer' is internalized.
- Continual process improvement is a basic premise of competitiveness.
- Continual process improvement is pursued company- and supplier- wide.
- Value added auditing facilitates continual improvement.
- Process capability, improvement, and risk mitigation require highly trained people.
- Risks are minimized.
- Benchmarks and goals are established in all activities.
- Everything is measured.
- Prevention and value addition are pursued throughout the supply chain.
- Flexibility is required to cope with change.
- All types of waste are eliminated.
- All employees are empowered to self manage and to improve processes.
- Employees are cross trained and compensated for learning.

COMPETITION IS GLOBAL AND FIERCE

The 1990s were characterized as the decade of competitiveness and this will continue into the new millennium. Why? Global power is shifting to those with the most vital, productive, and innovative economies, not particularly the biggest armies.

The value added auditor is an important element in the competitiveness equation. Value added auditing is used extensively in all businesses to improve competitiveness including the federal government. For example, with the massive shift of resources to homeland security, the federal government is urging that former and existing military contractors develop internal systems that focus on risk management and process management as the means to becoming competitive. This is difficult for military contractors that were weaned on guaranteed monies over the life of a military contract. Also, state and local governments realize that income through taxes is diminishing and that taxpayers want higher levels of service at the same cost. To meet this 'customer' demand, governments are using market incentives, such as privatizing operations to meet customer demands and auditing the value of contractor performance.

COMPETITIVELY PRICED. QUALITY PRODUCTS AND SERVICES SELL

Studies assert that quality products and services sell. Of the many elements in the buy decision including delivery, price, technology, promotion, and location, quality is one of the most important factors. Companies with strong reputations as producers of quality products and deliverer of quality services have a higher rate of return.

Quality products and services sell regardless of the industry. Health care costs are spiraling out of control while the quality of health care is stagnant, so hospitals are using value added auditing to monitor operations, control costs and improve the quality of health care.

SALES PRODUCE PROFITS AND JOBS

If quality products and services sell at sufficient margins, they guarantee continued profits for an organization. Profits ensure jobs. High-paying jobs are the major public policy issue of federal, state, and local governments. Because value added auditing provides stakeholder assurances of the lifeblood of the firm including market share, risks, sales, profits, and jobs, value added auditing is accepted as a key element of 'good corporate governance.'

TOTAL CUSTOMER SATISFACTION PRODUCES SALES

What satisfies customers? At one time, quality was defined as the ability to meet dimensional specifications or conform to quality standards. As mentioned, customer focused value is now considered the ability to anticipate and exceed customer expectations in all elements of the customer's experience with a product or service. The external customer orientation of auditing has assumed much more importance in value added auditing. So just having quality systems, such as ISO 9001 intensive documentation in place is no longer deemed sufficient unless they relate directly to risk management, operational effectiveness, process management, and final customer satisfaction.

TOTAL CUSTOMER SATISFACTION IS ACHIEVED THROUGH ADDING VALUE

Total customer satisfaction is achieved through value addition and business process improvement. Business process improvement includes all those activities that involve satisfying present requirements but also anticipating customer's expectations. For example, it may take two or more years to develop a new product. By the time the product is introduced on the market, it may be obsolete. Forward-looking companies try to anticipate customer's needs so that by the time a product or service is developed, it satisfies immediate needs. Otherwise, if a company tries to simply satisfy existing needs, then it may be too late because the market is saturated with emulators.

TOTAL CUSTOMER SATISFACTION IS THE GOAL OF ALL STAKEHOLDERS

Total customer satisfaction is the goal of all an organizational stakeholders, including employees, management, shareholders, community, and suppliers. A stakeholder is someone or a group that has an interest in the well-being and growth of the organization. Total customer satisfaction is the goal of each of the above stakeholders because only then can each stakeholder's interest be satisfied. Employees and management will maintain their jobs. Shareholders win as they see an increase in the worth of their stocks. Communities will be able to retain the company that pays taxes and helps to maintain the community. Suppliers will have continued business with their customers.

Until recently, the quality auditor may have been exposed to suppliers and to internal operations at a low level. Now, the value added auditor must interact with the many stakeholders or constituencies that are concerned with risk management and work improvement. This requires a high level of professionalism and communication abilities not simply checking compliance boxes.

Customer satisfaction results from preventing problems from occurring through continual improvement. To prevent problems and to continually improve, individuals must take ownership of processes. In other words to identify problems, individuals must develop solutions, eliminate the symptom, and the root cause. Then, internal control systems can be put in place to prevent problems from occurring and recurring.

'VOICE OF THE CUSTOMER' IS INTERNALIZED

Quality function deployment (QFD) is a practice for ensuring that the 'voice of the customer' is integrated into the organization. It means that final customer requirements and concerns are known, felt, and understood by everyone.

The value added auditor should know how customer requirements are operationalized into business processes and then know what types of management and controls are used to ensure that processes work properly to prevent deficiencies. In terms of statistical thinking, the value added auditor must know whether the proper process variables are in control and capable of meeting specifications. The value added auditor also needs to know how processes interact. This is known as 'managing the white spaces.'

CONTINUAL PROCESS IMPROVEMENT IS A BASIC PREMISE OF COMPETITIVENESS

Fundamental to all work is continual process improvement. Quality and customer requirements are not static. What was good enough yesterday is not good enough in the future? If complacency sets in, excellent quality and high customer satisfaction today lead to OK quality and low customer satisfaction tomorrow, which are a prescription for diminished sales. Because during the complacent interim, worldwide competitors will have improved existing products, developed new products, and increased customer service.

Continual process improvement consists of doing a little better in many things everywhere in the organization and with suppliers that provide key products and services. The many little things result in big things when all the little things are added together.

Having process management and internal controls that prevent defects and risks from occurring is fine but not good enough in a fickle marketplace that requires defect-free products produced quickly and delivered courteously. Since a project is also a process, this also applies to project management. The value added auditor needs to assess internal process controls, not, only in terms of prevention, control, and capability, but also in terms of continual improvement. All internal and key supplier operational systems and processes must continually improve. A customer may not want the auditor simply to report on the compliance to a policy, standard, or specification. The organization, whether it's top management or the line worker, wants the auditor to add value to the organization. This is done by improving the quality of work life, recommending operational improvement, mitigating risk, simplifying product flow, or creating process flexibility.

CONTINUAL IMPROVEMENT IS PURSUED COMPANY AND SUPPLIER WIDE

Continual improvement is pursued company and supplier wide. Companies realize that they cannot be the best in all functions, so companies are searching and certifying single-source suppliers. These suppliers are asked to supply higher-order products. For example, a supplier of a mechanical door assembly may be asked to design the entire door.

In some cases, 80% of the manufacturing dollar is outsourced to world-class suppliers. Value added auditors treat these suppliers the same way as internal suppliers. The value added auditor would assess suppliers' technology, service, delivery, quality, management, engineering, and manufacturing systems and controls.

VALUE ADDED AUDITING FACILITATES CONTINUAL IMPROVEMENT

Value added auditing facilitates the company's and its supplier's improvement. For example, the auditor may determine if controls or systems are in place ensuring that:

- Continual improvement is pursued company and supplier wide.
- People have a high level of authority to self manage their work.
- Process planning is integrated into all operations and processes.
- Work instructions are developed and followed.
- Extensive and intensive process records exist.
- Root cause corrective action is initiated.
- Costs are estimated and used extensively.

- Process drawings, flowcharts, documentation and changes are accurate, current, and understandable.
- Measuring and testing equipment is available and calibrated.
- Production tooling accurately reflects engineering prints.
- Advanced metrology is pursued.
- Key supply chain elements are monitored and required to improve.
- Statistical control and capability (Six Sigma improvement) are extensively used.
- Finished products are periodically inspected and tested.
- Handling, storage and delivery quality documentation is accurate and up to date.

PROCESS CAPABILITY, PROCESS IMPROVEMENT, AND RISK MITIGATION REQUIRE HIGHLY TRAINED PEOPLE

Value added auditing is evolving from an inspection/policing function to one that stresses collaborative problem solving. The value added auditor may lead an interdisciplinary team to conduct the audit. The perception that an expert can come in and tell a group what is wrong and how it will be fixed is history. The people working on a job daily know much more than any external auditor how the process works. In some cases, the value added auditor may bring some specific expertise that is not available to the audited. However, this expertise is used to fix one problem only. The permanent fix requires the daily involvement of those who work with the process each day.

The value added auditor's role continues to expand as process management and risk management are equated with competitiveness. This will continue throughout this decade. With this focus, the best and brightest people are now attracted to value added auditing who are adaptable and can assume different roles:

- Assessor.
- Communicator.
- Leader.
- Role Model.
- Initiative or Project Sponsor.
- Change Agent.
- Ombudsperson.
- Champion.
- Manager/Administrator.
- Coach.
- Organizer.
- Mentor.
- Enabler.
- Cheerleader.
- Delegator.
- Compliance officer.
- Translator.
- System Architect.
- Internal Consultant.

- Management Representative

RISKS ARE MINIMIZED

All types of risks are now higher – terrorists risks, business risks, and customer risks. Boards of directors and senior management don't want surprises. Senior management starts thinking: 'Do we have sufficient information, yes even knowledge, of what's happening with our operations and suppliers?'

This can be illustrated by ISO 9001 deployment. ISO 9001 has been accepted by thousands of companies as it offered compliance value. Companies are now asking: 'What other assessments can we use to assess risks and to evaluate end-to-end value chain processes?' The solution is a more comprehensive philosophy of control, assurance, and organizational governance based on identifying, controlling and managing risk.

Today, smart management is not being blindsided by unexpected events. Smart managers want to avoid crisis, firefighting, or reactive management. It's smarter decision making to have stable and capable processes. If there is a system or process spike, systems should be able to correct the problem and even anticipate the problem so that it doesn't spin out of control.

BENCHMARKS AND GOALS ARE ESTABLISHED IN ALL ACTIVITIES

Quality and process improvement can only be pursued if world-class benchmarks are established. Once benchmarks have been reached, new benchmarks are established in the pursuit of continual improvement. It's generally recognized that to survive and prosper in this competitive decade requires that a company be the best or near the best in a class of products or services. The value added auditor must assess risks and process controls in terms of how they compare against the best in the marketplace. It's no longer good enough that they exist and work properly, the value added auditor must be able to determine the effectiveness of risk management.

EVERYTHING IS MEASURED

To achieve benchmarks, all critical variables within the value chain are measured. This is sometimes called process metrics. The thinking behind this is that one can't get where one wants to go unless one knows the competitive milestones in the continual improvement journey. A list of the possible areas or benchmarks that can be measured would take pages. But, a partial list of metrics may include: project management, management performance, customer satisfaction, and risk management.

For example, process benchmarks and measures may be developed. The value added auditor may audit the level and the rate of improvement. Measurement is very important because the competition is moving forward as well. The goal is to stay ahead of the competition. This is accomplished not so much through quantum leaps but through incremental improvements throughout the organization and the supplier base.

PREVENTION AND VALUE ADDITION ARE PURSUED THROUGHOUT THE SUPPLY CHAIN

Waste and nonconformance prevention should be pursued company and supplier wide. However, before waste and nonconformance prevention can be pursued, risk management and internal controls are required. Prevention includes all the risk management, intervention and proactive activities required to prevent deficiencies or nonconformances from occurring. Process management and internal controls in managerial, administrative, professional, service, and manufacturing areas also prevent nonconformances from occurring. The value added auditor can assess prevention internally as well as throughout the supply chain. Process controls also can be used in manufacturing and administrative processes. Thus, the value added auditor needs a strong background in statistical techniques, such as Six Sigma analysis.

FLEXIBILITY IS REQUIRED TO COPE WITH CHANGE

Change is a fact of life in a very competitive and changing marketplace. For example, major changes may involve the customer. Customers expect more. For example, a quality product is no longer good enough. It must be cheap, easy to use, and look good. This already assumes that it is safe, reliable, and durable. As well, customer's expectations change quickly.

In general, what was good enough yesterday is not good enough today. Products must be adapted and modified quickly. That means that the entire organization must develop marketable products quickly and at a low cost. As well, the entire organization must adapt quickly.

Flexibility is also a requirement of value added auditors. Many quality and internal auditors have compliance mindsets and use a checklist approach to assess sophisticated problems. This is the road to professional obsolescence. The value added auditor must be able to adapt to many different types of conditions. Also, the value added auditor's challenge is to keep up with the major changes in technology, standards, and processes.

ALL TYPES OF WASTE ARE ELIMINATED

The elimination of waste is an important tenet of value added auditing. Waste can be inefficiencies, redundancies, nonconformances, ineffectiveness, cost, duplication, poor service, and low technology.

As mentioned, conformance or compliance to a specification is no longer sufficient in value added auditing. The value added auditor must also evaluate effectiveness, efficiency, economy, suitability, materiality, and many other factors of internal and supply operations. These factors often deal with the elimination of all types of waste and redundancies.

ALL EMPLOYEES ARE EMPOWERED TO SELF MANAGE AND TO IMPROVE PROCESSES

Responsibility and authority for work are pushed down to the lowest organizational level. Responsibility for improving quality and other related elements must be coupled with the authority to implement changes. Expressed in another way, all employees are responsible for their work and are empowered to improve. The individual or team has the power to make changes in their work processes, which may mean purchasing a new machine, instituting new training methods, or purchasing materials from a new supplier.

Value added auditing is often perceived as an enabling, facilitating, and problem solving function as much as an assessment function. Establishing and reinforcing a supply chain improvement environment is as important as determining whether stabilized and improving processes exist and are working properly. The latter verifies that control systems are in place, whereas the former states that if the organizational environment emphasizes participation, problem solving, and risk management, there is a lower probability that problems will arise. If they do exist, the people responsible for the process as well as its output will cooperatively solve the problem so that it doesn't recur.

'Command and control' is still an important management method in many organizations. The major difference is that in an empowered work environment, 'command and control' is being exercised over control systems and processes instead of over employees.

EMPLOYEES ARE CROSS-TRAINED AND COMPENSATED FOR LEARNING

To improve and to innovate continually requires continual training and development to keep up with trends and technology. This involves everyone: management, first line supervision, exempt employees, and non-exempt personnel. Training is not only job specific but also covers peripheral jobs. Employees are cross-trained in upstream and downstream processes.

Training involves human factors as well as technology-related areas. Human factors include problem solving, group dynamics, presentation skills, project management, and human relations. Technical skill building includes computer-aided design (CAD); computer aided manufacturing (CAM), metrology, computerized machinery, and computerized inspection. To encourage continual learning, employees may be induced through pay-for-knowledge programs. The more an employee learns and more skills that employee has, the more he or she makes. This ties the person's skill and learning to remuneration.

SUMMARY

Value added auditing is synonymous with excellent business management. Value added auditing started in the quality discipline to evaluate operational performance. As the quality issue expanded from inspection to process control, to quality assurance to continual process improvement, value added auditing has become an important tool to manage, evaluate and monitor long-term organizational performance.

CHAPTER 2
GOVERNANCE AND AUDITING

Corporate governance is all about regaining the confidence of investors and the financial community. This can only be done by developing processes for transparent reporting, creating an atmosphere of honesty and developing personal accountability throughout the organization. Good corporate governance comes down to personal accountability – 'do as you say, say as you do.' So, today's CEOs and CFOs of publicly traded companies must now certify the accuracy of all financial results and reports. Boards of directors, CEOs and CFOs must now develop processes that assess risks and assure operational effectiveness.

CORPORATE GOVERNANCE

The game rules have changed!
Anonymous

Sarbanes/Oxley Act was passed in 2002 and to a large extent created the need for value added auditing.

SARBANES/OXLEY ACT REPORTING REQUIREMENTS

Section 302 of the Sarbanes/Oxley Act requires that the CEO and CFO certify annual and quarterly financial statements. These senior executives must

- Have personally read, reviewed, and approved the report.
- Assure based on their knowledge that financial reports do not contain any untrue statements that can impact materiality or contain any misleading facts that can impact materiality.
- Assure based on their knowledge that financial statements fairly represent all material financial conditions, results of operations, and cash flow of the company.
- Have established and maintained 'disclosure controls and procedures so that material information is known to them.'
- Have developed effective disclosure controls and procedures.
- Have reported to the external financial auditors and to the board of director's audit committee the following: a. all significant deficiencies in the design and operation of internal controls which could adversely impact the company's ability to report financial information; b. any material weaknesses in internal controls; c. any fraud whether or not material, that may involve management or other employees who have a significant role in internal controls.

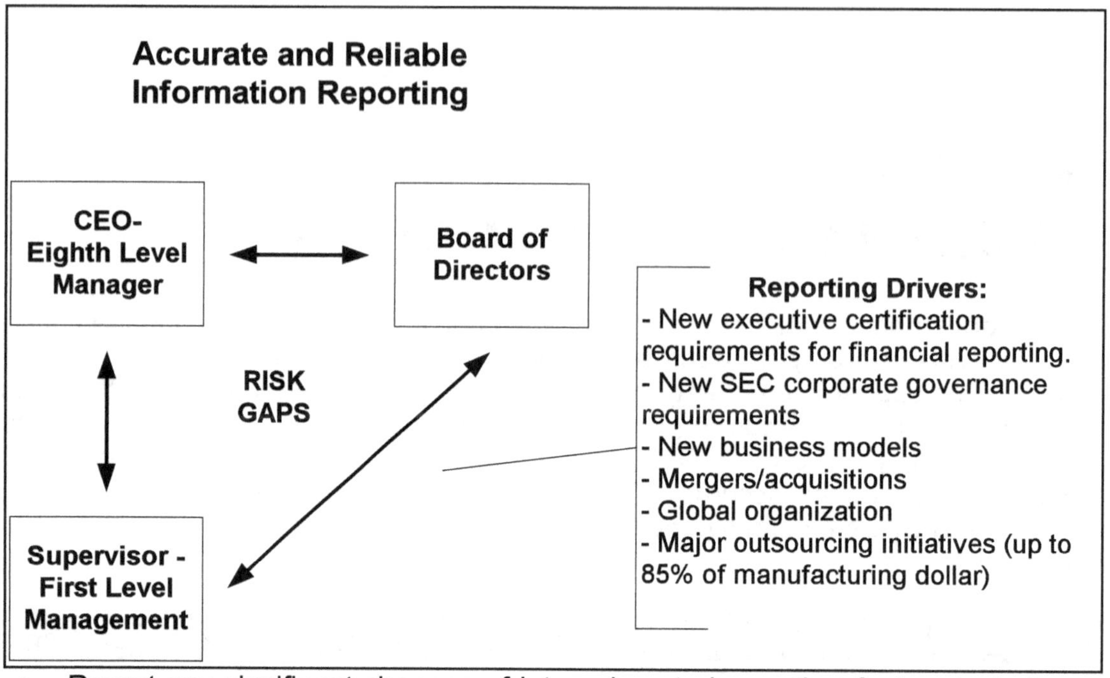

- Report any significant changes of internal controls or other factors including corrective actions with regard to significant deficiencies and material weaknesses.[1]

EXECUTIVE CERTIFICATION

Executive certification of financial statements is a critical element of today's corporate governance and directly impacts value added auditing. The critical issue for senior executives, some of whom may be 8th level managers within their organization, is how can they have assurance (peace of mind) that what they are personally certifying accurately reflects reality. In other words, they don't want surprises or 'risks.' The solution is internal controls over financial reporting and now risk - controls over operations, IT, product development, and supply management.

This is a huge challenge for today's multinational organizations. They must have transparent, accurate, and reliable reporting mechanisms. This becomes very difficult as the number of management layers increases from the CEO to first level management. Also, new business models, outsourcing, mergers, acquisitions, and other drivers have made created 'risk gaps' as illustrated in the above figure.

Bottom Line: Senior management wants assurance and ultimately peace of mind that all operations and suppliers are risk managed and operations are working effectively, efficiently, and economically. How is this assured? Through risk management and value added auditing. We believe this is the future of ISO 9001:2015 and quality auditing.

MORE VALUE

Over the last five years, there has been a growing awareness that more value must be realized from binary (comply/not comply) assessments. Traditional internal auditing has been financial and compliance based. Quality auditing has also been compliance based. Both disciplines understood the importance of compliance or conformance auditing to standards, policies, and procedures.

While internal auditing and quality auditing are still conducting compliance audits, they are moving to add more value to their clients, stakeholders, and organizations. The quality discipline is calling its audits 'effectiveness' 'performance audits' or 'management' audits. The internal auditing discipline is calling its audits 'risk' or 'process' assessments.

What does this mean to the future of these normally separate functions? I believe that we'll see their integration into one functional group responsible for conducting statutory, compliance, risk management, quality, contractual, compliance and internal audits. In order to fulfill their new expanded role, these functions will be matrixed into line and process based organizations.

WHAT IS AN AUDIT?

A small leak will sink a great ship.
Thomas Fuller

The GAGAS definition of a 'performance audit is defined in page 34. Additional definitions of auditing are found below:

- ISO 19011 definition.
- Malcolm Baldrige definition.
- Sawyer definition.
- Institute of Internal Auditors definition.

ISO 19011 DEFINITION

Let's look at the latest definition of quality auditing. The ISO 19011 doesn't even define 'quality audit.' The term 'quality audit' has been eliminated in favor of simply 'audit.' In the ISO 19011, audit is defined as:

> "systematic, independent and documented process for obtaining **audit evidence** (3.3) and evaluating it objectively to determine the extent to which the **audit criteria** (3.2) are fulfilled."

The above definition of auditing is largely compliance based. Question? 'Is this sufficient for most companies these days?' Many companies must submit to compliance audits for regulatory reasons. But, these 'yes or no' audits are frankly not sufficient for today's diverse stakeholders who want more transparent and reliable information from companies.

MALCOLM BALDRIGE AUDITING DEFINITION

The Malcolm Baldrige National Quality Award is considered 'world-class' criteria. The MBNQA defines auditing as:

> "A systematic process of objectively obtaining and evaluating evidence regarding assertions about economic actions and events to ascertain the degree of correspondence between those assertions and established criteria and communicating the results to interested users."[2]

This definition of auditing is more analytical and includes many elements of value added auditing. The emphasis is on gathering evidence regarding actions and events to determine the degree of correspondence (also called the degree of adherence) between actions and established criteria, which can range from 100 percent to 0 percent.

One hundred percent means that there is total correspondence or adherence with all the issues or audit objectives; 0 percent means there is no alignment or correspondence. In most value added audits, the degree of correspondence is somewhere between the two numbers. The information or evidence gathered can be qualitative or quantitative. The results are communicated to an interested party called the client or customer.

SAWYER DEFINITION

Early on, Lawrence Sawyer in his **The Practice of Modern Internal Auditing**, was one of the first to outline the general elements of value added auditing. Specifically, auditing is:

> "...a systematic, objective appraisal by internal auditors of the diverse operations and controls within an organization to determine whether (1) financial and operating information is accurate and reliable, (2) risks to the enterprise are identified and minimized, (3) external regulations and acceptable internal policies and procedures are followed, (4) satisfactory standards are met, (5) resources are used efficiently and economically, and (6) the organization's objectives are effectively achieved-all for the purpose of assisting members of the organization in the effective discharge of their responsibilities."[3]

This comprehensive definition addresses many of today's value added auditing concepts, specifically:

- Using organizational resources wisely.
- Obtaining operating information.
- Identifying and minimizing risks.
- Achieving organizational objectives effectively, efficiently, and economically.

In general, this also defines the value added auditor's role, purpose, opportunities, and responsibilities. Most importantly, this definition has a management orientation and establishes the foundation for today's standards of professional practice of value added

TIGHTENING RULES ON BUSINESS[4]

Accountants

- Create governance board to oversee enterprise risk management, corporate audits and discipline auditors.
- Limit auditor's consulting work.

Executives

- Require CEOs and CFOs to certify accuracy of financial reports.
- Force CEOs to give up gains from stock options, bonuses based on false reporting.
- Fine and imprison executives who lie to the SEC.
- Make easier criminal prosecution of executives who destroy evidence or defraud investors, and lengthen maximum jail terms.
- Require executives to disclose stock sales within two days. Ban personal loans to top executives of public companies.
- Require shareholder approval of option plans.

Corporate Boards

- Require that the majority be independent directors.
- Make it mandatory that board of director's audit committee approves auditors, auditor-consulting contracts, etc.

auditors. Much of this is due to the tightening rules of business as indicated in the above sidebar.

INSTITUTE OF INTERNAL AUDITORS DEFINITION

Until recently, the Institute of Internal Auditors (IIA) defined auditing as:

> "Internal auditing is an independent appraisal function established within an organization to examine and evaluate activities as a service to the organization. The audit objectives include promoting effective control at a reasonable cost." [5]

But, the IIA Board of Directors adopted the following expanded definition of internal auditing:

> "Internal auditing is an independent, objective assurance and consulting activity that **adds value** to and improves an organization's operations. It helps an organization accomplish its objectives by bringing a systematic, disciplined approach to evaluate and improve the effectiveness of risk management, control, and governance processes."[6]

We can infer a number of value adding best practices from the above definition. IIA value added audits aim to:

- Follow a systematic and disciplined approach in its assessment.
- Provide independent and objective operational analysis.

PERFORMANCE AUDITS

GAGAS 2.10 Performance audits are defined as audits that provide findings or conclusions based on an evaluation of sufficient, appropriate evidence against criteria. Performance audits provide objective analysis to assist management and those charged with governance and oversight in using the information to improve program performance and operations, reduce costs, facilitate decision making by parties with responsibility to oversee or initiate corrective action, and contribute to public accountability. The term "program" is used in GAGAS to include government entities, organizations, programs, activities, and functions.

- Examine every function, process, and activity of the organizational and external value chain.
- Help an organization achieve its business strategies and objectives.
- Evaluate and improve the effectiveness and efficiency of risk management, control, and governance processes.

James Roth In **Best Practices** contrasted the key differences between the two definitions:

> "Under the old definition, internal auditors provided arms length, compliance appraisals focusing on internal controls. Under the new definition, internal auditing has a larger scope of work providing assurance, risk management, internal control, governance, and consulting services."[7]

As can be seen from the above definition, internal auditing increased its purview and scope. Internal auditors may assess all business systems, processes, operations, and activities within the organization or in the supply based.

INTRODUCTION TO OPERATIONAL AUDITING

A new idea is delicate. It can be killed by a sneer or a yawn; it can be stabbed to death by a quip and worried to death by a frown on the right man's brow.
Charles Brower, executive

Operational audits are sometimes called different names. It is useful to explore different terms for auditing and understand how they add value. Generic terms for an audit include:

- Program.
- Product.
- Survey.
- Walkthrough.
- Surveillance.
- Regulatory.
- Functional.

GOVERNMENT AUDITING

GAGAS 1.01 The concept of accountability for use of public resources and government authority is key to our nation's governing processes. Management and officials entrusted with public resources are responsible for carrying out public functions and providing service to the public effectively, efficiently, economically, ethically, and equitably within the context of the statutory boundaries of the specific government program.

1.02 As reflected in applicable laws, regulations, agreements, and standards, management and officials of government programs are responsible for providing reliable, useful, and timely information for transparency and accountability of these programs and their operations. Legislators, oversight bodies, those charged with governance and the public need to know whether (1) management and officials manage government resources and use their authority properly and in compliance with laws and regulations; (2) government programs are achieving their objectives and desired outcomes; and (3) government services are provided effectively, efficiently, economically, ethically, and equitably.

1.03 Government auditing is essential in providing accountability to legislators, oversight bodies, those charged with governance, and the public. Audits provide an independent, objective, nonpartisan assessment of the stewardship, performance, or cost of government policies, programs, or operations, depending upon the type and scope of the audit.

PROGRAM AUDIT

A program audit is sometimes called a management or performance audit. An organization's chairman or chief executive officer authorizes the audit to evaluate organizational performance. This audit is an overview, determining whether operational goals, objectives, accountabilities, benchmarks, and timelines have been established and are being met. Customer satisfaction indices are checked. Market share analysis and financial tests are conducted. This audit has an overview perspective and is conducted at the highest levels of management.

For example, the Malcolm Baldrige National Quality Award (MBNQA) examines the relative quality and competitiveness of a firm by focusing on its ability to satisfy a customer, manage effectively, and plan strategically. The MBNQA is an excellent example of a management or business improvement assessment. Customer satisfaction and results are priorities in the award. The assumption is that if the customer is satisfied, then top management is committed to process improvement at the highest organizational levels.

PRODUCT AUDIT

A product audit evaluates a product's performance against a standard. A product audit may test or assess safety, maintainability, reliability, and durability. A product audit may be an inspection of a critical product quality attribute to determine compliance to a specification or standard. A product audit may assess the product's performance instead of inspecting a product characteristic for conformance to a dimensional specification. A product may be performance-tested under different conditions, or software may be tested against benchmark criteria. If a product fails a performance test, this often indicates that the product may have to be redesigned.

Product acceptance testing usually evaluates a product for compliance against a technical, performance, or safety standard. Testing and inspection are techniques or tools that are used in an audit to obtain data or information. Tests commonly used in an audit include:

- Statistical.
- Product testing.
- Reliability.
- Chemical/physical.

SURVEY

A survey is a brief process improvement assessment of an area or a supplier. Survey questions are often in a yes/no format or are open ended. Its purpose is basically informational, providing an overview.

WALKTHROUGH

A walkthrough, as the term states, is an informal visit of an area or facility. Its purpose is to obtain immediate impressions, inspire confidence, or solicit information. Several years ago, 'management by walking around' (MBWA) was popularized as a means by which top management could make an appearance, obtain information from people on the line, swap stories, and develop a sense of operational control. A walkthrough is very effective in terms of obtaining a broad overview or impressions, for example, in evaluating internal housekeeping.

SURVEILLANCE

In surveillance, a third party auditor or an internal auditor observes an operation to ensure that system or process practices, procedures, and controls exist and are being followed. The surveillance report goes to the customer stating that the client did or did not satisfy process requirements. Surveillance is more rigorous than a walkthrough, but less rigorous than a value added audit because the degree of adherence between process improvement criteria and observed conditions is not strictly determined.

REGULATORY AUDITS

There are many types of regulatory audits and all follow a similar format in checking for compliance. The Food and Drug Administration (FDA) conducts a Good Manufacturing Practice (GMP) audit on pharmaceutical companies and makers of health devices to ascertain compliance with regulated manufacturing practices. Its general purpose is to determine whether the manufacturer has internal controls to ensure safe and reliable products. If controls don't exist, the manufacturer is cited and corrective action request is initiated.

FUNCTIONAL AUDITS

Value added audits can be conducted in different functional areas of the organization including:

- Manufacturing.
- Quality.
- Engineering.
- Purchasing.

A functional audit may assess an area's:

- Quality systems.
- Existence of internal controls.
- Effectiveness of internal controls.
- Quality cost allocations.
- Internal customer satisfaction.
- Level and rate of continual improvement.

A functional audit is often conducted in two or more areas that must cooperate closely and coordinate effectively. For example, a multidisciplinary team may work together to develop a product quickly. A value added audit team may assess engineering, manufacturing, process improvement, research, and development to determine if product designs are periodically reviewed for calculation errors, peer review, and manufacturability.

INTRODUCTION TO QUALITY AUDITING

Let all things be done decently and in order.
Corinthians

Traditionally, quality and other audit departments spent most of their time conducting compliance audits. Quality departments checked compliance to ISO 9001 standards, specifications, and similar requirements. Internal auditing departments checked compliance against internal financial policies and budgetary procedures.

Compliance assessments usually followed a strict cycle. For ISO 9001 compliance, all systems are evaluated over a three-year period while internal groups checked compliance yearly or every two years.

ISO 9001 is not world-class system. It is a basic, entry-level quality system. It used to be documentation intensive. It now strives to establish a set of common sense, process based, quality management system that hopefully will provide some assurance of product quality.

COMPLIANCE DOCUMENTATION

Organizations usually have a compliance documentation structure to comply with regulations. The structure often consists of:

- Policies.
- Procedures.

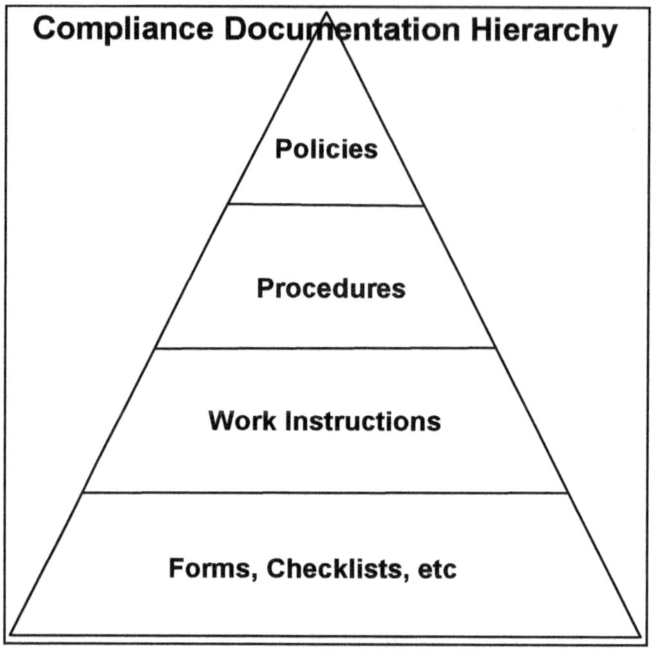

- Work instructions.
- Forms.

Policies are usually 'shall' documents. These documents specify overall organizational requirements. They may state the corporate vision or explain a business code of conduct. Policies are documents that all impacted stakeholders must comply.

Some companies are pursuing and developing world-class quality and business processes. However, many companies simply need a basic quality system to prevent deficiencies and to meet customer requirements. ISO 9001 - 2008 was a model for such a quality system. ISO 9001: 2015 is moving towards a world-class business management system. A major benefit of registering to ISO 9001 or implementing quality systems is that the standards can form the foundation for continual improvement and a more robust quality initiative in the future.

It is important when pursuing ISO 9001 that quality management educate senior executives in what they and the organization are getting into. Realistic expectations should be developed as to what resources are required, how long it will take, and what benefits will accrue. And, this information should be shared with senior management and the board of directors.

ISO 9001: 2015

ISO 9001: 2015 is a significant change from previous versions of the standard. Previously, ISO 9001 requirements were document intensive. The expression among quality auditors was 'do as you say and say as you do' and then prove it. The saying implied the existence of three or four levels of documentation, specifically quality policies, procedures, work instructions, and forms.

The result was that complying with the two previous iterations of ISO 9001 became an exercise in documentation. The assumption was that if the documentation was in order, then the work was being conducted according to the standard. And ISO 9001 auditors would conduct compliance audits where they checked that all the areas of the standard were addressed in the documentation and people were following the standard. The problem was that there was a lot of variation in how people wrote the documentation and followed the standard.

'GOOD MANAGEMENT' PRINCIPLES

ISO 9001: 2015 is a major revision from the two previous editions in 2008 and 2000. The standard developers reorganized the Quality Management System standard into a system of linked processes. The ISO 9001: 2015 process model was based on 7 'good management' principles and practices:

- Customer focused.
- Leadership.
- Engagement of people.
- Process approach.
- Improvement.
- Evidence based decision making.
- Relationship management.

ELEMENTS OF ISO 9001: 2015

ISO 9001: 2015 follows a Plan – Do – Check – Act as shown in the 'PDCA Cycle' figure on the next page. This is a closed-feedback loop that is common to most value added auditing. The basic element of the PDCA cycle is that improvement is continuous process. Major addition to ISO 9001:2015 is the inclusion of risk requirements.

The basic elements of a Quality Management System include:

- Context of the organization.
- Leadership.
- Planning.
- Support.
- Operations.
- Performance evaluation.

CHALLENGES FOR ISO 9001: 2015

More than 1,100,000 companies have been registered worldwide to the ISO 9001 standard. This is a phenomenal number and reflects its wide acceptance and value. However, many companies and quality gurus have insisted that it is a static quality system, does not emphasize quality improvement, and was needlessly documentation intensive. As all customer focused initiatives, the ISO 9001 developers heard these comments and implemented changes to the standards to emphasize risk assessment, continual improvement, and the appropriate level of documentation to achieve quality objectives.

PDCA CYCLE

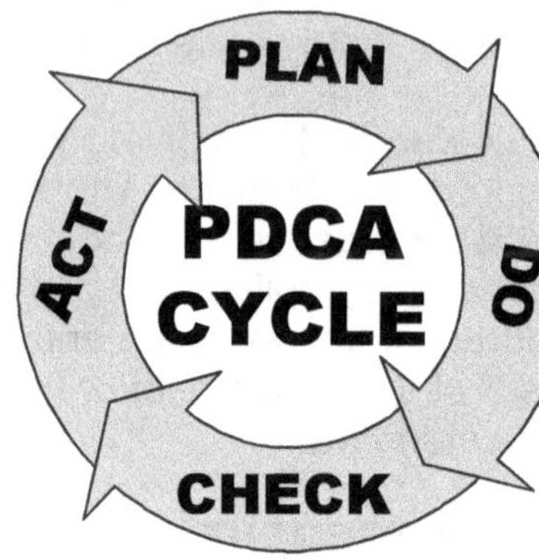

ISO 9001: 2015 in its Committee Draft (CD) does not require specific documentation, but requires evidence of compliance. For all other organizational areas, process owners can determine what type, level, and extent of the documentation they need to control their processes. In general, business and process risk determines the type and level of documentation that will be produced.

INTRODUCTION TO INTERNAL AUDITING

If standards are not formulated systematically at the top, they will be formulated haphazardly and impulsively in the field.
John Biegler

Recently, a *Fortune 500* company was on the financial ropes. It looked at its operations and asked: 'What doesn't add value?' This question was eventually posed to every senior manager. The VP of quality got together with the VP of internal auditing to discuss redundancies. ISO 9001 audits and internal audits were soon on the table for discussion because they were considered redundant. One had to go. What happened? Internal auditing won. ISO 9001 internal audits were outsourced while internal auditing became the dominant auditing function overseeing financial, operational, process, customer-supplier and ISO 9001 audits.[8]

This scenario seems to happen often? Internal, quality, operational, safety, environmental, compliance, and customer-supplier auditing are converging. This is going to be the dominant model for most organizations, publicly held and government organizations. This is something that all quality auditors, internal auditors, consultants, registrars, and professionals should be ready and prepared for.

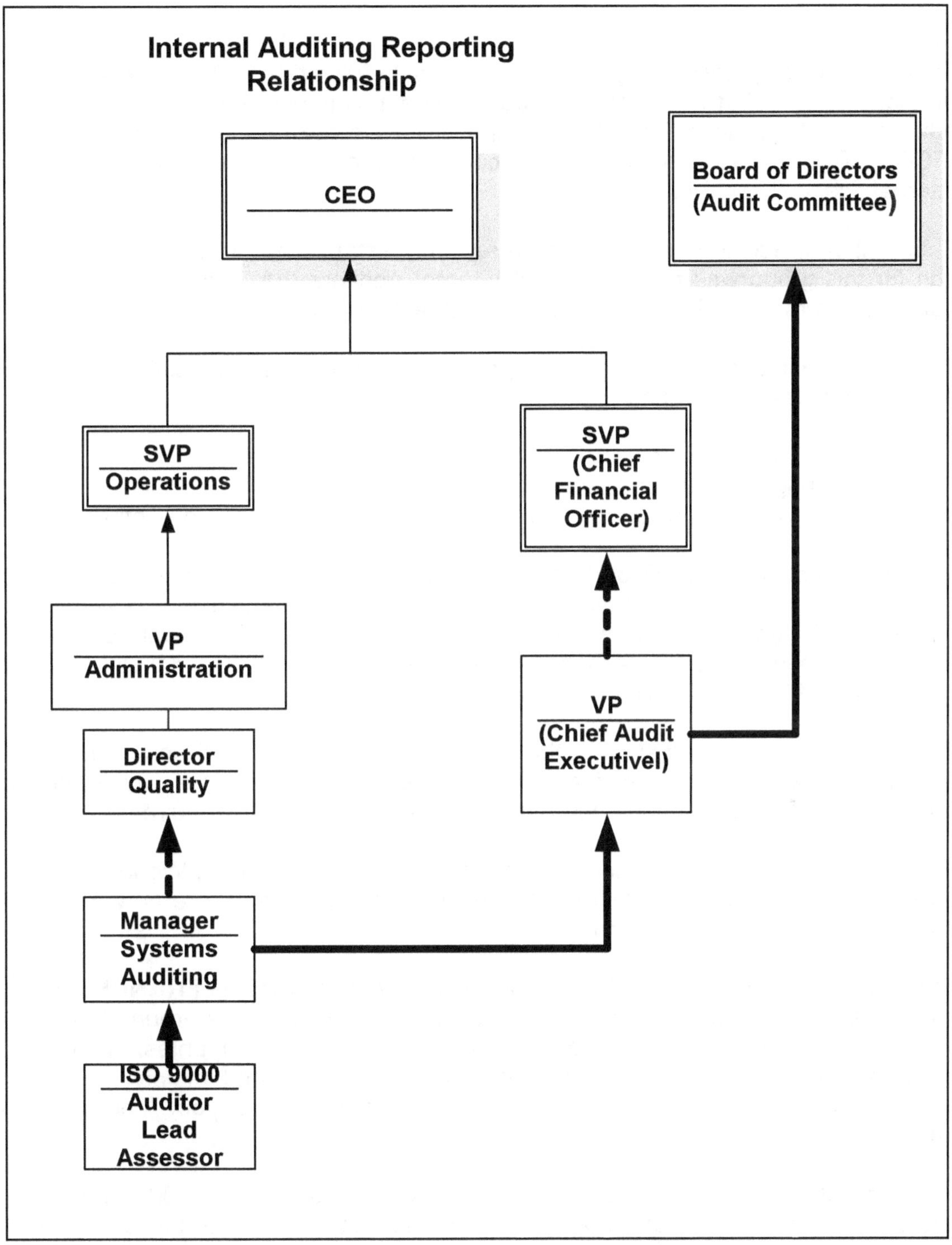
Internal Auditing Reporting Relationship
CEO
Board of Directors
(Audit Committee)
SVP
Operations
SVP
(Chief Financial Officer)
VP
Administration
VP
(Chief Audit Executivel)
Director
Quality
Manager
Systems Auditing
ISO 9000
Auditor Lead Assessor

FCPA

First a little history lesson may help. In the early 1970's and 1980's, there were a number of financial firms that went belly up due to illegal payments, embezzlements, fraud, and other criminal practices. The U.S. Congress passed the Foreign Corrupt Practices Act (FCPA). This act requires publicly held companies to maintain an appropriate system of internal financial control.

Thus, internal auditing departments were firmly established. The professional association for this group was the Institute of Internal Auditors (IIA). The equivalent group for quality auditors is the American Society for Quality (ASQ).

The FCPA institutionalized the concept of auditing internal financial controls. Internal auditing originally focused on:

- Safeguarding assets.
- Ensuring accurate and reliable accounting records.
- Encouraging compliance to company policies, procedures, and other standards.

Internal control is usually divided into administrative and accounting controls. Administrative controls include plans, policies, procedures to help managers achieve operational efficiency and effectiveness. Accounting controls are also plans, policies, and procedures to safeguard internal assets and ensure the accuracy of financial records. The overall goal of these is to eliminate redundancies and waste.

To assure effective internal controls, internal auditors conducted compliance audits of accounting and financial policies and procedures. In much the same way, internal quality auditors do the same thing to ensure compliance with ISO 9001 requirements.

Recently, the ethical, accounting, and auditing violations at Enron, WorldCom and many other companies have changed the business landscape. There are now more rules on auditors and on corporate boards.

CORPORATE GOVERNANCE AND THE INTERNAL AUDITING FRANCHISE

Internal auditing developed assessment tools and methodologies around the concept of internal control and these became institutionalized in many companies. Internal auditors became the owners of internal operational assessments and of internal compliance audits of financial controls. They also became the assurance of good 'corporate governance.'

In the quality arena, quality auditors were the owners of the ISO 9001 and system assessments. So who had the most power and visibility? Much of the authority and responsibility of internal auditing was based on the financial assurance that shareholders required. External financial auditors used internal audit findings to report financial compliance to GAAP (Generally Accepted Accounting Principles). External financial auditors relied strongly on internal auditors as can be seen in IBM's Annual Report:

> "IBM maintains an effective internal control structure. It consists, in part of organizational arrangements with clearly defined lines of responsibility and delegation of authority, and comprehensive systems and control procedures. .... To assure the effective administration of internal control, we carefully select and train our employees, develop and disseminate written policies, and procedures, provide appropriate communication channels, and foster an environment conducive to the effective functioning of controls."[9]

Internal control became the fundamental concept of accurate financial reporting and good 'corporate governance.' For example, IBM's Chairman of the Board and Chief Financial Officer would sign this statement. Recent congressional legislation now makes these senior managers personally liable for its accuracy.

Internal operational and supplier risk - controls and assessments will migrate to quality. This has important implications for ISO 9001, customer-supplier, and other operational audits, which will report risk-control operational effectiveness in a solid reporting line to the Chief Audit Executive. These consolidated reports will escalate in a solid line reporting relationship to the Board of Director's Audit Committee. This is a critical element of a 'line of accountability' to the board level.

CONVERGENCE AND CONSOLIDATED REPORTING

You never reach the promised land. You march towards it.
James Callaghan, statesman

What is happening to quality auditing? At the simplest level, the term 'quality audit' is fading from ISO 9001 quality vocabulary. Quality auditors are simply going to become business 'auditors' much like their counterparts in internal auditing.
The changes the IIA are advocating are critical to functional management including quality management. Under the old definition, internal auditors provided arms length, compliance appraisals focusing on internal controls. Under the new definition, internal auditing has a larger scope of work providing assurance, risk management, control, and maybe even internal performance improvement consulting services. And most importantly, this group more often reports directly to the company's board of directors. This is the future of quality auditing.

The changes mean that internal auditing may provide services that were often the purview of industrial engineering, manufacturing engineering, quality assurance/control, purchasing, and other functional departments. How the new IIA standards will be operationalized will be interesting in light of these competing interests of other departments.

INTERNAL AUDITING REPORTING STRUCTURE

Internal auditing usually has two reporting relationships. One reporting relationship is to the audit committee of the board of directors. The head of internal auditing is called the chief audit executive. This person communicates to the board the effectiveness of risk management and internal controls. The chief audit executive develops an audit plan that ensures sufficient evidence is obtained to evaluate the effectiveness of the control processes.

The continuing importance of internal auditing is reflected by its high corporate profile and position. In most organizations, the head of internal auditing is a senior vice president (SVP), vice president (VP) or at least a director level. Post Enron debacle, this person more often reports to the Chief Financial Officer in a dotted line relationship and to the Board of Directors in a solid line relationship as indicated by the 'Internal Auditing Reporting Relationship' sidebar on page 41.

QUALITY AUDITING

Quality auditing started from the need to ensure that suppliers and contractors produced and delivered products that consistently met quality standards and specifications. Incoming material inspection based on statistical sampling plans was widely used to check material for conformance to an engineering specification. This was costly and ineffective. There was a risk that good products would be rejected and bad products accepted.

The Department of Defense (DOD) developed a quality standard called MIL-Q 9858A, which became the highest quality standard in the military. It was later adopted by many commercial organizations as their quality standard and became a condition of business for suppliers wanting to supply products to government contractors. This standard later evolved into ISO 9001.

The scope of quality auditing was specific quality requirements that were spelled out in quality standards such as ISO 9001. Anything outside the scope of the quality standard was not addressed. Depending on the organizational culture and mission, ISO 9001 became a lowest common denominator or compliance-based, quality standard. ISO 9001 matured so that in its latest revision, ISO 9001: 2015, it is a performance and risk based standard. The new ISO 9001: 2015 is customer focused, risk based, and process based. It is moving to a value added standard focusing on business performance.

VALUE ADDED AUDITING

Why is 'value added audits' the preferred term for management, internal, and performance audits? The critical question for all assessments and audits is: 'Do they add value to specific stakeholders?' This is the question organizational stakeholders, especially investors, today want answered. More often compliance audits, whether internal audits or quality audits, don't add the value and deliver the critical decision making information that critical stakeholders today want.

Let's look at what audit stakeholders want. Senior management wants to know that processes are stable and capable. They want to know that risks are identified and mitigated. Shareholders and market analysts want to know that overall business risks are controlled, senior management is ethical, and there are no major surprises. Customers want a safe, consistent, and value priced product. Question: 'do compliance audits answer these questions?" No is the common answer. So what can be done about this situation?

CONVERGENCE OF OPERATIONAL, QUALITY, AND INTERNAL AUDITING

What is the major difference between the internal auditing and quality auditing definitions discussed early in this chapter? The quality definition still seems to be compliance based while the internal auditing is more business driven. We are seeing a much more business oriented focus with ISO 9001: 2015.

We're also seeing the disappearance of the term 'quality auditor' to simply 'auditor.' Also in much the same way, the terms 'accounting controls' or 'financial controls' are disappearing in favor simply 'internal controls.' Previously, internal control was viewed as an auditing consideration to get a favorable financial GAAP report. Now, the responsibility for the integrity and reliability of internal controls has shifted to the Board of Directors.

Bottom line: we're going to see the convergence of internal auditing, quality auditing, and other assurance assessments over the next three years. As mentioned, internal auditing, quality auditing, and other internal assurance/control functions are being reinvented. We're seeing more cross functionality, where perceived operational redundancies are being eliminated and functions are being integrated. Many assurance functions including cybersecurity, safety, environment, quality, customer-supplier evaluations, and internal auditing will be subsumed into a new assurance organization. This group will consist of multidisciplinary professionals that provide best-practice guidance, internal consulting, risk management, and regulatory auditing.

Instead of fighting these developments, quality professionals, quality auditors, environmental auditors, and others should understand what's occurring within organizations and be prepared. If compliance auditors resist or fight the integration of value added assessments, they may be dismissed. The smart option is consolidate assurance assessments so they get senior management and board level exposure so that ultimately everyone wins.

SUMMARY

The environment in which value added audits are conducted is also changing. Originally, quality auditing was considered a compliance service to an organization's management and ISO 9001 quality systems. Quality was considered an operational, technical issue.

The problem was that these compliance audits, such as ISO 9001 were static and after the fact. They usually fixed the documentation symptom but did not address the removal of the root cause of the problem, which was often work or process related. Even if the root cause was removed, people in a work area would wait until something else failed or would not initiate actions to improve existing systems. However, the problem would recur maybe in some other form and the cycle would start again. The cycle of conducting the documentation audit and fixing the problem became chronic and systemic. Although documentation problems may have been fixed or corrected, the underlying nature of the problem - which was often work, process or organizational - would erupt in some other deficiency in another part of the organization. This was called the 'don't fix it if it isn't broke' attitude.

A major goal of quality auditing has been to get more and higher level of exposure for their reports. Now, many compliance audits end up at a first or second level manager for subsequent action. Internal auditing reports go to the Chief Financial Officer and ultimately to the Audit Committee of the Board of Directors. This is where value added audit reports should end up.

CHAPTER 3
VALUE ADDED AUDITING 101

Auditing is a widely used assessment and assurance technique. Generally speaking, auditing is an independent and objective assessment of some area or activity. Auditing originally was used by accountants and tax investigators to assess financial and accounting control systems. Because it was such an effective tool, it moved beyond accounting into other areas. Instead of analyzing just financial, accounting, and budgetary controls, new auditing methodologies were developed to assess risks, evaluate controls and evaluate efficiency and effectiveness of an organization's general processes and operations.

While auditing is used in many organizational areas for different purposes, this book focuses on value added auditing. Value added auditing is an emerging discipline that is integrating internal auditing, environmental, and quality auditing methodologies.

FROM DETECTION TO ANALYTICAL AUDITING

Situations are rarely ever black or white; they are usually varying shades of gray.
Professor Ray Brown

Compliance auditing is 'after the fact', specifically meaning that it is done after a product has been produced or a service has been conducted. Some call compliance auditing, a form of detection or inspection. As quality auditors have known for years, detection is the equivalent to inspection. The problem is that inspection locates nonconforming products, but does not resolve the root cause of the problem. The root cause of nonconforming products may be the lack of a robust design. Assurance and other control activities should have occurred where the problem originated and the risk identified. Maybe, calculation reviews, peer reviews, or coordination checks should have occurred during the design of the product. In much the same way, compliance or systems auditing is losing favor to analytical auditing or value added auditing.

In many organizations, compliance auditing is perceived as the equivalent to inspection. As mentioned, companies are asking: 'Where's the value in a compliance audit?' Only a few years ago, the inspection group inspected materials coming in from suppliers, materials that were being processed, and materials just before they were sent to suppliers. Inspection is losing favor to process management.

A line person usually conducted inspection and usually reported results to manufacturing management. Inspection was not independent and was considered a line activity to stop defective material going to the next operation. In this atmosphere, the inspector was considered a policeman looking over the shoulder of the people doing work. Roving inspectors were sometimes called auditors. In much the same way, compliance auditors must now provide higher value by understanding core business processes, risks, and mitigating controls.

MOVING AWAY FROM PUNISHMENT

This occurred internally as well. Quality auditing and internal auditing would assess documentation looking for compliance discrepancies, mistakes, flaws, or irregularities. If one was found, it was written up as a finding for subsequent corrective or preventive action. The report went to line supervision or to an auditing supervisor.

Management perceived the discrepancy as someone's inability to manage or to get work done properly. The supervisor of the area or a worker was told to fix the problem, which the supervisor eventually did. This became a black mark on the supervisor's record and was reflected on the supervisor's evaluation at the end of the year.

Compliance auditing was perceived as a tool for policing and punishing. The client resented the intrusion and was sometimes adversarial. The result was that the auditor did the assigned task, the audited suffered through the audit, and in some cases even tried to cover up flaws and discrepancies.

FROM DETECTION MONITORING TO CONTINUAL IMPROVEMENT AND RISK BASED PROBLEM SOLVING

Another problem developed. The client, the area supervisor, or supplier would receive an audit report with many nonconformances. This report card was perceived as the work of someone outside the organization who did not know how things really worked. The auditor was process clueless. After the audit, the supervisor would say 'Sure, I'll fix the problem,' which usually meant doing the minimum. This established or hardened existing adversarial feelings.

The supervisor would then tell employees to take care of it and then walk away. The employees didn't care if a solution worked because they weren't consulted as to the nature of the problem; the purpose, timing, and scope of the audit; or possible solutions. The symptom would be fixed, usually as a quick fix. The problem would recur. The problem would be fixed again. The result was a chronic, self-perpetuating cycle. A preventive, participatory, problem-solving approach needed to be developed.

EVOLVING TO A COLLABORATIVE, PROBLEM-SOLVING ASSESSMENT

The goal of value adding auditors is to evolve where process owners may request or even conduct their own value added audits, test systems, reach findings, and implement solutions. There are major benefits to this. The people on the line have bought into risk management, process control and process improvement. Internal customers are satisfied. Attitudes change. Continual improvement is pursued.

The people on the line are not told by supervision what to do and how to do it. Their active involvement provides an immediate payback. Teams participate in improving their quality of work life. Line personnel have more process knowledge of what occurs, so if there are problems, they have immediate process knowledge to pinpoint causes and to implement lasting solutions that fix the symptom but also the root cause. Finally, line personnel as process owners are in the best position over time to improve the process.

Line personnel may want to enlist the specialized help of the value added auditing group. The audit group brings knowledge, skills, and expertise that the line area doesn't have. For example, a complicated chemical process may consist of multiple process steps. Process chemistry can be very complex. Problems on the line may be intractable and beyond the technical abilities of line personnel, the process engineer, and the line supervisor. An audit team of Ph.D. research chemists may be called to audit the process and offer solutions. The R&D team may use sophisticated testing equipment to understand the problem and use design of experiments to isolate the prime causes of the problem.

VALUE ADDED AUDITING ⇔ ANALYTICAL AUDITING

Also, quality, system, and internal auditing are focusing on risk management, process control, process capability, and organizational effectiveness. These value added audits require more information to be collected and analyzed to evaluate value, risk, waste, effectiveness, and efficiency. As an organization moves toward value added auditing, the auditor or audit team will collect more information, conduct in-depth analysis, and obtain sufficient evidence in order to reach conclusions.

What does this mean? Value added auditing for effectiveness, risk and efficiency requires more analysis. This may mean:

- Flowcharting critical processes.
- Identifying critical process variables.
- Deciphering critical process variation.
- Uncovering unexpected gaps between business objectives and process metrics.
- Discovering gaps or white spaces that can result in higher risks.
- Discovering potential risks and errors.
- Uncovering the absence or differences from what is expected.
- Identifying chronic or systemic issues.
- Uncovering fraud or illegal acts.

VALUE ADDED AUDITING CHARACTERISTICS

The biggest challenge and opportunity lies in involving everyone of our people in what amounts to a quality obsession. Everyone needs to say by his or her actions that quality is a way of life; to affirm that they apply the same high standards of performance to their jobs as they do in their personal lives.
Douglas D. Danforth, Chairman, Westinghouse Electric

Many professional groups are developing value added auditing standards. The Institute of Internal Auditors (IIA), American Institute of Certified Public Accountants (AICPA), and the American Society for Quality (ASQ) are developing standards for conducting value added audits. What will these look like? There has been a lot of discussion of this, which we'll discuss throughout this book.

WHAT IS VALUE?
First, we need to define value. All organizations exist to add value to their stakeholders. Value is added when products and services satisfy critical organizational stakeholders. Value means different things to different stakeholders. Value to shareholders means raising stock value. Value to management means no surprises. Value to regulatory authorities is compliance to laws and regulations.

To assess or assure value, value added auditors may assess specific economic attributes, which may include:

- Effectiveness.
- Efficiency.
- Cost reduction.
- Waste elimination.
- Risk management.
- Business controls.
- Process controls.

The Institute of Internal Auditors offers a more comprehensive definition of 'adding value', specifically:

> "Organizations exist to create value or benefit to their owners, other stakeholders, customers, and clients. This concept provides purpose for their existence. Value is provided through their development of products and services and the use of resources to promote those products and services. In the process of gathering data to understand assess risk, internal auditors develop significant insight into operations and opportunities for improvement that can be extremely beneficial to their organization. This valuable information can be in the form of consultation, advice, written communications, or through other products all of which should be properly communicated to the appropriate management or operating personnel."[1]

VALUE ADDED AUDITING 101

CHARACTERISTICS OF VALUE ADDED AUDITING

Auditing is evolving from compliance to standards to enhancing corporate governance through evaluating operational effectiveness, risk management, and internal control. Value added auditing is considered the major assurance, assessment, validation, verification, measurement, control, and monitoring tool in the pursuit of ethical corporate governance, continual improvement and customer satisfaction.

These are the key characteristics of value added auditing:

- External customer satisfaction.
- Internal customer orientation.
- Alignment with achievable organizational objectives.
- Systematic project approach.
- Performance objectives and measures.
- Risk management orientation.
- Knowledge of diverse operations and functions.
- Compliance to standards.
- Degree of correspondence.
- Process management and control.
- Prevention emphasis.
- Corrective and preventive actions.
- Value chain process improvement.
- Technical and statistical orientation.

EXTERNAL CUSTOMER SATISFACTION

It is universally acknowledged that external customer satisfaction is a requisite to competitiveness. This means that companies must design aesthetic, defect-free products and deliver them quickly and courteously, and this requires everyone's participation and cooperation. Compliance auditing as a policing tool does not promote this ethic, but value added auditing as a participatory, problem-solving, risk management tool does.

INTERNAL CUSTOMER ORIENTATION

Originally, the objective of quality audits was to satisfy the client who authorized the audit and the client was often top management. As a result, the organization perceived the compliance audit as a policing, gotcha, and punishing tool. When compliance auditors appeared, people within the organization would sometimes shy away, doubting and sometimes resenting the auditor's intrusion. In this atmosphere, compliance auditing was often adversarial and confrontational. But now in many organizations, value added auditing emphasizes external and internal customer satisfaction, fact-finding, measurement, value, and improvement.

ALIGNMENT WITH ACHIEVABLE ORGANIZATIONAL OBJECTIVES

Value added audit results should be communicated so they satisfy specific customer audit objectives but also are aligned with the organizational mission-critical objectives. The value added auditor may have been retained to provide management with specific information or used as an internal consultant. Whatever the function, the value added auditor always should maintain the perspective of the organization's goals and objectives.

For example, the customer may want to know whether the client's processes are in statistical control. However, if the organization is promoting continual improvement, then the auditor should also address the client's process stability, capability and continual process improvement efforts. In general, the value added auditor should know the audit customer's process improvement policies, goals, objectives, accountabilities, responsibilities, and measures so they can be addressed in the value added audit.

SYSTEMATIC PROJECT APPROACH

Value added audits are structured so they follow a logical sequence of assurance and appraisal from planning, to conducting, to reporting results in a project sequence. Each stage of the project sequence has sub elements, which are discussed in this book. This sequence is called the framework of value added auditing. Commonly used synonyms are structure, methodology, outline, and foundation.

The critical project steps in a value added audit are:

- Managing the audit.
- Planning the audit.
- Conducting the audit.
- Reporting the audit.

PERFORMANCE OBJECTIVES AND MEASURES

Value added audits should report on performance measures critical to the organization. Traditional internal audits had distinct performance measures, such as number of audits conducted, number of audit reports issued, organizational areas audited during the year, and so on. Quality audits used similar performance measures or metrics including number of audits, number of findings, and status of outstanding corrective actions. Bottom line: what did these measures do for what matters to critical organizational stakeholders? Was decision-making improved? Was information accurate and reliable? Was information used to improve operations? Did information mitigate risks?

ENTERPRISE RISK MANAGEMENT ORIENTATION

There is seldom 100 percent assurance at a low cost. The problem is that the value added auditor attempts to reach the highest level of assurance so that facts, information, and evidence can support a reliable audit decision. In most cases, the level of assurance and risk are complementary. A higher level of audit assurance requires that more data must be gathered and analyzed to lower the risk of making an incorrect decision, consuming time and resources, which adds cost.

In terms of value added auditing, there are a number of definitions of risk. One definition of risk is variation from a desired target. The target or objective can be a specification, standard, drawing, policy, procedure, work instruction, or flowchart. The target or objective is what the process or product aims at, adheres to, corresponds with, complies with, or a number of other criteria.

The possible risks in a value added audit may include outright fraud, lying, conflict of interest, unrepresentative samples, mistakes, abnormal variation, strikes, misunderstandings, and so on. To reduce these audit risks, the value added auditor should follow a systematic project audit methodology in planning, conducting, and reporting audit results, thereby identifying and minimizing audit risks at each stage of the audit.

COMPLIANCE TO STANDARDS

Compliance or conformance to international, national, industry, or organizational standards, regulations and specifications can still be a critical value added auditing objective. As such, compliance audits add value to regulatory stakeholders. Standards can also consist of organizational policies, procedures, work instructions, and flowcharts. The value added auditor must be able to identify system, process, and product requirements that are applicable to the situation, and then determine whether systems or processes comply with the applicable standards.

DEGREE OF CORRESPONDENCE

The degree of correspondence between the auditee's assertions and established audit objectives is the degree of adherence between evidence gathered in the client's operations and the audit objectives. In a compliance audit, the degree of correspondence is a binary decision. It is a matter of determining conformance in terms of acceptable and not acceptable compliance with a regulation or standard. Compliance or conformance audits are fairly straightforward.

Value added audits evaluate performance, effectiveness, risk, or other factors. These audits are more analytical, difficult, time consuming, and knowledge based. The process starts when the customer requires information and authorizes an audit. The value added auditor develops audit objectives based on the customer's requirements. The value added auditor obtains evidence that supports these objectives. Evidence and information can be gathered by testing, interviewing, sampling, and observing. The value added auditor compares this information against business objectives, risk/controls, process requirements, quality standards, engineering drawings, procedures, policies, purchase orders, contracts, or similar documentation. The value added auditor then determines the level of correlation between evidence and objectives, which forms the basis for findings and an audit determination.

PROCESS MANAGEMENT AND CONTROL

Process management is the means by which to monitor and regulate a process to prevent defects or nonconformances from occurring. Process analysis can be used to understand any process, from manufacturing to a service process. For example in manufacturing, a nonconforming product has dimensions that are outside specification limits shown on an engineering print or does not perform what is spelled out in a performance specification, or, it may be discourteous or tardy customer service. The result of any nonconforming product is a dissatisfied customer. If processes are properly managed, the number of nonconforming products is minimized. The customer is then assured that process improvement is being pursued, monitored, and controlled.

Value added auditing could be used to assure the effectiveness of process management and control. These can be managerial, operational, administrative, service, manufacturing, or engineering controls, internal or external to the organization. For example, with the need to satisfy the customer, companies are integrating single-source, world-class suppliers into product development and other processes. Since suppliers are a critical part of the supply process chain (or simply the supply chain), they must have the similar process controls as the customer. How is this assured? Prospective suppliers are extensively and intensively audited to ensure they have internal process risk = controls.

PREVENTION EMPHASIS

Prevention is a major tenet of the process improvement ethic. Prevention means that organizational systems and processes are managed and controlled to prevent nonconformances, deficiencies, or flaws from occurring. Prevention is stressed in all managerial, line, administrative, service, and professional areas.

Inspection usually does not add value. Inspection during production, warrantee returns or at worst a recall are very expensive. Inspection is ineffective because:

- It is after the fact.
- It is inaccurate.
- It is costly.
- It sends the wrong signals, that nonconformances are acceptable.

Once defective products are in dissatisfied customers hands, the horse is already out the barn door. Procedures are not in compliance with statutes or standards. These are after-the-fact detection decisions.

Prevention focuses on upstream issues, on doing the right things right in the first place. Prevention focuses on eliminating risks. Prevention also focuses on adding value and eliminating non-value adding activities. An error found and corrected as early as possible can cost ten times or more if the error is found downstream the process. For example, if a customer wants a customer service representative to fix a product problem, it will cost $150 or more to send a technician to find the problem and to fix it. If the problem is design or production related, then it is replicated in each product and the recall can cost many millions.

CORRECTIVE AND PREVENTIVE ACTIONS

If there are significant differences between the results of the value added audit and the established criteria, the value added auditor should report these to the customer. The value added auditor or the customer may request and/or suggest the type of corrective or preventive action. Instead of specifying the type of corrective action, the value added auditor may defer to the client, because he or she knows better what must be done to correct the problem. Corrective action focuses on the fix and the cure. The fix is removing the symptom or the immediate cause. The cure is removing the root cause. Preventive action ensures that a possible nonconformance can be detected sufficiently early to prevent its occurrence or recurrence. Preventive actions can be thought in terms of managing and mitigating risks. In much the same way, we advocate the 4P's: being proactive, preventive, predictive, and preemptive™.

VALUE CHAIN PROCESS IMPROVEMENT

The value chain can include internal processes or can include the entire supply chain's core processes. Both involve a process that adds value to a product or service. More often, the value chain is called the supply chain. Supply chain implies supply processes. 'World class' companies now strive to facilitate the continual process improvement internally as well as with suppliers. Traditional financial and internal auditing focused on the efficiency and effectiveness of the status quo. At present, the status quo in most organizations is in a rapid state of flux. Product life cycles have been reduced by half. World-class suppliers are integrated into customer's product development processes. This is required to develop and deliver new products that satisfy the customer. An important element of value added auditing is to ensure that continual improvement is pursued throughout the supply chain.

TECHNICAL AND STATISTICAL ORIENTATION

The Malcolm Baldrige National Quality Award (MBNQA) and top management consultants advocate 'managing by fact' and 'managing by evidence.' This means that management and all decision makers including those on the line should have sufficient hard data to make reliable and accurate decisions.

Quality in general has a strong technical and statistical orientation, the heritage of Walter Stewart and W. Edwards Deming, two seminal authorities on process management and improvement. The statistical concepts of minimizing variation, process control, capability, and improvement are fundamental to all organizational processes. We're seeing this influence in the wide application of Six Sigma.

VALUE ADDED AUDITING EXAMPLES

Let all things be done decently and in order.
Corinthians

To understand value added auditing, it's useful to look at various types of internal and operational auditing. It is helpful to understand similarities, differences, and uses among value added auditing methodologies and techniques.

In this section, we discuss the following types of value added audits:

- Compliance audits.

- Operational audits.
- Process audits.
- Self audits.
- System audits.
- Risk audits.
- Attestation.
- Internal control audits.

COMPLIANCE AUDITS

A compliance assessment is conducted to determine whether federal, state, local, agency, industry, or company guidelines are understood and are consistently followed. A compliance or conformance is a binary assessment. Policies and procedures are evaluated to determine they are followed. A compliance audit is often performed under a regulatory, contractual, or legal authorization. The assessment is formal and follows a prescribed methodology.

Compliance audits are fundamentally documentation reviews. The result is a binary decision, compliance or noncompliance. If there is noncompliance then the auditor will issue a Corrective Action Request (CAR) or a Preventive Action Request (PAR).

Compliance auditor usually conduct a 'gap analysis' between what is specified or required in a standard and what is presently done. This gap reveals a binary situation. A particular activity is in compliance or it is not. If it isn't, there is a corrective action request or recommendation. Compliance audits are black and white.

Compliance audits add value to governmental agencies and to commercial organizations that mandate contractual or regulatory compliance. Compliance audits are probably the easiest to conduct because requirements are written and less auditor discretion is required.

For years, quality, internal, and financial auditors were tainted with the clipboard mentality. The auditor would come in, look at something and check 'yes or no' on the checklist. Most compliance auditing still follows this procedure. The problem is there is little analysis and evaluation of core business processes. Hence, the perception is these assessments add little value. On the other hand, value added audits require more analysis and evaluation especially if auditors must evaluate risk, controls, efficiency, and effectiveness. Auditors, in other words, must be more analytical.

OPERATIONAL AUDITS

Operational auditing evolved from industrial engineering to address organizational performance. As with many internal assessments, we're seeing an integration of operational, quality, and internal auditing.

Operational auditing focuses primarily on nonfinancial performance analysis. It is used to assess internal systems, resource flows, and managerial performance. Operational auditing has attracted industrial engineers, whose knowledge and skills fit well.
Operational auditors primarily focus on the three Es - economy, efficiency, and effectiveness of operations. They emphasize optimizing these three factors within the constraints of a project, operation, or contract. Process improvement is a secondary matter, and sometimes not even addressed. Efficiency and economy factors involve 'doing the right thing'; effectiveness on 'doing the thing right.' Economy and efficiency therefore deal with maximizing resources while effectiveness deals with ensuring that the organization's mission, objectives, and mandates are addressed.

Operational auditing emphasizes administrative controls, and often has a stopwatch, time/motion, and productivity orientation. This can lead to problems when the organizational mandate is to satisfy external and internal customers by improving continually, because in order to accomplish this, employees must participate and initiate change. They may not buy into the process if they know that there is a staff-auditing group monitoring performance and reporting results to management. Typical examples of operational reports include:

- Production increases studies.
- Efficiency, effectiveness, and economy recommendation studies.
- Redundancy analysis.
- Workflow analysis.
- Time studies.
- Value analysis.
- Costing reports.
- Training and development studies.

Customer Focus
Operational auditing has an industrial engineering, sometimes a 'stop-watch' orientation. Value added auditing has more of an external and internal customer focus. External customer focus is required to maintain business effectiveness and improve competitiveness. It is generally acknowledged that competitiveness is enhanced through customer satisfaction. An internal customer focus is also required to improve continually and to ensure the external customer is satisfied.

Similarities of Operational and Value Added Auditing
Both operational and value added auditing:

- Use internal employees as auditors.
- Stress objectivity and/or independence.
- Review operations and risk - controls.
- Follow a systematic methodology.
- Stress economy, efficiency, and effectiveness.
- Emphasize root cause corrective action.

PROCESS AUDITS

The process approach is fundamental to all value added auditing. Traditional management focused more on the person than on the process. If there was a product nonconformance, blame was immediately placed on the person who it was assumed made a mistake. Emphasis has shifted from the person to the process. The individual, the process owner, is now responsible for understanding, controlling, and improving his or her process which may be answering the phone, designing products, or delivering products.

In the horizontal organization, hierarchal levels disappear, work is simplified, and information is universally available. Responsibility, authority, resources, information, and decision-making are downloaded to the lowest organizational level. Hierarchal - functional workers evolve into process owners.

A process is an activity consisting of three elements: 1. inputs, 2. the process, and 3. outputs. A value added audit ensures that inputs are consistent and conform to requirements, the process is in control and capable of meeting requirements, and finally, the output satisfies internal and external customer requirements. A process can be a machine or a service operation. An operation can process information, paper, raw material, parts, or money. Process audit ensures that the process is monitored, controlled, and meets business objectives.

A critical requirement in ISO 9001: 2015 is demonstrating Quality Management System 'effectiveness' and 'risk – control'. The major question is how to audit for 'effectiveness?' Most quality and ISO pundits think that an effectiveness audit will be some type of process audit. There is still much confusion and little standardization on how to conduct a process audit, however the following are commonsensical steps for conducting a process audit:

- Identify business objectives.
- Identify high risk processes.
- Flowchart processes.
- Identify process measures.
- Evaluate process performance against measures.
- Recommend improvements.

The power of process audits is that they can go beyond evaluating effectiveness of ISO 9001 quality management system clauses to evaluate value chain processes against internal business objectives and external business benchmarks.

SELF AUDITS

The workplace is galloping towards self-managed work teams who self assess their work. Self managed teams are also composed of self directed individuals who accept responsibility for developing schedules, managing quality, controlling costs, upgrading worker skills, assigning work, improving process performance, focusing on results, and ensuring stakeholders are satisfied. Many job classifications are replaced by one work classification. The work environment is open and friendly. Time clocks are eliminated. Compensation is based on pay-for-knowledge so people are paid on the basis of training, experience, knowledge, and value-addition.

Self-managed process or project teams do work simultaneously or concurrently. Specifications are jointly developed so everyone is familiar with them. Processes are stabilized, documented, and capable. If there are process deficiencies, the team has the skills to solve them. When team and self managed teams work, results are stunning. Self-managed, process teams can be up to 30-50% more productive than conventional ways of doing work.[2]

Self managed teams and individuals can assess themselves through:

- Balanced scorecards.
- Checklists with ratings.
- Internal control questionnaires.
- Team written procedures and instructions.
- Process control information, such as SPC.
- Flexible and reinforcing work environment.

SYSTEM AUDITS

ISO 9001 system audits traditionally were compliance assessments. The ISO standard or similar standards consist of 'shall' requirements. The auditor assesses a business system, process, or product against a standard. The auditor in a compliance audit verifies that documentation complies with the standard's requirements and verifies implementation to 'say what you do and do as you say' criteria.

In general, a system is a set of interrelated parts that work together to achieve a desired objective. ISO 9001 in the 1987 through 2015 revisions are considered system standards. A system is therefore a set of interrelated parts consisting of subsystems and elements that work together to satisfy a customer with a product or service.

An ISO 9001 assessment is a systems audit. A systems audit is used to determine the existence and effectiveness of an organization's documentation system involving quality policies, procedures, and work instructions. A basic system audit is used to investigate whether organizational elements, activities, functions, and departments have process controls. Examples of an organization's quality systems include:

- Quality manual.
- Quality organization.
- Training and certification.
- Quality planning.
- Quality documentation.
- Incoming material control.

- In-process product control.
- Final product test and evaluation.
- Measuring equipment and calibration.
- Storage and delivery.

For example, a manufacturing operation may consist of 100 or more steps and each step may be critical. One process step can be a lathe that turns and forms a piece of metal. The lathe is one machine in a work area producing a component used in a subassembly. A system audit would not check the lathe and machines. It would check the management, operational, and administrative elements of the process improvement system including policies, procedures, manuals, and internal controls.

RISK AUDITS

The American Management Association recently said the following about risk:

> "For virtually every business in the United States, the implications of economic change are enormous. The rapidly changing and more uncertain environment not only has made corporate decision-making and planning more difficult, but also has significantly increased business risks. Operating successfully in this new environment will require a very different approach to business management. It involves more, rather than less, attention to external factors, as well as new priorities and strategies and a sharply increased focus on risk management."[3]

All auditors and professionals are seeing more emphasis on risk analysis and new forms of risk assurance. Up to ten years ago, quality was the primary filter through which American senior management reached decisions. Customer satisfaction was the critical quality attribute. Well things changed. Cost and schedule overshadowed quality as the primary senior management decision filters. First to market, first to critical mass, and other time elements became critical to senior management as they competed with other companies. September 11, 2001 and corporate failures changed all that. Risk and its management are now the primary filters by which management makes its decisions.

ORCA is a common organizational risk assessment methodology. Its principal elements are:

- Identify business and process **Objectives.**
- Identify operational and other **Risks.**
- Define business or other **Controls.**
- **Analyze and Assure** the effectiveness of the business process to satisfy objectives.

This book follows a modified ORCA model. There are a number of advantages to this model. It is well accepted and adopted. It incorporates elements of other types of assessments including process, internal control, and system audits. Also, it fits neatly into today's corporate governance focus on risk management and operational effectiveness. This is explained more fully in the next chapter.

ATTESTATION AUDITS

Attestation audits require complete independence. Audits and assurance services are usually defined as professional services that improve the quality and reliability of information to third parties and/or decision makers. There are several forms of attestations such as those defined below by GAGAS:

> "Attestation engagements can cover a broad range of financial or nonfinancial objectives about the subject matter under review. Attestation engagements can result in an examination, a review, or an agreed-upon procedures report on a subject matter or on an assertion about a subject. Three types of attestation engagements are:
>
> **a.** Examination: Consists of obtaining sufficient, appropriate evidence to express an opinion on whether the subject matter is based on (or in conformity with) the criteria in all material respects or the assertion is presented or fairly state in all material respects, based on the criteria.
>
> **b.** Review: Consists of sufficient testing to express a conclusion about whether any information came to the auditors' attention on the basis of the work performed that indicates the subject matter is not based on (or not in conformity with) the criteria or the assertion is not presented (or not fairly stated) in all material respects based on the criteria. Auditors should not perform review-level work for reporting on internal control or compliance with provisions of laws and regulations.[18]
>
> **c.** Agreed-Upon Procedures: Consists of auditors performing specific procedures on the subject matter and issuing a report of findings based on the agreed-upon procedures. In an agreed-upon procedures engagement, the auditor does not express an opinion or conclusion, but only reports on agreed-upon procedures in the form of procedures and findings related to the specific procedures applied."[4]

INTERNAL CONTROL AUDITS

This book, **Value Added Auditing**, is written around the concept of internal and external controls. Internal control audits are usually conducted as part of an overall risk assessment. First risks are identified and then internal controls are identified to mitigate the risks. Internal control assessments usually evaluate these 5 interrelated elements of effectiveness:

- **Control environment.** Senior management sets the tone for vision, mission, quality, ethics, goals, and controls. Daily operational control defers to the people who know the process or a product – the process owners.
- **Risk assessment.** Risk management is the fundamental objective of all managers in the next few years. The precondition to effective risk management is identi-

fied core processes, stabilized processes, capable processes, and control of process variation.

- **Control activities.** Control activities are the people, policies, suppliers and other factors that ensure risks are identified, monitored, and mitigated throughout the project, product, or contract lifecycle. Controls may include approvals, authorizations, validation, verification, reconciliation, and segregation of authorities.
- **Information and communication.** No information and no communication result in no control. It's that simple.
- **Monitoring.** Internal control systems and processes must be monitored. It's not enough to have a process out of control or worse that is noncompliant with a specification or standard. Ongoing monitoring should ensure corrective and preventive actions.[5]

The nature and purpose of internal control auditing can be understood by examining what is done and produced. The main products of internal auditing are reports detailing the efficiency, economy, and effectiveness of management and financial controls, policies, plans, and procedures. Typical products and outputs of these audits may include:

- Risk management and assurance analysis.
- Corporate governance assessment.
- Management control effectiveness.
- Internal financial control analysis.
- Budget reports and analysis.
- Fiscal policy compliance.
- Statutory compliance.
- Contract analysis.

INTERNAL CONTROL FUNDAMENTALS

Internal control is the foundation of good corporate governance.
Anonymous

Much of this book is based on the concept of risk – controls. This is the age of risk. Executives want a better way to control the risks within their enterprise as well as the extended enterprise of the supply chain. How should this be done? The development of internal and external controls is becoming the preferred method to achieve continued profitability as well as manage risks.

Internal and external risk - controls allow an organization to deal with rapidly changing competitive environments, shifting customer requirements, restructuring for growth, and managing the supply chain risks. Controls offer a number of mission critical benefits to an organization, specifically to:

- Improve corporate governance.
- Promote operational efficiency and effectiveness.
- Manage enterprise risks.

REASONABLE ASSURANCE

GAGAS 6.03 In performance audits that comply with GAGAS, auditors obtain reasonable assurance that evidence is sufficient and appropriate to support the auditors' findings and conclusions in relation to the audit objectives. Thus, the sufficiency and appropriateness of evidence needed and tests of evidence will vary based on the audit objectives, findings, and conclusions. Objectives for performance audits range from narrow to broad and involve varying types and quality of evidence. In some engagements, sufficient, appropriate evidence is available, but in others, information may have limitations. Professional judgment assists auditors in determining the audit scope and methodology needed to address the audit objectives, and in evaluating whether sufficient, appropriate evidence has been obtained to address the audit objectives.

- Ensure reliability of financial statements.
- Ensure compliance with regulations and laws.

COSO INTEGRATED FRAMEWORK

COSO is an acronym for 'Committee of Sponsoring Organizations.' The sponsoring organizations are the American Institute of Certified Public Accountants (AICPA), the Institute of Internal Auditors (IIA), Financial Executive International (FEI), Institute of Management Accountants (IMA), and the American Accounting Association (AAA).

Value added auditing incorporates the key elements of the COSO Integrated Framework. Why? The statutory authorities and the regulatory authorities, such as the SEC, recognize the framework as a consistent and rigorous methodology for controlling risks. And, more companies have or are adopting the COSO framework. The framework is an excellent methodology for conducting risk management and operational effectiveness assessments.

The value added auditing approach in this book is harmonized with COSO. For example, there are three critical elements for conducting a COSO internal control assessment. The first is to determine business objectives. As discussed, internal controls are designed and deployed to provide reasonable assurance that business metrics and objectives are achieved in the effectiveness and efficiency of operations, reliability of financial reporting and compliance with applicable laws and regulations. In this book, we focus on evaluating the effectiveness and efficiency of operational and compliance with applicable laws and regulations. Reasonable assurance is defined in the above sidebar.

In the 'Planning Value Added Audits' section of this book, we discuss the importance of determining business objectives and then ensuring there is alignment with business objectives to a business strategy.

The second element of conducting a COSO internal control assessment is to evaluate an enterprise or entity based on key control attributes. Attributes are covered in the 'Planning the Value Added Audit' and 'Conducting the Value Added Audit' sections of this book. Generic questions are supplied in each chapter to facilitate conducting the audit as well as providing tips for the auditor to develop more specific or granular questions addressing the specific area being audited.

The third element of conducting an internal control assessment is to assess processes and activities. These are covered specifically in the 'Conducting the Value Added Audit' section of this book.

WHAT ARE INTERNAL CONTROLS?

As discussed, the concept of internal risk - control is woven and integrated throughout this book. The concept is abbreviated to simply control in order to include both internal and external controls. The concept of internal control means different things to different people. Regulatory authorities as well as lay people use the concept. For example, the concept of internal control has been written into law, such as Sarbanes-Oxley Act.

In terms of this book, we use the COSO Integrated Framework of Integrated Control, specifically to mean:

> "Internal control is broadly defined as a process, effected by an entity's board of directors, management and other personnel, designed to provide reasonable assurance regarding the achievement of objectives in the following categories:

- Effectiveness and efficiency of operations.
- Compliance with applicable laws and regulations.
- Reliability of financial reporting.[6]"

This book covers and emphasizes the first and second bullets. The first category addresses the enterprise's ability to meet its business critical objectives, including performance, profitability, reliability and other goals and metrics. The second category deals with complying with applicable laws, regulations, and standards.

This book does not address the third bullet of 'reliability of financial reporting.' This category focuses on assuring that published financial statements are prepared accurately, are transparent, and comply with applicable standards. Also, these assessments more often require a CPA and knowledge of accounting and finance.

It's important to mention that the goal of value added auditing is not absolute assurance, but reasonable assurance. We will explain this concept in greater detail throughout this book.

VALUE ADDED AUDITOR RESPONSIBILITIES

This book, **Value Added Auditing**, is structured around the concept of risk - control, specifically internal and external controls. What's the role of the value added auditor in internal control? Internal control systems operate at different levels effectiveness, efficiently and economy.

The value added auditor is responsible for providing senior management and even the board of directors 'reasonable assurance' that:

- Objectives had been deployed down the organization and into the supply chain.
- Mission critical objectives are being achieved.
- Applicable laws are regulations are being complied with.

COMPONENTS OF INTERNAL CONTROL

Internal control consisted of five interrelated components:

- Control environment.
- Risk assessment.
- Control activities.
- Information and communication.
- Monitoring.[7]

Control Environment

The control environment is basically the culture of the organization. The environment establishes the ethic of the organization. Senior management sets the 'tone at the top,' which permeates the organization; guides, role models, and reinforces behaviors; and influences the control ethic of all stakeholders. The control environment is the foundation of all elements of the control system

The control environment includes:

- Core values.
- Oversight by the board of directors.
- Credibility of the board of directors and senior management.
- Integrity of the organization.
- Ethical values.
- Senior management's operating style and philosophy.
- Management deployment of authority and responsibility.

Risk Assessment

All organizations have mission-critical strategies, objectives, tactics, and plans, which are deployed down the organization and into the supply chain. One definition of risk is the ability to meet these objectives consistently. In other words, the ability to assess and ultimately manage risks reflects on the ability of an organization to meet its business objectives.

Risk is the key filter for senior management decision-making. An organization faces risk from many sources; from within and outside the organization. How it identifies, monitors, controls, mitigates, and ultimately manages overall risk indicates how successful and profitable it will be.

Risk assessment includes:

- Determining critical business objectives.
- Identifying risks that impact the ability to meet objectives.
- Developing a system to manage the risks.
- Developing mechanisms for managing change.

Control Activities
Controls activities occur through the organization and into the supply chain. There are basically two types of controls: 1. soft controls and 2. hard controls. Soft controls deal with the messages and reinforces that the board of directors and senior management want to communicate. This is sometimes called 'tone at the top.' Hard controls include policies, procedures, and work instructions that detail how management directives and work is carried out. These help ensure that the necessary actions are anticipated and taken to address the risks of not meeting an organization's objectives.

Information and Communication
Reliable data and accurate information are required to control processes and activities. Without them, there is no control. So critical control information must be identified, captured, and communicated to the right parties so it's relevant for informed decision making and external reporting. The information must also be in a form and timeframe so process owners and primes can meet their responsibilities.

Information should be captured based on critical needs of the organization. Risk points are identified throughout the organizational value chain and externally into the supply chain. Communication is also reported externally to customers, suppliers, regulators, and shareholders. Risk points become organizational points of control. Information from these points, nodes, or areas may be communicated up, across and down the organization.

Monitoring
Once processes are stable, capable, and improving, these processes must be monitored. Monitoring may mean first party assessments; real time monitoring; second party evaluations such as internal auditing; or third party audits such as by regulatory authorities.

Monitoring ensures critical system, process, and product performance improves over time. Management should Pareto (80 – 20 rule) critical risk-control points within the organization. The scope and frequency of monitoring depends on the evaluation of the control effectiveness to manage critical risks. Then, control deficiencies are reported to process owners, senior management, or the board of directors depending on the risk, materiality, or exposure to the organization.[8]

BENEFITS AND LIMITS OF INTEGRATED FRAMEWORK OF INTERNAL CONTROL

This framework is an integrated system of controls and should be flexible to react dynamically to changing business conditions. These 'built in' controls are the essence of good management practices. For example all 5 integrated control components should exist, be documented, and functioning effectively for the value added auditor to conclude that internal controls are working properly to "provide reasonable assurance regarding the achievement of objectives" in the "effectiveness and efficiency of operations." Again, the goal is reasonable not absolute assurance.

The internal control framework is also an excellent framework for achieving performance and profitability targets. Internal control is a transparent model that can be used in many industries and businesses. Controls can be used to manage risks throughout the organization and into the supply chain. Internal controls are also fundamental to many corporate governance laws and regulations.

However, even an integrated system of internal control:

- Cannot guarantee absolute or one hundred percent (100%) assurance.
- Cannot ensure success or profitability.
- Cannot guarantee meeting business objectives.
- Cannot change improve systemic or chronic problems.
- Cannot evaluate or change the fundamental business model.
- Cannot ensure customer satisfaction.
- Cannot modify or change people or management behaviors, fundamentally the biggest source of variation or risk.
- Cannot impact external variables, such as changing customer requirements, regulatory mandates, and macro economic conditions.

Finally, the threshold of professionalism for value added auditors is to provide 'reasonable assurance' to the board, senior management, process primes, process owners, or other report stakeholders. The best conceived, deployed, and operating control system can still be circumvented, be misused, not understood, overridden, waived, colluded, tainted, or have diminished effectiveness for any number of reasons.

SUMMARY

Internal auditing is primarily a risk management, monitoring, and control tool. Internal audits are usually conducted by staff employees to assess operations and processes as a service to senior management and to the company's board of directors. Internal auditors originally had accounting backgrounds. Now, more internal auditors are coming out of operations, engineering, quality, and information technology. They more often submit reports to the audit committee of the board of directors.

The value added auditor must be able to assess risks and controls. Also, the value added auditor must use his or her analytical skills to add value to the work data so that it becomes information and knowledge useful for decision-making.

VALUE ADDED AUDITING 101

Value added auditing has expanded beyond a simple yes/no format evaluating compliance to a process effectiveness and risk mitigation assessment. Its general value is becoming more recognized and accepted. As the importance of process management and control spreads, the uses of value added auditing as an assurance assessment are increasing. In general, the value added audit can be used to:

- Verify compliance with known standards.
- Comply with contractual requirements.
- Evaluate manufacturability of engineering drawings.
- Determine suitability of specifications.
- Evaluate risk - control effectiveness.
- Establish process improvement baselines.
- Measure improvements.
- Maintain competitiveness by benchmarking world-class competitors.
- Identify and reduce liability exposure.
- Ensure external and internal customer satisfaction.
- Improve internal operations.
- Measure and monitor process improvement program effectiveness.
- Measure and report on processes (i.e., manufacturing, engineering).
- Monitor internal and external customer satisfaction.
- Monitor continual improvement.
- Set and measure benchmark attainment.
- Monitor product process improvement.
- Monitor personnel effectiveness.
- Monitor and measure supplier improvement.
- Establish performance baselines.
- Evaluate and measure service process improvement.
- Identify areas for corrective action.
- Identify areas for process improvement.
- Improve financial performance.
- Confirm effectiveness of management action.

CHAPTER 4
ENTERPRISE RISK MANAGEMENT 101

Federal agencies, state governments, and companies are facing increased uncertainty and complexity due to cybersecurity risks, new regulatory requirements, offshore competitive pressures, and new governance regulations. Inability to anticipate, identify, mitigate, and manage risks can result in catastrophic failure. As a result, risk management has become critical to all organizational executives and process owners.

"We are all paid to take risks," says Lawrence A. Bossily, the former leader of aerospace company AlliedSignal. Andrew Grove, chair of Intel Corporation, refers to risk as "points of inflection" or "the moment that massive change occurs and all bets are off." For these reasons, senior management, and the boards of directors want to know how organizational, systems, suppliers, processes, and value chain risks are managed. We call this Enterprise Risk Management (ERM). In this chapter, we cover event-based risks and in the next chapter, we cover process-based risks.

WHAT IS RISK?

All business proceeds on beliefs or judgments of probabilities, and not on certainties.
Unknown

The difference between successful and less successful companies today is how they manage risks in terms of mitigating downside threats and taking advantage of upside opportunities.

Executives and managers for all organizations are developing processes to identify management risks across and down the enterprise, hence the term Enterprise Risk Management (ERM). From an enterprise point of view, ERM is the means to architect and deploy good governance practices, manage uncertainty, and mitigate risks.

WHAT CAN GO WRONG?

Enterprise risk management is a systematic process of managing unwanted events or unwanted change. The business view of risk is determining 'what can go wrong?' Risks are inherent in any project, process, or work. Smart enterprise risk management is the ability of not being blindsided by unexpected events. This is illustrated in the figure on the next page, 'Managing Risk Gaps.' Uncertainty is increasing for all organizations. Existing organizational processes cannot keep up. The result is 1. Enterprise, 2. Pro-

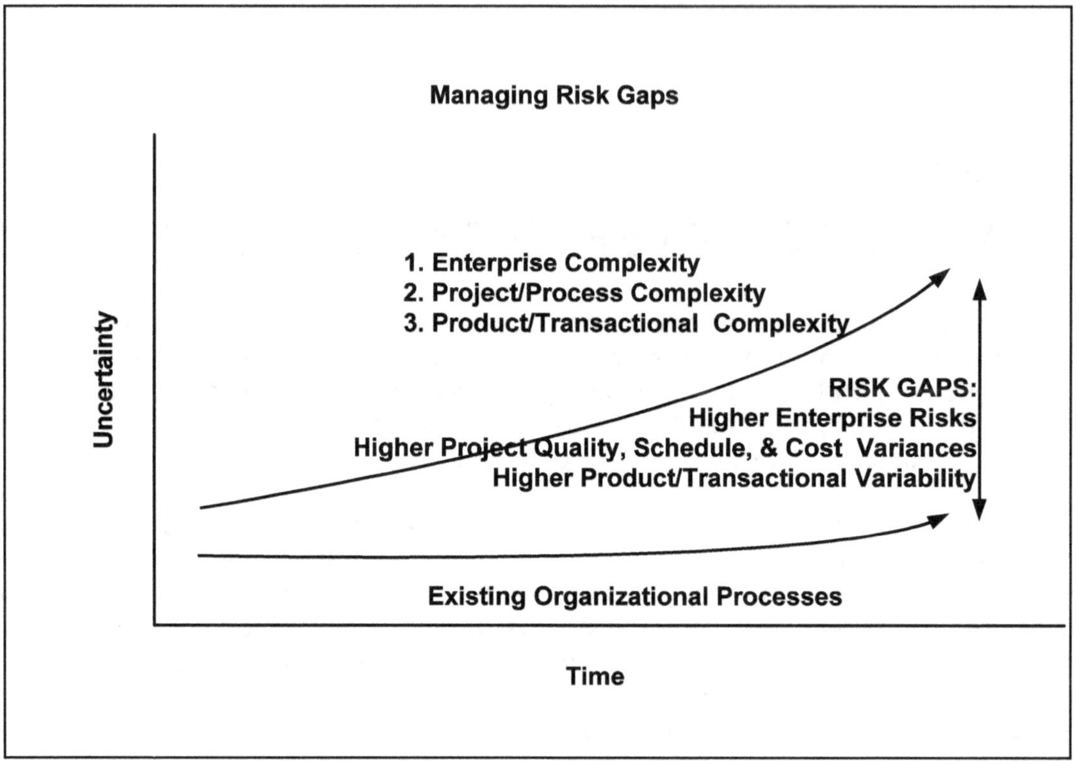

grammatic/project/process, and 3. Product/transactional risk gaps, which must be mitigated.

Colloquially speaking, risk has a negative connotation. It is mainly about an adverse event or loss. Risk is often thought in terms of a black cat crossing one's path or breaking a mirror. However today, risk may be the most critical element of management decision-making.

WHAT IS RISK?

Before we start an in-depth discussion of risk and enterprise risk management, it is critical that we define risk. Below are common risk perspectives and definitions:

- "The chance of something happening that will have an impact upon (business) objectives. It is measured in terms of consequences and likelihood."[1]
- "A situation or circumstance, which creates uncertainties about achieving program objectives."[2]
- "The possibility that an event will occur and adversely affect the achievement of objectives."[3]
- "The risk of loss due to deficiencies in information systems, business processes, or internal controls as a result of internal or external events."[4]
- "Risk is the probability that an event or action may adversely affect the organization or activity under review."[5]

Risk is an interesting concept. Two elements to risk can be seen in the above definitions. There is upside risk and there is downside risk. Upside risk is opportunity risk, the reward that can be achieved in undertaking a new endeavor. Downside risk is the risk of loss. For example, project risk is the inability to achieve project objectives within cost, schedule, and quality constraints. Process risks are the risks of producing nonconforming products or services due to unstable and incapable processes.

Elements of most risk definitions include the following:

- Defined process targets and business objectives exist.
- Expected and unexpected variation and variances result in higher risks.
- Possibility of harm or loss can occur as a result of variances.
- Probability of an undesirable event with critical consequences can be estimated.

MANAGING UNCERTAINTY AND RISK GAPS

WHAT IS RISK MANAGEMENT?

Risk management is the ability to identify, mitigate, and manage risk gaps. This is illustrated in the figure on the previous page, 'Managing Risk Gaps.' Existing organizational processes are often static. They do not change, unless there is an external driver for change. Why? Most folks support the status quo, because change is difficult and uncomfortable. The challenge is that enterprise, programmatic/process/project, and transactional uncertainty continues to increase. In addition, current processes are not capable of responding appropriately, which result in risk gaps.

Risk management is considered the ability to make smart choices in uncertain times and manage risk gaps. Below are two common definitions of risk management:

- **Risk management:** "Culture, processes, and structures that are directed towards the effective management of potential opportunities and adverse effects."[6]
- **Risk Management:** "Identification, assessment, and response to risk to a specific objective."[7]

The definitions of risk and risk management offer several lessons learned:

- Satisfy the organization's risk appetite or tolerance.
- Please the customer by delivering projects and products on time and on budget.
- Encourage process owners to achieve desired business outcomes.
- Meet programmatic, project, and process objectives.
- Assure external (supplier) and internal processes are stable (controlled), capable of meeting business objectives, and improving.
- Assure programs and projects are controlling scope, quality, schedule, and costs.
- Anticipate and control undesirable events.
- Avoid and mitigate negative consequences.

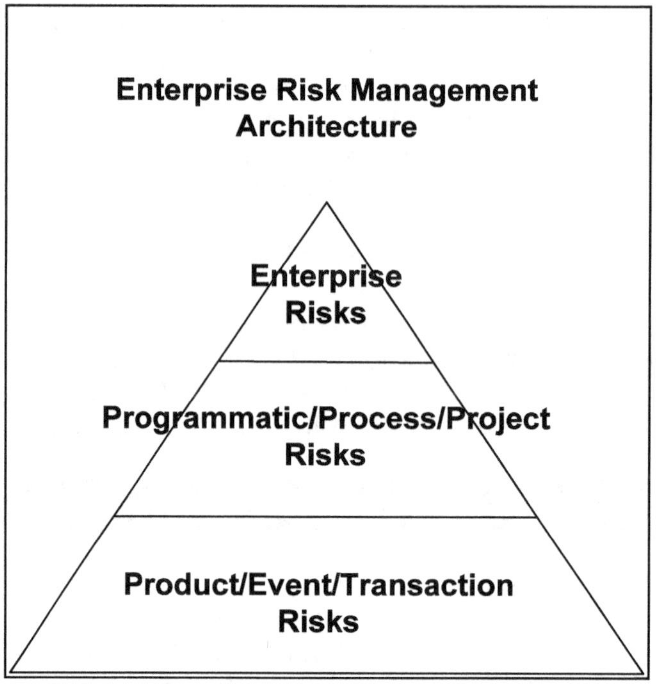

- Control process or project variation.
- Support smarter decision-making.

ENTERPRISE RISK MANAGEMENT ARCHITECTURE

Enterprise and operational risk can be evaluated many ways, for example from an enterprise, program/project/process, or transactional/product perspective. We will analyze risk and conduct the value added audit according to the structure illustrated in the above figure. The differences among the levels are shown below

- **Enterprise risk:** Risks related to the inability to meet business objectives, strategic goals, 'tone at the top' requirements, cultural norms, ethical standards etc.
- **Project risk:** Risks related to not being able to meet a program's or project's 'triple constraints' – 1. Quality/scope; 2. Schedule; and 3. Cost.
- **Process risk:** Risks related to a production or service process not being able to demonstrate 1. Stability (control), 2. Capability, 3. Improvement objectives; inability to achieve consistent outcomes or conforming products.
- **Product/Event/Transactional risk:** Risks related to a product not being able to comply with form/fit/function requirements; risks related to a service not meeting stakeholder requirements; an event that may result in injury, illness, or death to people.

ENTERPRISE RISK MANAGEMENT

Value is created, preserved, or eroded by companies when resources are deployed to increase shareholder value. This concept is fundamental to Value Added Auditing. For governmental agencies, value is realized when stakeholders and constituents recognize that services are provided at an acceptable cost to stakeholders.

WHAT IS ERM?

As illustrated in the 'ERM Architecture' figure on the previous page, ERM can also be defined as the assessment and management of 1. Enterprise risks; 2. Programmatic/Project/Process risks; and 3. Program/transaction/event risks. Sarbanes-Oxley Act of 2002 accelerated its adoption for companies, especially in managing financial risks.

Enterprise risk management is defined as the following:

> "Enterprise risk management is a process, effected by an entity's board of directors, management, other personnel, applied in strategy setting and across the enterprise, designed to identify potential events that may affect the entity, and manage risks to be within its risk appetitive, to provide reasonable assurance regarding the achievement of entity objectives."[8]

ERM is a flexible framework and is sometimes conceptualized in terms of:

- **Governance:** Overarching purpose of ERM is to form a proactive, preventive, predictive, and ultimately preemptive structure of enterprise risk management.
- **Assurance:** Assurance is independent and objective monitoring, verifying, and assuring that risks are controlled within the risk appetite of the organization.
- **Architecture:** Architecture defines the risk and control relationships at the enterprise, programmatic/project/process, product/event/transactional levels.
- **Documentation:** Documentation structure consists of enterprise policies, program/project/process procedures, and product/transactional work instructions.

ERM MODELS/METHODOLOGIES

There are a number of ERM models or methodologies. In this book, we use elements of the COSO and GAO ERM models. COSO is an acronym representing the Committee of Sponsoring Organizations of the Treadway Commission. COSO developed a conceptual framework that integrated risk principles, common risk terminology, and practical implementation guidance for ERM processes. The Government Accountability Office (GAO) developed a robust and mature risk management methodology shown on page 75.

Most companies have adopted an Enterprise Risk Management model for the following reasons:

- Demonstrates good governance.

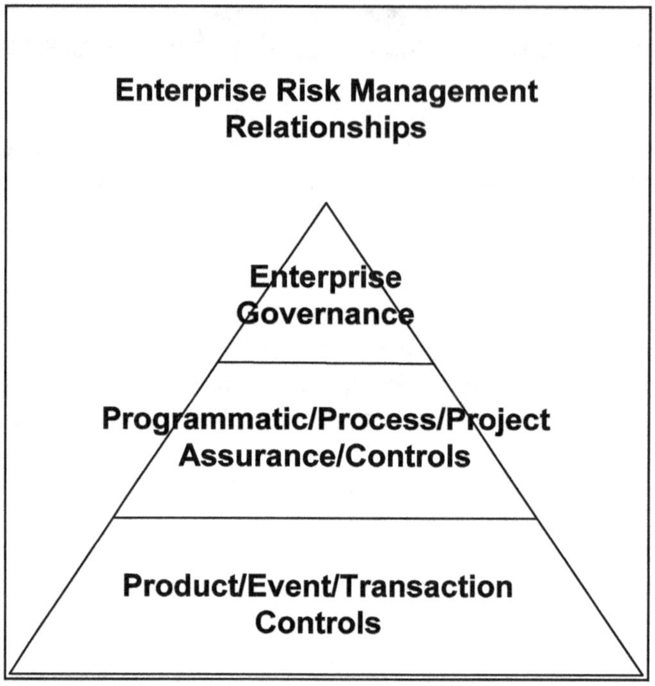

- Encourages an organization/person to ask the right questions, such as determining organizational risk sensitivity.
- Creates a consistent decision-making process.
- Ensures compliance with regulatory requirements.
- Results in process owner and individual accountability.
- Results in lower programmatic and project variance risks.
- Results in consistent operational processes.
- Creates a decision-making audit trail.
- Results in lower costs.
- Is process based.
- Emphasizes the appropriate level of controls.

COSO ENTERPRISE RISK MANAGEMENT (ERM)

Publicly held companies must have internal financial controls, which are often based on COSO standards and guidelines. Many companies use the COSO Enterprise Risk Management model because it assists in complying with the Sarbanes-Oxley Act of 2002. As well, the COSO ERM model fits in well with federal agencies and governance.

The COSO ERM is divided into the following components:

- Internal Environment.
- Objective Setting.
- Event Identification.
- Risk Assessment

Government Accountability Office (GAO) Risk Assessment Process/Criteria

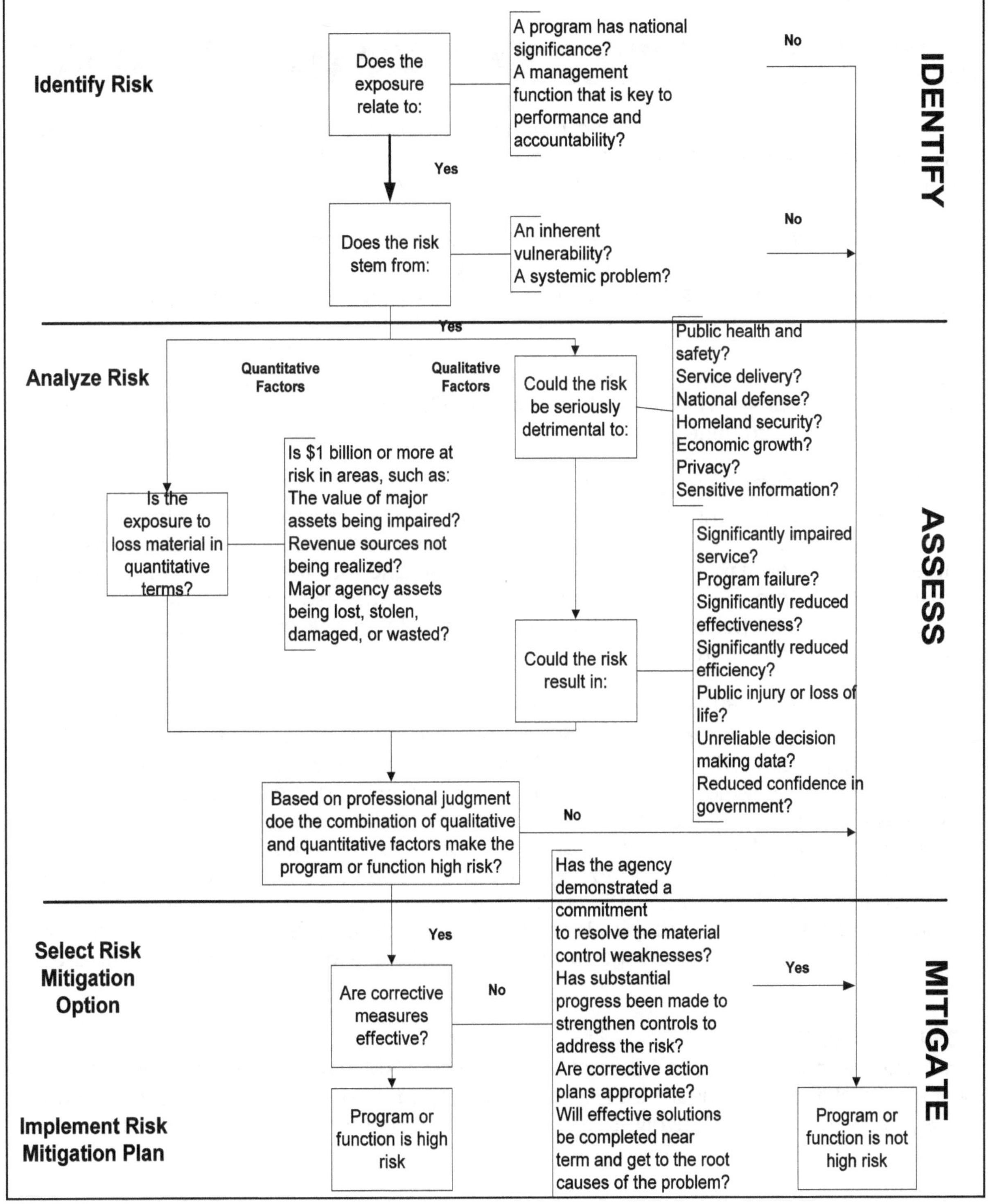

- Risk Response.
- Control Activities.
- Information and Communication.
- Monitoring.

INTERNAL ENVIRONMENT

The board of directors and executive management set the tone and philosophy regarding enterprise risk and risk appetite. The enterprise point of view and internal environment are the foundations for how risks are ultimately viewed, treated, and addressed.

One of the first issues to address in risk management is to determine the enterprise risk appetite, also known as risk sensitivity. Risk appetite determines the level of risk critical organizational stakeholders can or are willing to assume. Risk appetite can be visually divided into risk averse on the left side and risk taking on the right side of a spectrum. Somewhere in the middle an organization or person is risk sensitive. These concepts relate to the amount of risk an organization is willing to assume and the amount of controls it want to impose. This is illustrated in the 'Organizational Environments and Risk Sensitivity' sidebar on page 78.

ENTERPRISE PORTFOLIO VIEW OF RISK

Also, ERM can be thought as an umbrella approach to risk management. Mature ERM is an integrated collection of capable processes, policies, and procedures used to assess, define, and manage risk throughout an organization and into the value stream involving suppliers. ERM provides a common framework to assess the risk resulting from changes. Also, ERM helps reduce the number of isolated risk events, which may result in wasted time and resources.

ERM by definition is an overall or entity approach to risk. Product, transaction, process, and project risks may be within the tolerance band of the organization. However when these are aggregated, the composite risk may well be outside the risk tolerance of the organization. Why? When an individual risk analysis is conducted, then interdependent risks, dependent risks, white space risks, hand-off risks, and other risks accumulate. 'White space' risk refers to the white spaces in an organizational chart, where roles, responsibilities and work rules tend to be defused and sometimes unknown. Handoff risks occur when material, communications, or other resources are shared among business entities and slips occur. Interdependency risks are those that relate to how one risk can exacerbate another. The challenge is that many risks are often unknown and unknowable.

TONE AT THE TOP

ERM culture is both structural and attitudinal. The ERM structure must be in place as well as the attitudes to ensure that risks are managed. Risk culture goes beyond adherence to procedures. ERM involves the active participation of all critical stakeholders and

RISK APPETITE FACTORS

A number of factors affect risk appetite, such as the following:

- Senior executive risk perception.
- Organizational culture (environment).
- Personal risk perception.
- Previous successes/failures.
- Timing of potential event.
- Amount of information available to decision makers.

process owners. Duties must be carried out correctly; diligence ensures that actions are thought out, good judgments are applied; and responsible parties are personally accountable for actions.

RISK SENSITIVITY

Every organization has a risk sensitivity. The discussion of risk sensitivity should start with senior management and the board of directors. The risk sensitivity discussion determines organizational strategies, tactics, controls, mergers, acquisitions, revenue, exposure, and other critical business decisions.

The challenge is that risk sensitivity is often not discussed and understood within an organization. Risk sensitivity is a function of the industry, processes, and products. Regulated industries, such as medical devices, have a higher risk threshold or tolerance than a commodity manufacturer of tissue paper. Once risk tolerance is understood, then operational managers can make decisions on capital expenditures, product development, controls, and other risk related ventures.

AMOUNT/QUALITY/ACCURACY OF DECISION-MAKING INFORMATION ABOUT RISK.

As discussed, risk has an upside and a downside. There should be explicit discussions about potential upside opportunities and down side implications of business risk exposure. This risk discussion frames what is expected of senior management, employees, suppliers, and other stakeholders. If there needs to be an expansion or rescoring in the acceptance of risk, then this would require further executive discussion and authorization. Ultimately, risk tolerance defines risk management and controls throughout the organization.

Risk tolerance forms the basis for business unit, process, and individual accountability. While everyone is responsible for monitoring and managing risk in their operations, ultimate ERM responsibility rests with executive management and even the board of directors.

Process owners are accountable for managing risk in their areas of responsibility. Clear lines of risk responsibility and accountability are necessary to foster a risk culture in

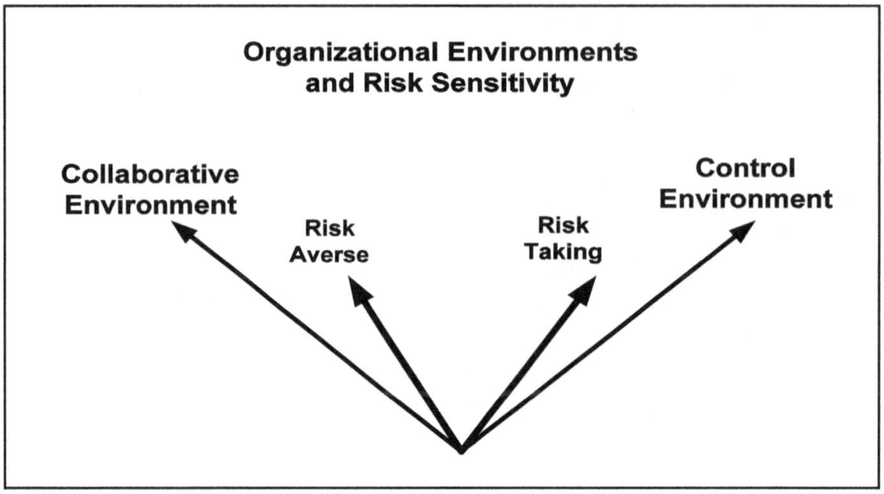

which all employees feel personally responsible for the risk under their control. The importance of accountability can be understood in terms of risk gaps. If there is no accountability for controlling risk, then the probability of a risk event increases as risk gaps increase. The solution is to determine risk gaps, identify responsibilities for controlling risks, and then assign responsibility to closing the risk gaps.

OBJECTIVE SETTING

Objectives are the yardsticks or metrics for organizational profitability and performance. All risk management activities should be grounded and mapped to enterprise, programmatic/process/product, and product/transactional. This is what adds value. Period.

When adopting, implementing, and adapting ERM to an organization, risks can be thought as roadblocks, constraints, or inhibitors to reaching objectives. So expressed another way, if an organization is to meet its business objectives, it must remove, control, or mitigate risks that impede or inhibit reaching objectives.

Enterprise risk management ensures that management has a process to set objectives, that support and align with the entity's mission/vision and are consistent with the entity's risk appetite.

Objectives can be found at the enterprise, programmatic/process/project, and transactional/product levels. Typical ERM objectives include:

- Business objectives.
- Balanced scorecard objectives.
- Statutory/regulatory compliance reporting.
- Efficiency/effectively/economic objectives.
- Intervention/correction/prevention action objectives.

An organization faces many risks. It is critical that risks are matched to objectives. When designing and architecting the ERM, risks that impede or hinder the achievement of an enterprise or programmatic objective should be addressed and managed. Enterprise risks may include:

- Toxic culture.
- Organizational risk appetite not mapped to controls.
- High exposure level and types of enterprise risks.
- Level, location, and types of enterprise risks.
- Level, location, and types of enterprise controls.
- Level of required enterprise assurance.
- Enterprise risk consequences.
- Enterprise risk likelihoods.
- History of risk management success.

Process risks may include:

- Lack of process stability (control).
- Lack of process capability (ability to meet requirements).
- Lack of process improvement (less variation).

Programmatic and project risks may include:

- Scope variance.
- Quality variance.
- Schedule variance.
- Cost variance.

Product and service risks may include:

- Nonconforming product attributes.
- Service complaints.
- Product development lapses.

There is a simple linear relationship regarding risks: risk deltas/gaps > higher risks > more controls > more risk mitigation > more assurance. There are few universal solutions for managing risks, except they usually involve additional controls, standardization, stable/capable processes, and consistency.

EVENT IDENTIFICATION

Events are internal or external activities or occurrences that may impact the enterprise and the achievement of critical objectives. Events can be either positive or negative.

Positive events are opportunities for business development. Negative events are hazards that can adversely affect the enterprise, processes, programs, projects, transactions, and products.

This step requires an enterprise point of view. The board of directors and executive management should take a portfolio perspective at the enterprise to identify events that can impact the organization. The challenge at this step is that events can be difficult to identify. There are always risks and hazards that may be unknown and unknowable.

The following are critical to consider at this step:

- Ensuring there is an enterprise and common definition of risk appetite and tolerance.
- Understanding the project/process lifecycle risks and controls.
- Understanding risk consequences and likelihoods at different stages of a program or project lifecycle.
- Understanding 'white space' hazards and risks.
- Understanding interdependent risk.
- Understanding risk dependencies.
- Identifying cascading risks and consequences.
- Ensuring critical stakeholders have same view of risk.

RISK ASSESSMENT

Once risk is detected, it is assessed in terms of consequences and likelihood. There are a number of definitions of risk assessment. One definition of risk assessment is:

> "A systematic process for assessing and integrating professional judgments about probably adverse conditions and/or events. The risk assessment process should provide a means of organizing and integrating professional judgments for the development of the organization's work schedule."[9]

Another perspective of risk assessment follows:

> "The company faces a variety of risks from external and internal sources that must be assessed. A precondition of risk assessment is establishment of objectives, linked at different levels and internally consistent. Risk assessment is the identification and analysis of relevant risk to the achievement of the objectives, forming a basis for determining how that risk should be managed. Because economic, industry regulatory and operation conditions will continue to change, mechanisms are needed to identify and deal with risks associated with change."[10]

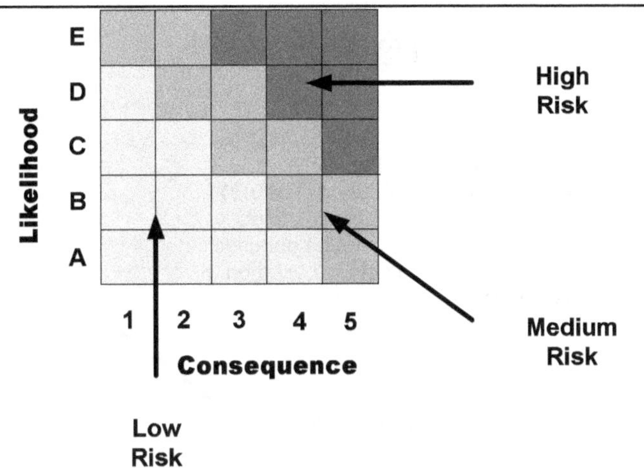

Risk Profile Matrix

Risks are assessed to form the basis for determining how they should be managed and if required mitigated. Again, risk assessments are always associated with being able to meet objectives. Risks are assessed on both a present and a residual basis. As well, risk is assessed in terms of likelihood and consequence. A range of possible outcomes may be associated with a potential risk event.

Risks can be identified by:

- **Consequence:** This is sometimes called impact and is evaluated from a one (1) to five (5).
- **Likelihood:** This is sometimes called frequency or probability. It is evaluated from an one (1) to five (5). Both consequence and likelihood are then multiplied to determine total risk. The data is then mapped to a risk profile matrix as illustrated in the above 'Risk Profile Matrix' figure and in the 'Event Based Risk figure on the next page.
- **Source:** This can be either the hazard or the root cause of the risk or a pattern of risks.
- **Uncertainty:** Risk is all about uncertainty, variability, variance, confidence level, probability, and statistics.

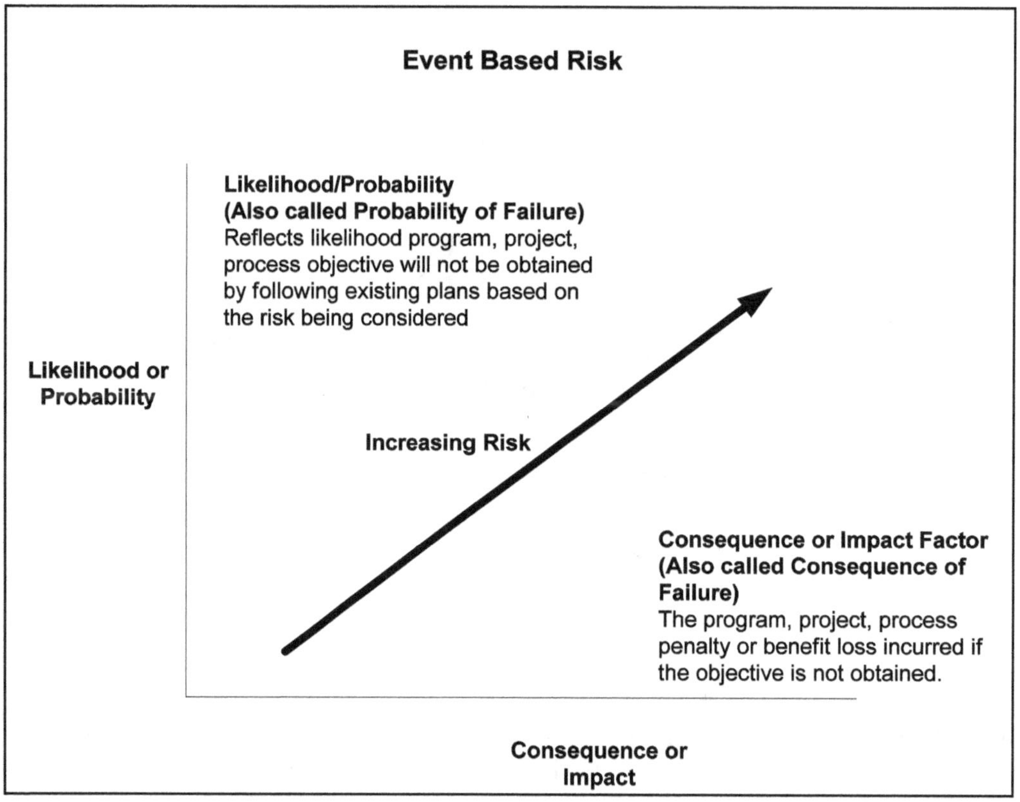

ERM as it is practiced by most organizations is not a science. Most ERM analyses are qualitative and subject to uncertainty and differing confidence levels. Uncertainty is: "The estimated amount or percentage by which an observed or calculated value may differ from the true value."[11]

Identifying critical organizational risks because of uncertainty poses challenges for an organization as well as the auditor. The challenge for the process owners is to separate 'trivial many' risks (acceptable risks) from the 'critical few' risks (unacceptable risks)?

RISK RESPONSE

Executive management along with the board of directors identify risk response options based on event likelihood and consequence. These are then balanced and assessed in terms of the organizational risk tolerances, costs/benefits, and organizational objectives. Effective, efficient, and economic enterprise risk management requires that executive management and process owners select responses that are expected to bring risks within the organizational risk tolerance band.

RISK APPETITE

Residual risk and risk appetite are among the most confusing and frustrating concepts in Value Added Auditing. Usually, this is a qualitative decision to determine whether an organization is risk averse, risk sensitive, or risk taking. This decision is critical because it determines how much risk the organization is willing to take as well as determines the level of controls, risk mitigation, and residual risk, the organization is willing to assume based on risk appetite.

Who determines the risk appetite or tolerance of an organization? From the enterprise point of view, the board of directors with discussions with executive management determines risk appetite. One challenge is that different business units, programs, projects, processes, and product areas may have different risk profiles. Usually, the process owners or executive management over the area determines the organizational risk acceptance. The determination of the above issues have a substantial impact on the value added auditor as well as process owners.

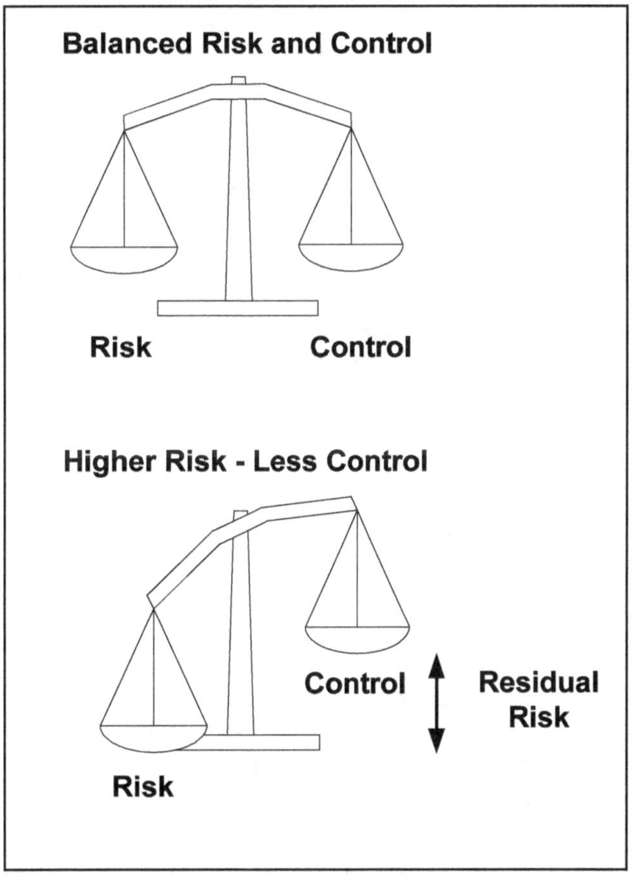

Let us discuss this in light of the figure on the right. The preferred option is have risks and controls be balanced. If risks outweigh controls, then there is an imbalance between risks and controls. The amount of imbalance is called residual risk. If risks outweigh controls by an acceptable level, then additional risk mitigation or treatment has to be architected into the process at the point of risk or at the point of control.

The lower figure illustrates where there is higher risk and less control. The difference between the amount of risk and level of control is called residual risk. This is a critical concept to the enterprise, process owners, and the value added auditor. Let us look at each:

- **Enterprise residual risk:** Residual risk from an enterprise point of view is what is acceptable to the organization's risk appetite. If the residual risk is higher than the enterprise risk tolerance, then the nature of the risk (left side) will have to be lessened or more controls added (right side of the figure) will have to be architected into the system. This is a critical concept. On the left side of the scale, risks can be decreased or on the right side of the scale, controls can be added, so the result is to move residual risks within the acceptable risk band of the enterprise or organization.

- **Programmatic/Process owner residual risk:** As discussed, process owners may determine risk appetite based on their specialized knowledge of the program, process, project, or product. The risk appetite decision at the local or process level is based on the enterprise risk appetite. If the enterprise or organization is risk averse, then process level acceptable residual risk must be within the enterprise risk tolerance band. This means the local process owners must have sufficient controls or risk mitigation/treatment tactics to harmonize with the organizational risk appetite.

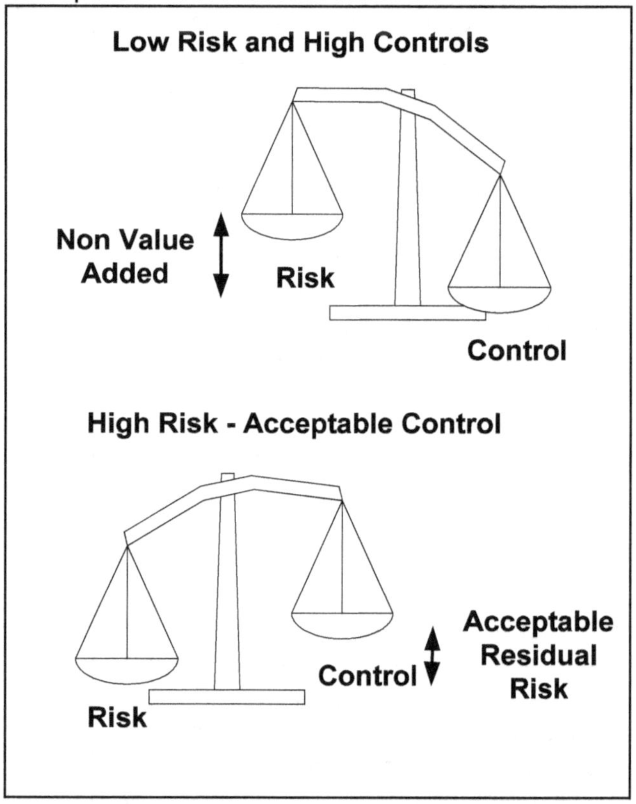

- **Auditor residual risk:** As discussed, the auditor does not own enterprise or local risks. The process owner, executives, and board of directors own and control risks. The auditor simply evaluates if critical risks have been identified, controls are appropriate to the risks, and residual risks are within the risk appetite of the organization. If there are too many controls over programmatic, processes, or product risks, then this is a non-value added. If the controls are inappropriate, again, this is non-value added and the auditor should notify process management if they can be root cause fixed or remediated. If the controls are unsuitable and result in high and unacceptable risks, then executive management should be informed to mitigate risks at the enterprise, systemic, and root cause levels, which may be beyond the abilities of local process management.

RISK RESPONSE STRATEGIES

When evaluating risk response strategies, executive management along with process owners must align risk responses with the organization's risk appetite, business objectives, costs/benefits, and overall risk strategy/tactics. Risk response strategies and tactics may involve:

- **Avoidance:** Action is taken to depart from the risk situation or remove the activities giving rise to the risk i.e. exiting a product line, triaging a project, declining service to an area, using stable designs, using less complex IT methodology, standardizing processes, centralizing facilities, and moving up the capability and maturity curve.

- **Reduction:** Action is taken to reduce the risk likelihood or consequence, or both, which may involve any of a myriad of everyday business and process control decisions, i.e. lowering level of risk, increasing level or breath of controls, using redundant systems, using joint application design, using design-build, adding more float to project schedules, and using enterprise, process, and project risk management methodologies.

- **Sharing:** Action is taken to reduce risk likelihood or consequence by transferring or otherwise sharing a portion of the risk, i.e. purchasing insurance, pooling risks, hedging transactions, outsourcing an activity, using QA/QC service provider with suppliers, retaining a third party project manager, retaining prime and acceptable alternate suppliers, and purchasing errors/omission insurance.

- **Acceptance.** No action is taken to affect likelihood or consequence, i.e. accepting quality level of commodity, falling within risk sensitivity threshold for enterprise, using commodity products, using products from ISO 9001 suppliers, and lowering level of assurance.[12]

COSO identifies 4 risk treatments. However, each of the above risk management strategies and treatments can be subdivided further. In this book, we use the following risk management strategies: avoid risk, mitigate risk, accept risk, share risk, diversify risk, control risk, and increase risk.

Once the appropriate risk response is determined, then management assigns responsibilities for deploying the risk management plan. The risk management and response plan should address how to mitigate the consequences and likelihood of critical risks. Some risks are more critical than others. Therefore, the process owners should separate the critical few risks from the trivial many risks, also called Pareto risks. A cost benefit analysis is always helpful in implementing a risk response plan.

CONTROL ACTIVITIES

Risk controls come in many forms from procedural documentation to machine controls to Information Technology controls. Controls are designed and implemented to assure that risk management and risk responses are efficiently, effectively, and economically implemented.

Controls ensure that business objectives previously defined are consistently met, which may involve controlling and ensuring:

- Effectiveness (doing the right thing).
- Efficiency (doing the thing right).
- Economy (doing the thing at right cost).
- Accuracy (doing the thing right).
- Reliability (doing the thing the same).
- Verifiability/measurability (doing the measured thing).
- Completeness (doing the same thing from beginning to end).
- Authorization (doing the approved thing).

In general, the type, level, scope, and duration of the controls are based on the risk appetite and risk tolerance of the organization. Architecting, designing, and deploying controls should be tailored to the following:

- Organizational culture.
- Enterprise risk appetite.
- Present control environment.
- Desired control environment.
- Risk magnitude and likelihood.
- Level of assurance (risk appetite).
- Complexity of project/process/product.
- Operating environment.
- Cultural sensitivity.

CONTROL STRATEGIES/TACTICS

As discussed, the board of directors and executive management set the tone at the top for risk appetite, which define the control architecture. There are a number of control decisions, including:

- Decentralized and/or centralized risk decision making.
- Collaborative and/or unilateral controls.
- Highly capable and mature or less capable mature control processes.
- Ad hoc and/or standardized controls.
- Informal and/or formal controls.
- Decentralized risk authority and/or centralized risk authority.
- Empowered (collaborative) and/or accountability based controls.
- Hard and/or soft controls.
- Self verification and/or oversight validation.

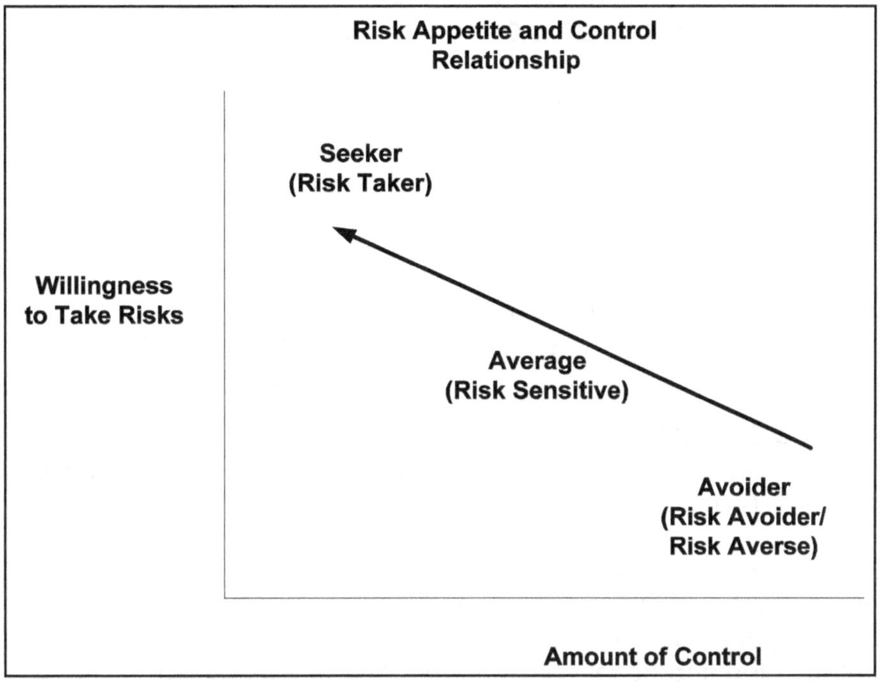

INFORMATION AND COMMUNICATION

Relevant and accurate information regarding risks and controls are identified, captured, and communicated so process owners, sponsors, and executives can improve decision-making. Information and communications is a top-down and bottom-up process from the enterprise, program/project/process to the product/transaction levels. The desired result is that critical stakeholders are fully informed for smart decisions making, which may involve:

- **Closed lop information stakeholders:** Information can be passed feed backward or feed forward. The purpose is to ensure than once an unmitigated risk is identified, then the appropriate person is notified to respond appropriately.
- **For your information (FYI) stakeholders:** Communication is passed across the enterprise or process to inform or status critical stakeholders.
- **Escalation (approval) stakeholders:** Bottom-up communications is passed to appropriate stakeholders for approval, additional resources, etc.
- **Command and control stakeholders.** Top-down communication is passed to appropriate stakeholders to intervene, correct, prevent, predict, or preempt risks and to architect, design, and deploy appropriate controls.

RISK-BASED DECISION MAKING TIPS

- Work on the right problem.
- Specify your objectives.
- Create imaginative solutions.
- Understand the consequences of decisions.
- Grapple with your tradeoffs.
- Clarify your uncertainties.
- Think hard about your risk tolerance.
- Consider linked decisions.
- Focus on 'Doing the Right Things' and 'Doing Things Right.'

RISK BASED DECISION-MAKING

Risk based, decision-making is the smart way to manage an enterprise, program, project, or process. The following are hard lessons learned of risk management:

- Uncertainty is increasing.
- Determining risk appetite is often difficult.
- Following standard methodology ensures consistency.
- Determining appropriate risk/controls is time consuming.
- Risk management is cultural.
- Risk management requires new decision making and operating styles, which requires cultural adaptation.
- There is no such thing as perfect information to determine risks and to manage them.
- There are always unknown or unknowable risks.
- There is no such thing as 0 risk.
- Threshold for assuring and/or determining risk is 'reasonableness.'
- Technology solutions are easy – People adaptation is difficult.
- Irony: Goal is to reduce risks but change requires new behaviors, which can result in increased variation and increased risks if not handled properly.

MONITORING

The Monitoring step can be considered a form of enterprise risk oversight, assurance, and control. The enterprise risk management process must be monitored because risks, controls, and people change; uncertainty increases; and variability/variances increase. Enterprise, programmatic/project/process, and transactional/product systems are dynamic and uncertain. The enterprise may have new management, new business model, new supplier, new tone at the top, new culture and a number of other factors that can increase or decrease risks. Processes also may become unstable and incapable of meeting business objectives. Programs and projects may have additional scope, quality, schedule, and cost variances, each of which is a risk hazard. Also, products and trans-

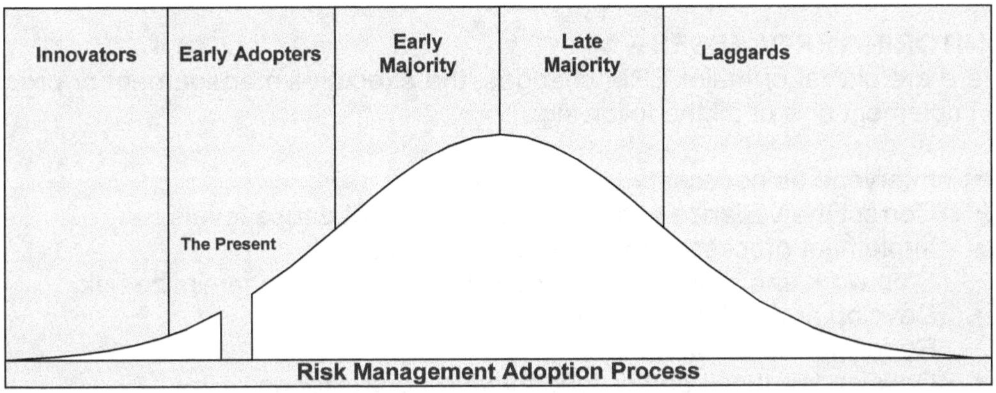

actional systems may be redesigned, which may have added failure mode and effects risks.

So as systems change and act dynamically in new and unexpected way, the ERM system needs to be monitored, continually evaluated, and if deviating from objectives should be modified. The following types of monitoring are possible:

- Board level governance and oversight.
- Executive management fiscal, schedule, scope, and quality monitoring.
- Third party objective and independent attestation and assurance.
- Second party, internal audit objective assurance.
- First, party, self-monitoring.

When organizations adopt Enterprise Risk Management, it diffuses at different rates as shown in the above figure 'Risk Management Adoption Process.' The further ERM diffuses in the organization, the higher the risk maturity and the higher the process stability, which implies higher controls. In general, an organization that has extensive controls with widespread adoption of ERM requires less intensive monitoring but more oversight assurance.

WHEN TO RE-EVALUATE RISKS

Risks change and controls degrade. The risk management and response plans have to be periodically evaluated. So, when should risk responses be reevaluated? The following are several tips for evaluating risk management:

- New business objective.
- Risk appetite/sensitivity changes.
- New risk is identified.
- Existing risk changes.
- Key input variables change.
- Key internal/external variables change.
- New information sources.

MONITORING PROCESSES

If there are critical or major ERM changes, the executive management or process owner can implement one or all the following:

- Intervene as necessary.
- Correct the variance at the symptom and root cause levels.
- Implement processes to prevent recurrences.
- Stop work and redesign the process, product etc. to minimize risk.
- Develop new risk responses.
- Redesign the ERM system.
- Develop feedback and/or feed forward control loops.

Monitoring should be appropriate and suitable to the recipient, business objectives, type of risks, risk appetite, and control effectiveness. For example, executive oversight may focus on:

- Develop standardized risk processes.
- Ensure they are followed consistently.
- Develop appropriate controls.
- Identify interdependencies and 'white space' risks.
- Present risk information in a consistent and timely manner.
- Promote proactive, preventive, predictive, and preemptive risk management.

VERIFICATION AND VALIDATION OF RISK AND CONTROLS

The control of risk can be at the 1. Enterprise level; 2. Programmatic/project/process level; and 3. Product/transaction level. Control can be anything that mitigates real and potential conditions that can cause injury, illness, or death to people; degradation of the environment; damage to or loss of a system, equipment, or property.

Control efficiency, effectiveness, and economic can be assessed in the following manner:

- **Validated:** Controls and requirements that are unambiguous, correct, complete, and verifiable.
- **Verified:** Controls and requirements that are objectively determined to have been met by the design solution.
- **Recommended:** Controls that have the potential to mitigate a hazard or risk but have not yet been validated as part of the system or its requirements.[13]

SUMMARY

Today's biggest challenge is not quality, cost, or schedule. It is uncertainty surrounding competitiveness, globalization, technology, regulation, mergers, and acquisition. Uncertainty presents both risk and opportunity. ERM provides a framework for critical organizational stakeholders to deal with uncertainty and associated risks.

ENTERPRISE RISK MANAGEMENT 101

All companies and governmental organizations face uncertainty and risk. The basic premise of ERM is that every company and organization exists to provide value to its stakeholders. Today's fundamental challenges is to determine how much uncertainty a business and government entity is willing to accept as it strives to grow and mitigate unexpected events.

ERM properly architected, designed, and deployed can provide the following value to an organization:

- Align risk appetite to enterprise strategy.
- Ensure deployment of strategy to operational tactics.
- Identify high-risk areas within the organization.
- Identify controls to mitigate these risks.
- Minimize operational surprises and losses.
- Identify 'white space' and interdependent risks.
- Provide integrated responses to multiple risks.
- Seize opportunities.
- Rationalize deployment of capital.

CHAPTER 5
PROCESS MANAGEMENT 101

Value Added Auditing is based on process management and process control principles. Process management is the control of variability and ultimately operations risk. The value added auditor needs to know process management fundamentals to evaluate business, operational, and supplier risks.

PROCESS FUNDAMENTALS

People support what they help to create.
Anonymous

It is essential the value added auditor is familiar with the auditee's industry, company, business units, organizational structure, systems, processes, and products. This information may be obtained during the planning stage, specifically during a preliminary interview or survey. This background information says who does what, where, when, and why. It also saves time and reduces frustration because the value added auditor can focus on conducting the audit, not on researching the basics, which should have been done at headquarters or in the auditor's office.

WHAT IS A PROCESS?

There are a number of definitions of a process. However, these are the common elements to most definitions including:

- Series of actions with a start point and an end.
- Sequence of related activities or tasks that lead to an outcome.
- Sequence of steps that transforms inputs into value added outputs.
- Inputs involving people, machines, material, environment, and methods that add value.

Nortel Networks, a global and integrated telecommunications company defined a process as:

> "an organized group of related activities that together create an outcome of value. A holistic view of a process includes multiple components that must be effectively defined and integrated.

'ADVERTISE AND AWARD' FLOWCHART

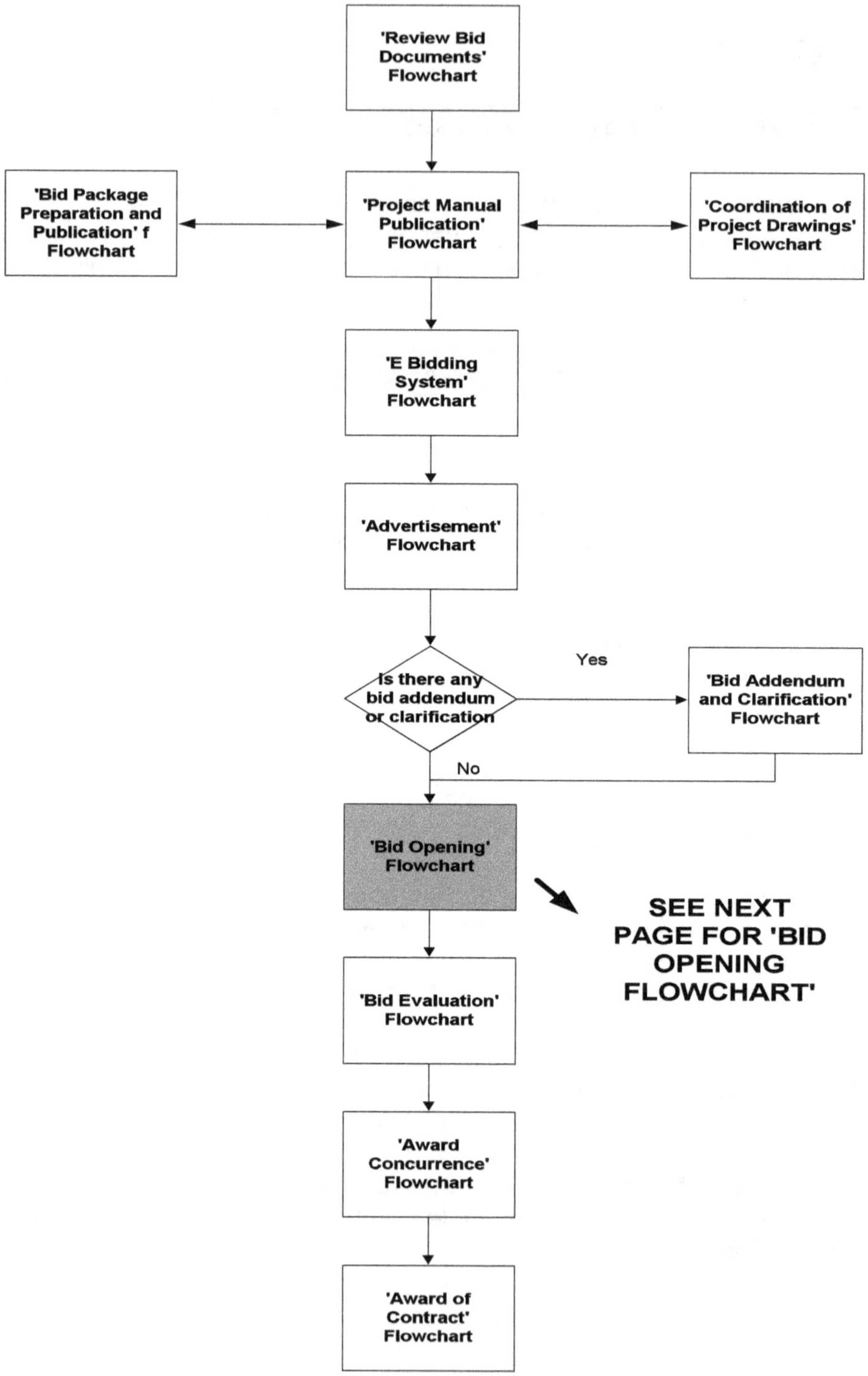

SEE NEXT PAGE FOR 'BID OPENING FLOWCHART'

❑

BID OPENING FLOWCHART

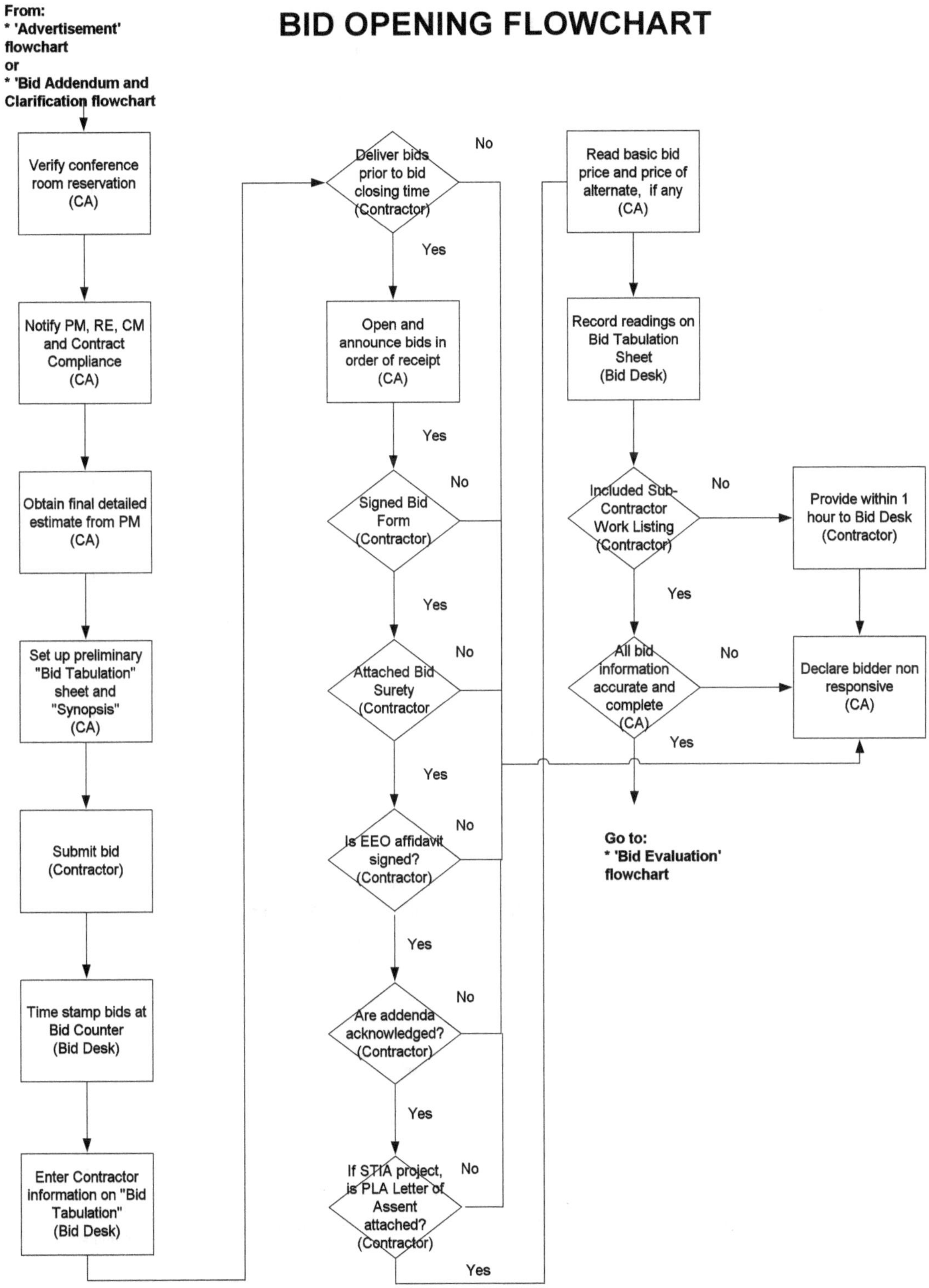

TUBE/PIPE PRODUCTION FLOWCHART

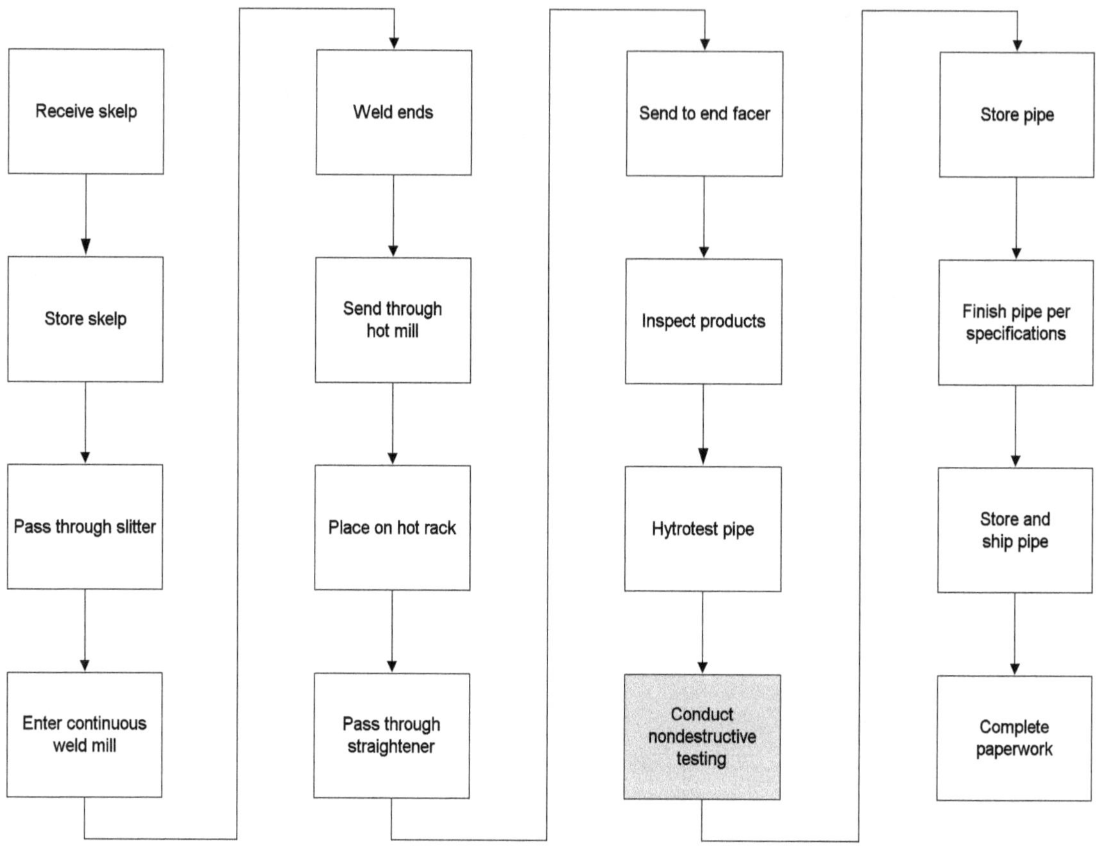

SUB FLOWCHART FOR 'CONDUCT NONDESTRUCTIVE TESTING

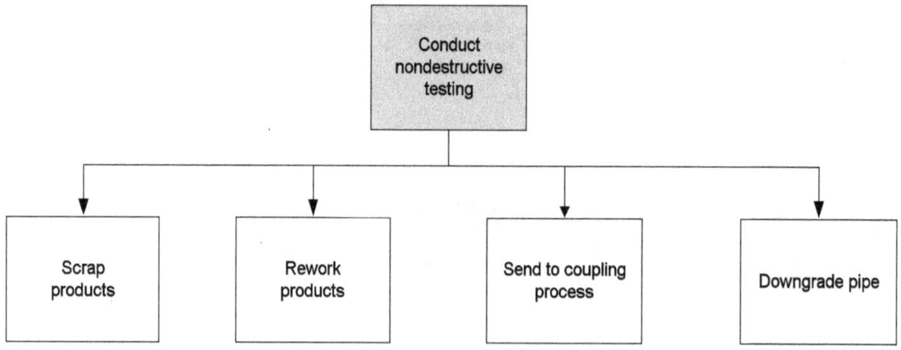

Inspection Versus Process Control

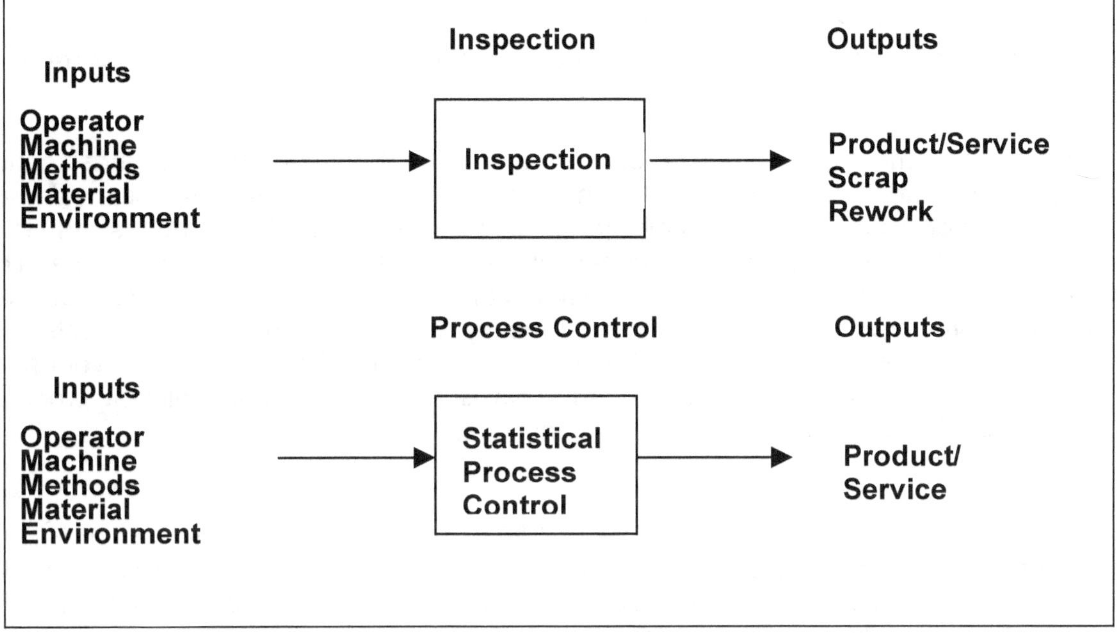

- Vertical integration: all components within the process.
- Horizontal integration: between two or more processes".[1]

The previous pages show flowcharts of a service process, specifically contract administration and of a manufacturing process, specifically producing pipes. See 'Advertise and Award Flowchart Process' and 'Bid Opening' figures on the following pages. Note that the 'Bid Opening' flowchart is a sub flowchart. Also, 'Tube/Pipe Production Flowchart' and 'Sub Process Flowchart for Conduct Nondestructive Testing' figure) are shown in the following pages.

PROCESS MANAGEMENT FUNDAMENTALS

Process management and control is fundamental to value added auditing, risk management, and internal control. The following are the assumption to most business process management:

- All work follows a process.
- Each process adds value.
- Each process has inputs and outputs.
- Each process has a supplier(s) and customer (s).
- Each process can be controlled, made consistent, and improved.
- Wastes are eliminated.
- Risk is managed and controlled.

- Quality, service, delivery, service, performance, efficiency and economics are measured.
- Each process can be documented and measured.
- If process works as intended, the outputs should conform to customer requirements.

Process management and process control add value because they emphasize risk management, defect prevention and consistency of output. See 'Inspection Versus Process Control' figure in the previous page. In contrast, inspection emphasizes 'after the fact' control. The inputs to both inspection and process control are the same. The difference is that inspection output hopefully is conforming products, not scrap, and rework. Using process controls, output should mainly consist of conforming products. If a process is continually being monitored, controlled, and adjusted, it should prevent nonconforming products from being produced. This is an effective and efficient process. Also, the risk of a nonconformance is also minimized.

POWER OF STANDARD DEVIATION

Standard deviation (SD or σ) is a measure of variation, the spreading tendency of data and is fundamental to process management. process decision-making, and value added auditing. A large standard deviation indicates greater variation in data than a small standard deviation. The symbol for standard deviation is a sigma, σ. The study of six sigma and variation is beyond the scope of this book, however the following is a brief introduction to the topic.

Most six sigma models are based on DMAIC, which stands for Define – Measure – Analyze –Improve – Control. Essentially, statistics is fundamental to six sigma analysis and process decision-making. For example, six sigma statistics provides evidence that a shipment of incoming products conforms to specifications and the process producing the products is running naturally.

Six sigma is based on the science and study of variation. Standard deviation affects the shape of a normal curve. A large standard deviation results in a flat and broad normal curve. A smaller standard deviation results in a peaked and narrower normal curve. However, the relative shapes of the two normal curves are the same, both bell shaped, symmetrical around the measurements, and diminishing toward the tails of the curves.

Six sigma is an objective organizational improvement system for evaluating business and process performance. Six sigma tools help eliminate subjective evaluations and allow the team or operator to monitor and control a machine or process. It provides assurance to the value added auditor of process management and control.

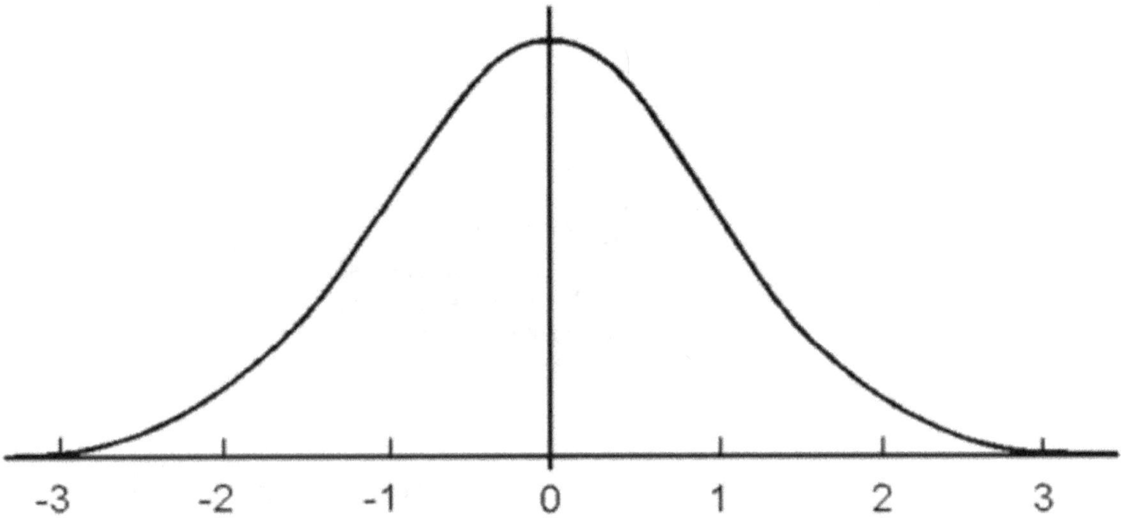

Bell Curve With Standard Deviations 1

Statistics and six sigma consist of powerful decision-making tools. However as the above quote illustrates, they should be used carefully. Six sigma involves the collection, analysis, and interpretation of data from a machine, service, or process being monitored. Six sigma incorporates process control, capability, stability, prevention, and improvement.

The above 'Standard Bell Curve' figure illustrates the shape, symmetry, average, and standard deviations of a generalized normal curve. As seen in the figure, the laws of probability state that a specific percentage of the measurable population will fit within certain standard deviation (S.D. or σ) limits:

- Plus or minus 1 standard deviation has 68.27% of the data.
- Plus or minus 2 standard deviations have 95.45% of the data.
- Plus or minus 3 standard deviations have 99.73% of the data.

In terms of risk, the bell curve should be centered in the middle of the specification spread, between the upper specification and lower specification. Statistically if a standard bell curve, as shown in the above 'Less Risk – On Target with Less Variation,' is in the middle of the curve, there is less probability of risk events or nonconformances occurring.

Understanding fundamental statistics is the basis for conducting value added audits. While value added auditors don't need to be statisticians and don't own process varia-

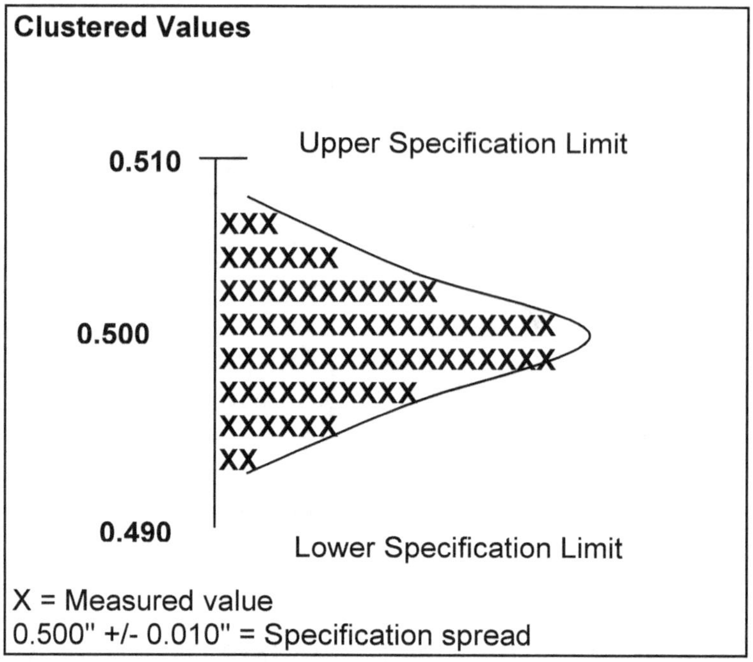

tion, they should be able to determine if system, process, or product variation is normal or abnormal. Then, the auditors need to be able to evaluate if the process owners are controlling it appropriately and suitably.

PROCESS MANAGEMENT DECISION MAKING

Statistician: A person who can go directly from an unwarranted assumption to a preconceived conclusion.
Anonymous

Process owners are now responsible for managing and controlling their work. For example, personnel would perform process control studies to evaluate whether a machine or process was capable of making products within the specification limits.

The following are the fundamentals to process based, decision-making:

- Value is defined in specifications or other documentation.
- Processes and products vary.
- Variation in processes and products can be measured.
- Normal variation follows a bell shaped curve.
- Abnormal variation distorts the bell shape.
- Causes of variation can be isolated and identified.
- Process variation can be controlled.[2]

VALUE IS DEFINED IN SPECIFICATIONS OR OTHER DOCUMENTATION

Each business process does something that adds value. Value is defined in terms of being able to meet a business objective, metric, or standard. For example, manufacturing, quality, and engineering may jointly establish standards for a product, which are then deployed to a process. Important characteristics for a threaded fastener may be pitch depth, number of threads per inch, major pitch diameter, and minor pitch diameter.

An organization first determines the requirements and expectations of the final customer. Several methods commonly used are:

- Customer interviews.
- Customer surveys.
- Field service reports.
- Quality studies.
- Warranty reports.
- Field return reports.
- Nonconforming material reports.
- Product audits.
- Quality function deployment.

PROCESSES AND PRODUCTS VARY

All processes vary, whether they are service or production processes. Help desk personnel may answer a phone differently, have different level of knowledge, or have different attitudes. In a manufacturing process, no two machines are the same. Even two parts made from the same machine are not identical. There is always some variation, which may only be detected at the microscopic level.

As was mentioned, process management and internal control are the study of variation. In a production process, it is the study of the dispersion of measurements in a product characteristic or dimension. Many statistical studies in manufacturing and business have shown that measured data have a tendency to cluster around a measurement point and be dispersed around this point, as shown in the 'Cluster Values' figure in page 100.

VARIATION IN PROCESSES AND PRODUCTS CAN BE MEASURED

Process and product variation can be detected and measured by critical measuring instruments, some of which can measure a dimension to a millionth of an inch until variation is detected. To ensure that external factors don't influence the measurement, critical measurements may be performed in special rooms where humidity, particulates, and temperature are carefully controlled.

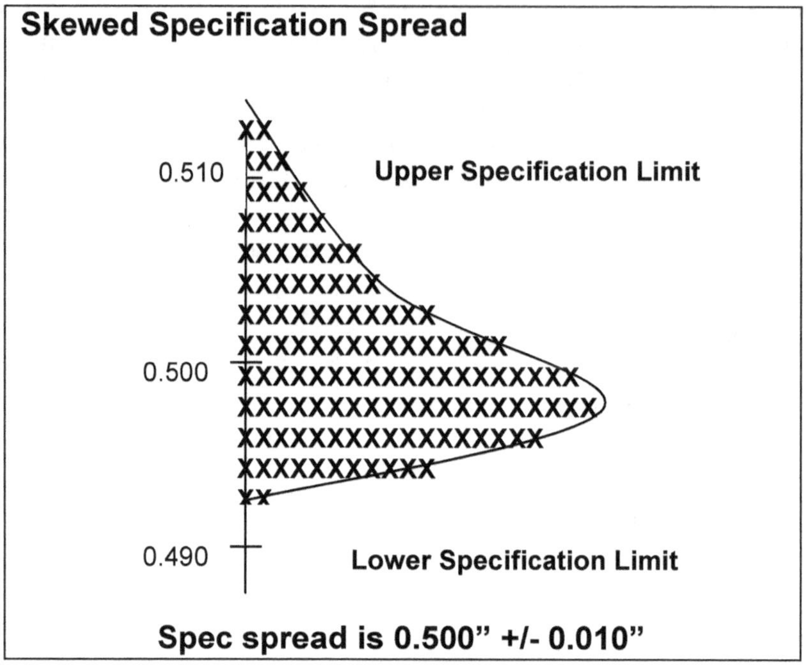

NORMAL VARIATION IN MEASUREMENTS FOLLOWS A BELL SHAPED CURVE

After many observations of similar patterns of variation, mathematicians saw that variations followed a regular, symmetrical, bell shaped pattern. In 'Clustered Values' figure on page 100, a bell-shaped curve is superimposed over the measured points. The bell pattern is a distribution one would expect if there were no abnormal factors influencing the process. The bell shaped pattern is called the normal curve. As mentioned, variation is measured in terms of standard deviation.

ABNORMAL VARIATION DISTORTS THE BELL SHAPE

Variation due to unnatural, abnormal factors distorts the bell curve. For example, measuring instruments that read high would move the distribution toward the top of the specification spread, as can be seen in 'Skewed Specification Spread' figure on the next page. The challenge is to determine whether process variation is due to normal or abnormal causes.

CAUSES OF VARIATION CAN BE ISOLATED AND IDENTIFIED

Variation is detected and controlled through statistical process control (SPC) and other types of internal control. This method, based on statistical principles, allows an operator to detect deviations in a measured value and then adjust the process back to the target dimension.

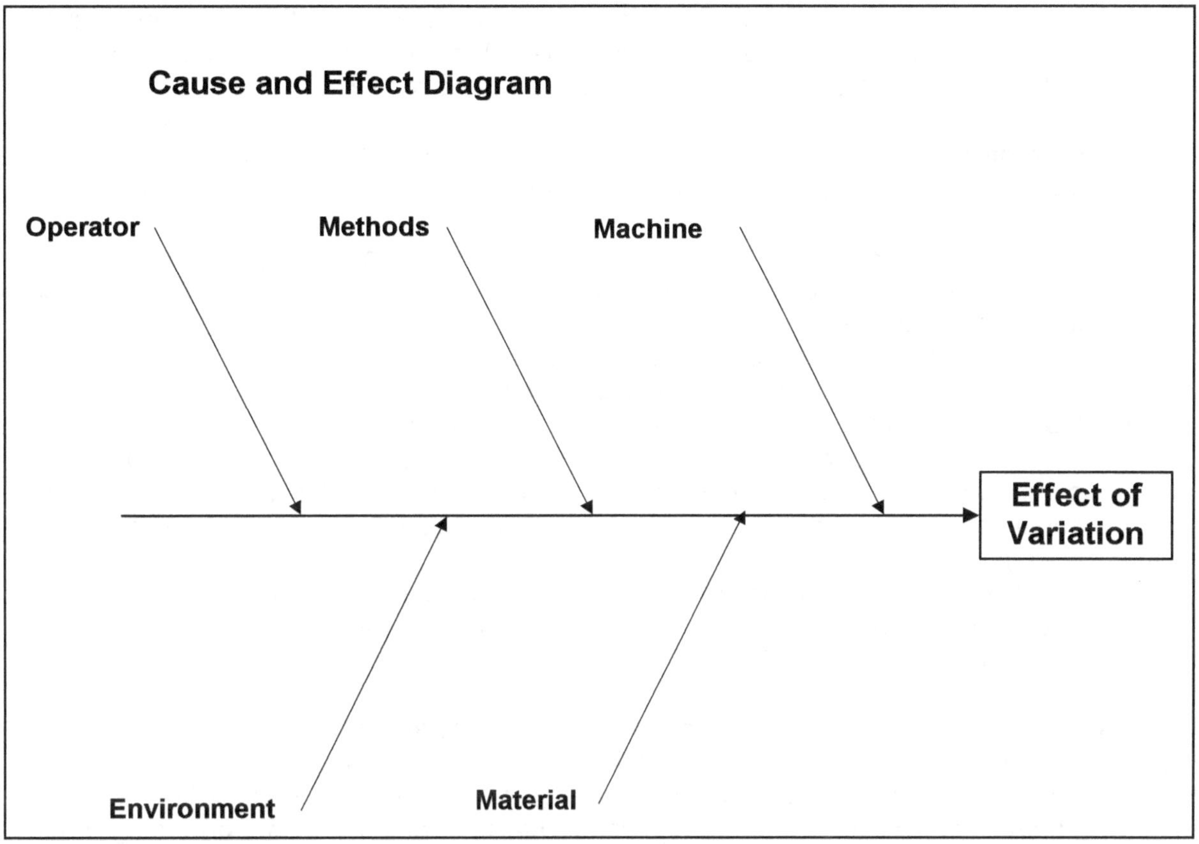

Using control charts as shown in 'Skewed Specification Spread' figure on the previous page, the operator of a machine process can determine if the variation is normal or abnormal. These charts tell operators and process owners whether to intervene and remove abnormal causes or whether to leave the process alone.

So, process variables as measurable characteristics can be understood if variation and its causes are understood. Variation can be caused by any of the factors identified in a cause and effect diagram, such as materials, machines/process, methods, operator, and environment. In this way, a team or process operator can work on ensuring the system or machine is stable. If the machine or process shows unexpected variation, then the operator examines the cause, eliminates the symptom and root cause, and stabilizes the process again. This becomes a form of management by exception where the supervisor, team or operator focuses on the unexpected process variation.

PROCESS VARIATION CAN BE CONTROLLED

Let's look at what process control may look like at the activity level for something like a machining operation. It all starts with internal control or critical process variables.

Internal controls check for unwanted, abnormal, or excessive variation. Internal controls check that there is uniformity, meaning little variation in the inputs, the process, and the outputs. This may be better understood by examining the following:

- **Operator:** Operators use similar and approved procedures to produce defect-free parts. To ensure uniformity, operators are trained to follow similar procedures.
- **Methods.** The methods of moving, storing, inspecting, sampling, segregating, and machining are completely and accurately spelled out in standards, specifications, or procedures.
- **Machine.** Machines are statistically controlled and capable of meeting specifications. Statistical process control is discussed later in the chapter.
- **Material.** The material used in the process has uniform physical, chemical, and metallurgical properties. Material can include raw, partially finished, supplied, or components. This is controlled ensuring that the raw material suppliers have a continual process improvement program and use Statistical Process Control (SPC).
- **Environment.** The environment in which processes are located is safe, clean, and stable. Stable means that humidity, particulates, temperature, and airflow are monitored and controlled.

Any one of the above inputs is a source of variation that can affect the level of assurance and confidence. Depending on the criticality of the audit, the value added auditor may have to check each of the above controls to ensure that the total process is being sufficiently controlled. This is a commonly used quality tool called fishbone diagram or cause-effect diagram, as shown in the 'Cause and Effect Diagram.'

Depending on risks and the required level of assurance from the audit, the value added auditor may have to assess process controls and the inputs and outputs of a process. Specifically, the value added auditor may determine that process inputs are a major cause of abnormal variation. For example, methods between operators may change, machines may have different characteristics, and raw materials may differ in chemical and physical properties. Each is a source of variation that may have to be controlled.

PROCESS ANALYSIS

Statistics are no substitute for judgment.
Henry Clay

The concepts of statistics and variation were not on management's radar screen until recently and only until Walter Stewart and W. Edwards Deming introduced the business benefits of process analysis. In particular, W. Edwards Deming popularized the concept that variation is an indispensable part of all business decisions.

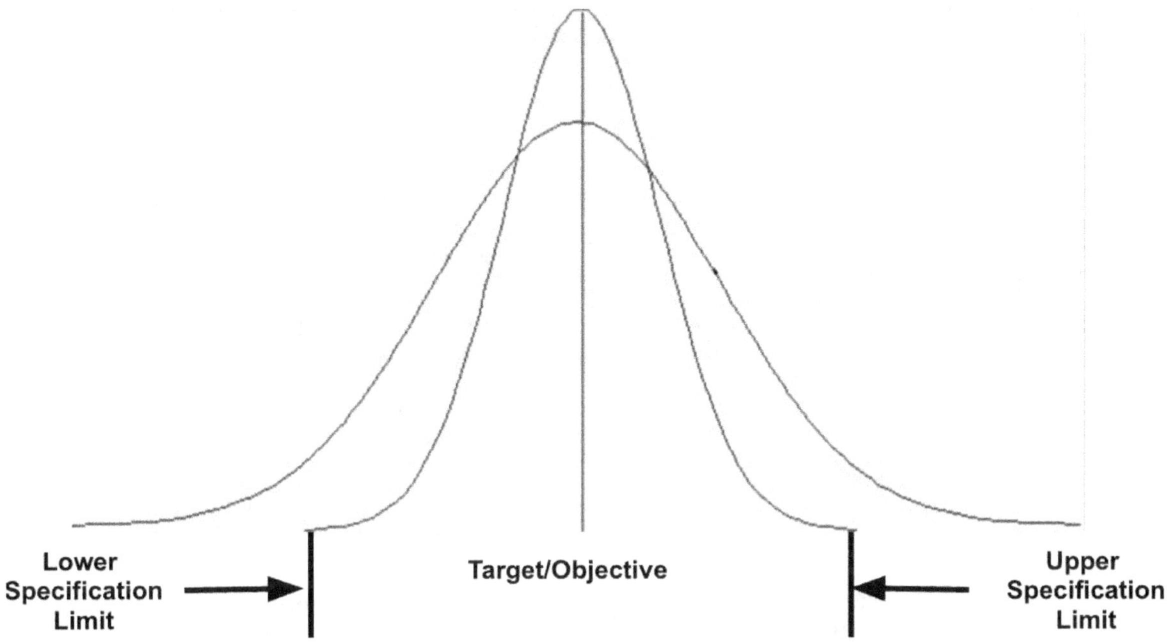

Less Risk - On Target
With Less Variation

STATISTICS AND VARIATION

Value determinations are only as good as the information and data generated. Conducting value added audits requires that large amounts of data are collected, analyzed, deciphered, and acted upon. Decision-making and value added auditing are only as good as the quality of the process information and data generated.

Data and information therefore must be accessible and understandable to management, teams, and all employees. Statistics can be used to make data and information understandable for value added decision-making.

Statistics involves collecting, processing, and then presenting data in an understandable form. Statistical analysis provides techniques and tools for studying variation and patterns by examining data samples to estimate characteristics of the larger population.

Statistics by themselves should be used and interpreted carefully. For example, some data even though it is reported to six significant decimal places, may be suspect. Data may be generated by a measurement instrument to 0.00001 inch, but if the instrument has not been calibrated properly or recently, measurements may be inaccurate. Or, a worker or team may not have been instructed in its use. Also, data may not represent typical conditions or circumstances.

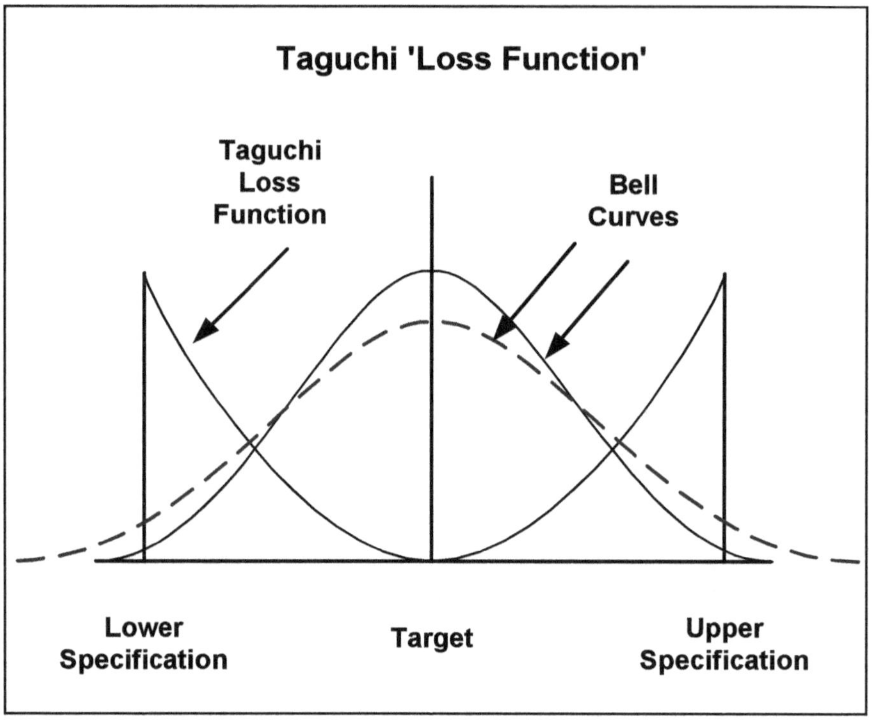

VALUE ADDED AUDITING AND THE STUDY OF VARIATION

A business objective may be defined as a variance or distance from a business objective, metric, or standard, all of which indicate risk. Value added auditing is the study of variation and risk, how to detect, measure, and control it. For example, quality that can be specified in terms of a dimensional tolerance or a surface finish is a variable that can be controlled and ensured. If a target product dimension can be kept in the middle of the specification spread and the variation of measurements are distributed inside the specification limits and process control limits, then dimensional quality can be controlled as illustrated in the 'Skewed Specification Spread' figure on page 102.

In much the same way, the relationship of variation and risk is illustrated in the following sidebars: 'Less Risk – On Target with Less Variation' and 'Higher Risk – On Target with More Variation.' The business target or objective is in the middle of the bell curve. There are upper and lower specification limits. When the bell curve is narrow, tight, and centered at the business target, then there is less risk of a nonconforming product and less variation. If the bell curve becomes flatter and has more spread, then there is higher risk and more variation resulting in more nonconforming products.

Taguchi also developed the concept of the 'loss function', which is illustrated in the above 'Taguchi Loss Function' figure. The 'loss function' is a parabolic curve that is centered in the middle of the curve. There is no monetary cost or loss at the middle of the curve, which is also located in the middle of the specification spread. Loss, cost, and risk increase as products are produced that are beyond the target value of the specification.

CONTROL OF THE 'CRITICAL FEW' PROCESS VARIABLES

Value added auditors should also be able to focus their attention on the 'critical few' items that can create the highest risk or greatest variance as opposed to the 'trivial many' variables in a process. This is called the Pareto principle or the 80 – 20 rule. The following may explain this concept. Service deficiencies or product nonconformances can be classified in terms of variation away from a business objective, standard, or target. The following three categories are often used:

- **Critical.** Critical risks and nonconformances can adversely affect product function and can result in a loss of life. A material crack that compromises structural integrity would be a critical defect.
- **Major.** Major nonconformances can cause a product to malfunction without a loss of life. A major nonconformance lowers performance or decreases product marketability.
- **Minor.** A minor nonconformance does not meet a specification but does not affect performance, 'fitness for use' or customer acceptance. A surface blemish on the inside of a product is an example of a minor defect.

As can be seen, the value added auditor should have sufficient product and process knowledge to detect significant variances in a project or the critical features of a product. For example, a product may have different types of defects. A product with 100 characteristics may have 10 critical, 10 major, and 80 minor nonconforming classifications. One nonconformance in a critical product characteristic always rejects the product. One major or critical defect is usually sufficient for a product to be rejected.

COMMON OR SPECIAL CAUSES OF VARIATION

Facts speak louder than statistics.
Justice Streatfield

There are basically two types of variation:

- Common causes.
- Special causes.

COMMON OR CHANCE CAUSES

Common cause variation, also called chance cause variation, is due to the natural and random changes in a process. These points are also inside the curve's control limits. See 'Out of Control Conditions' sidebar.

Chance causes are an inherent part of the process and usually cannot be identified or eliminated by an operator simply adjusting a machine. Fluctuations of material composi-

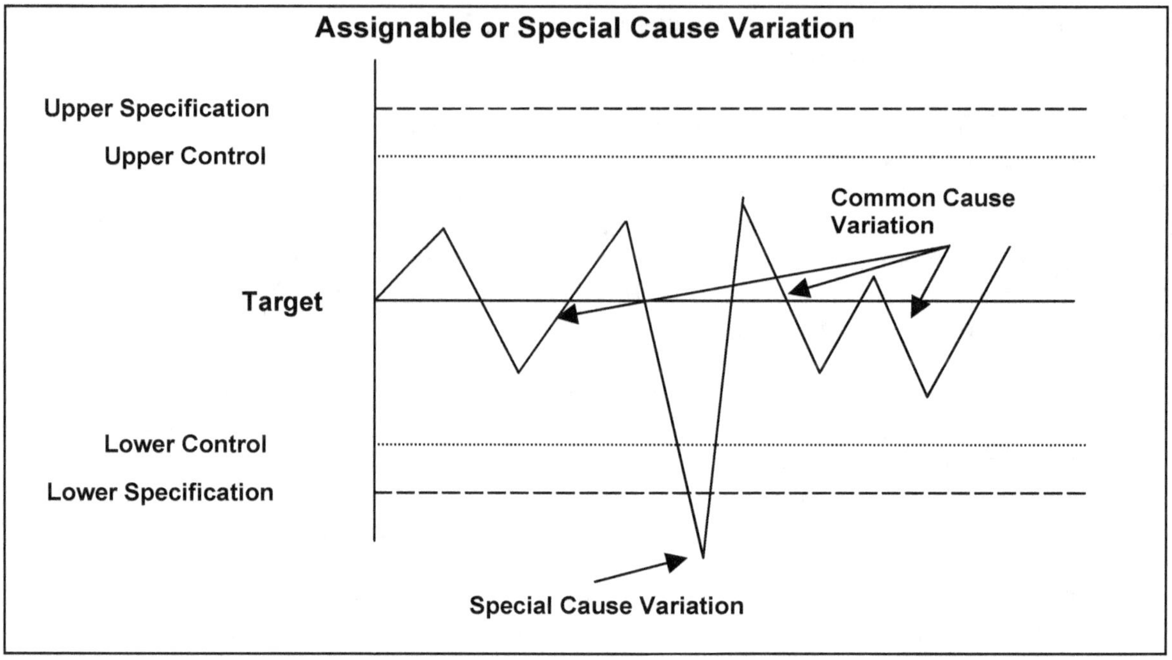

tion within a piece of raw material, changes in tool deflection, or variations in weld penetration are examples of chance causes.

W. Edwards Deming and Joseph Juran maintained that 85% of the chance or system causes are the responsibility of management. These problems can only be reduced by management action. For example, an operator cannot consistently machine products so they conform to specifications if the machine is old and shows inherent vibration. The solution is to buy a new machine or widen the specification spread. Both actions require management attention and action.

SPECIAL OR ASSIGNABLE CAUSES

Special cause variation, also called assignable cause variation, is due to unusual or identifiable causes in a system. These causes of variation can be detected and usually can be identified as a point outside of the specification limits or as an abnormal trend in a process. See above 'Assignable or Special Cause Variation' figure.

Assignable causes of variation are introduced from outside a process. These causes either increase or shift variation in unpredictable ways. Assignable causes result from untrained operators, poor instructions, or poor equipment.

VARIABLE AND ATTRIBUTE DATA

Statistical process control (SPC) charts are based on the type of data obtained from a process, which can be either variable data or attribute data. The diameter of a bore, thickness of this book, or today's temperature are examples of variable data. A variable is any product characteristic that can be measured over a range of values.

Variable data and information are obtained by tracking a manufacturing process over time. An operator or inspector continually samples, measures and compares dimensions of parts to ensure they comply with specification data shown on the engineering drawing.

Conforming or non-conforming, go or no/go, good or bad, and acceptable or not acceptable are examples of attribute data. An attribute is a characteristic that the product has or does not have. A surface blemish is an example of an attribute type of data. An inspector will determine whether the blemish affects the product's marketability or function and will make an accept/reject decision based on the examination.

CRITICAL PROCESS QUESTIONS

Hindsight is good, foresight is better; but second sight is best of all.
Evan Esar

The value added auditor wants to know that a process is monitored, controlled, capable and improving. Each of these terms has a specific definition that value added auditors should be familiar with. The problem is that a process can remain stable for years and show little indication of improvement.

So, the value added auditor wants to know:

- Are the major processes in control?
- Are the major processes capable?
- Are the processes improving?

ARE THE MAJOR PROCESSES IN CONTROL?

A process in control is stable. This is usually called statistical control, where its process measurements are plotted inside calculated action limits such as on a variable control chart. If measured points are inside the control limits and follow a random pattern, variation in the process is natural, as shown in 'Out of Control Conditions' sidebar. If the process is running naturally and is stable, the process owners leave the process alone.

When a change in a process occurs, results can be seen on the variable control chart. So, if points follow a nonrandom pattern such as a rising trend or are outside the control limits, an abnormal condition exists which should be immediately eliminated, as shown in the 'Out of Control Conditions figure on page 111.

If the operator recognizes a nonrandom pattern of measurements or a point outside control limits, the operator identifies the cause, adjusts the process, and eliminates the immediate cause. The immediate cause of the unnatural variation may not be the root cause. Eliminating the root cause usually requires more time.

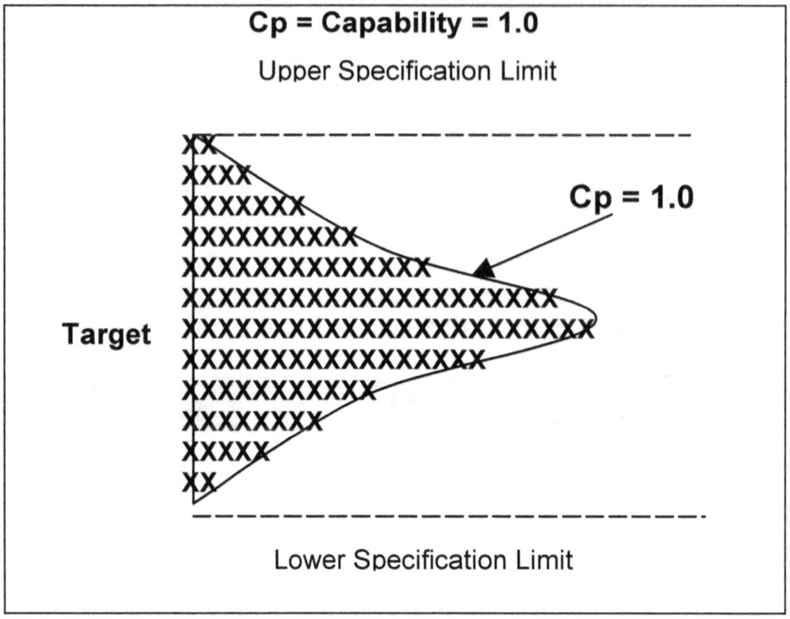

Out of control or abnormal conditions follow recognizable trends on the chart as shown in the above 'Out of Control Conditions' figures on page 111. These trends shown in the figure are:

- Changes or jumps in level.
- Trends or steady changes in level.
- Recurring cycles.
- Two distinct populations.

To understand and identify the causes of the abnormal conditions, thorough knowledge of statistics and of the process are required. A complete analysis of statistical tools is beyond the scope of this book. Its value however, can be illustrated by an example of changes or jumps in a chart. For example, if there is a sudden change or jump in level on a variable chart, possible causes can be:

- **Operator.** A new or inexperienced operator over adjusted a process.
- **Material.** Raw material varies in physical or chemical properties.
- **Method.** Procedures changed between shifts.
- **Machine.** A machine has a new fixture or tooling that changed the basic settings.

ARE THE MAJOR PROCESSES CAPABLE?
Process capability is determined once a process is in statistical control. Abnormal causes have been identified, monitored, and eliminated. Then process spread can be

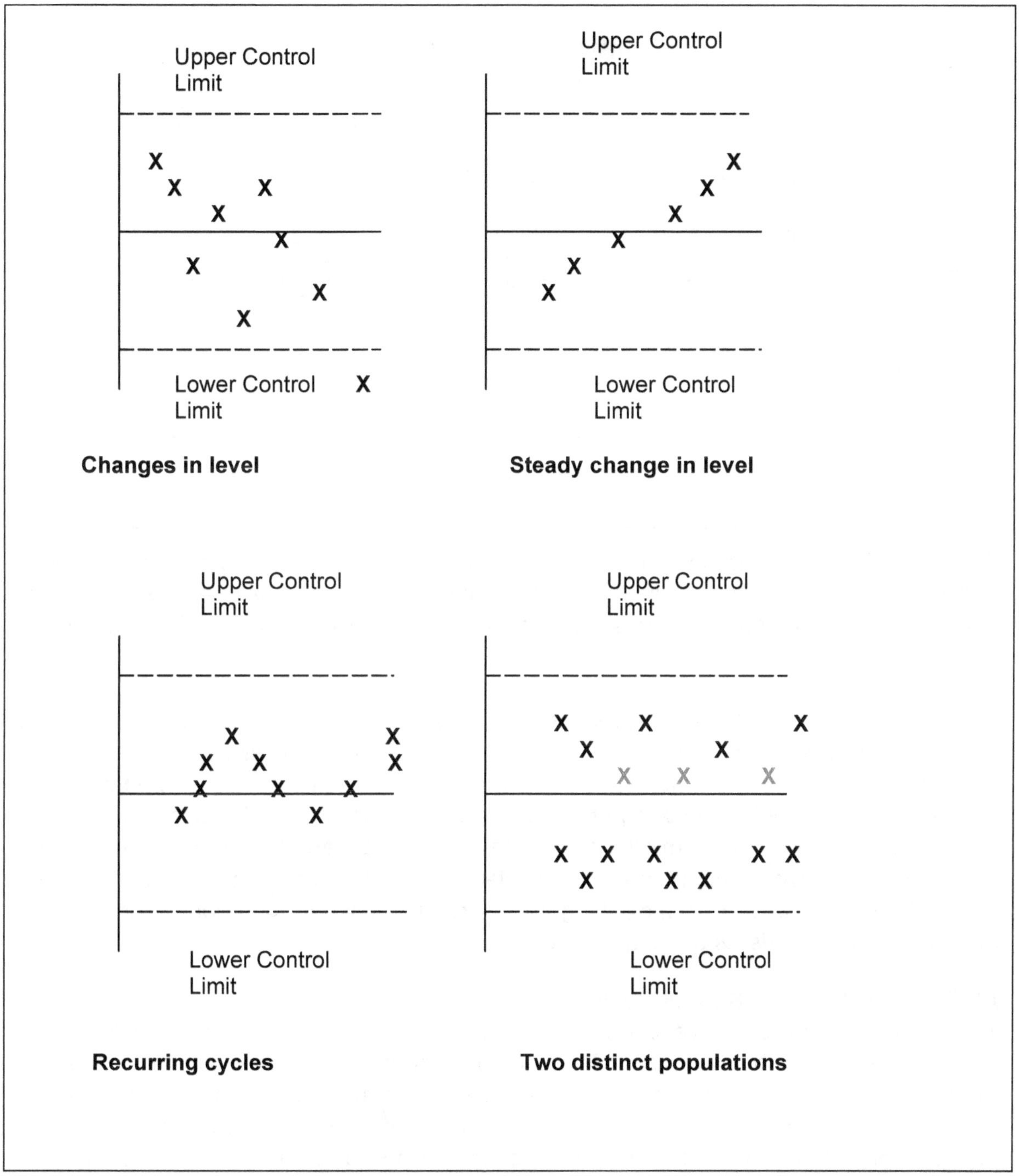

compared with specification spread to determine whether process specifications or objectives can be achieved and consistently met.

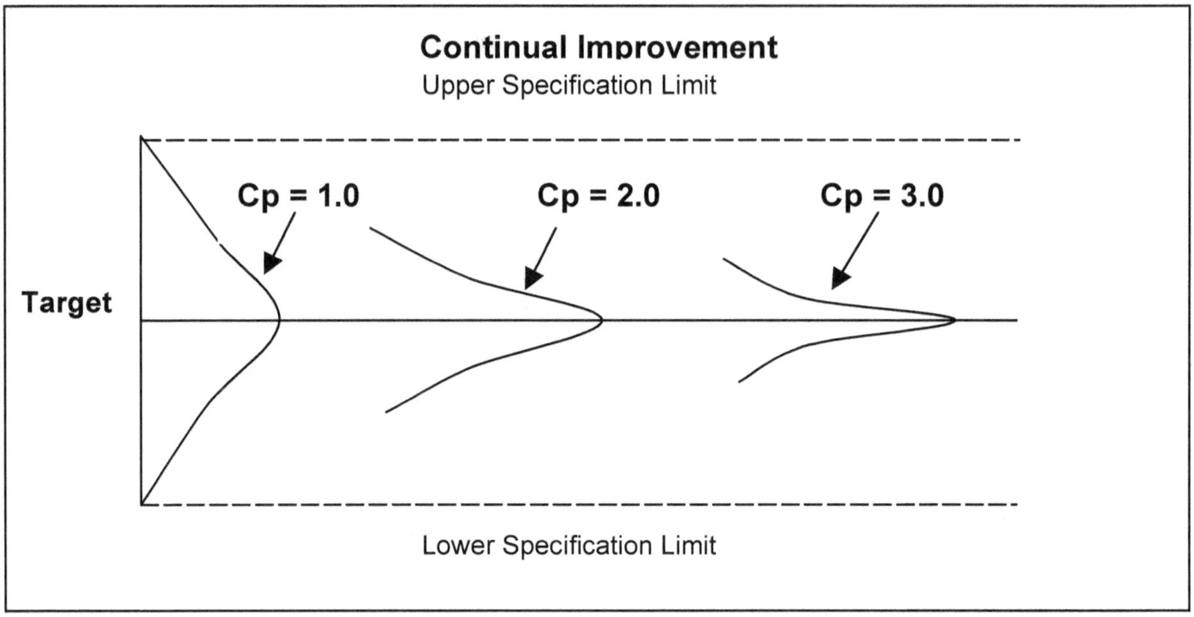

Cp and Cpk are two common measures of capability. Both compare specification spread with process spread. Cpk, however, specifically measures how well centered the process spread is compared with the specification spread. It also measures the tightness of the variation spread.

If Cp = 1.0 and the bell curve representing process spread is centered in the middle of the specification spread, then by definition the process has 99.7 percent of its products within the specification spread, as shown in the above 'Continual Improvement' figure. This is, however, still 3,000 parts per million defectives, which is still unacceptable. If the bell curve is centered in the middle of the specification spread, then there are only parts per million defectives. This is the fundamental element to all six sigma analysis, which implies that the process is sufficiently good to product only 3.4 parts per million nonconformances. This leads us to the next question.

ARE THE PROCESSES IMPROVING?

To improve the process, variation around the specification target, the middle of the specification, has to be minimized. The capability index will increase, which means that the bell curve should be centered in the middle of the specification spread and is narrower, with a higher peak, as indicated in 'Continual Improvement' figure. If this process of improvement continues, the bell curve becomes so narrow that the probability of a point appearing outside the specification limits is in the realm of parts-per-million (PPM).

As discussed, often management action is required to improve a process once it is in statistical control. Management action means focusing on eliminating or minimizing the natural causes of variation, which is a significant conceptual contribution of W. Edwards Deming. Deming, an eminent quality authority, proposed that management controls 85 percent of the process while the operator only controls 15 percent of the improvement. For example, if a machine cannot consistently produce products that conform to a specification, management may have to authorize the purchase of a new machine. Another management option may be to relax or broaden the specification limits.

SUMMARY

If measurements tend to form a bell shape and if the shape is distorted, then the variation is caused by special factors. Statistical decision-making is used to detect, isolate, and eliminate special causes of problems. This is fundamental to all process management and value added auditing.

Good management is all about understanding customer requirements and providing consistent products, then deploying these into a process. Much has been written about statistics and Six Sigma being the panacea for all business ills. This may exaggerate the accomplishments of this prevention tool. Properly used, however process management and statistical decision-making can result in the following benefits:

- Provides objective and measurable feedback on supplier quality performance.
- Improves product flow.
- Pushes control to process owners.
- Eliminates sorting or incoming inspection.
- Improves product quality.
- Reduces need to accept off-standard parts.
- Reduces costs.
- Lowers scrap and rework.

PART II: MANAGING VALUE ADDED AUDITING

VALUE ADDED AUDITING LIFECYCLE(TM)

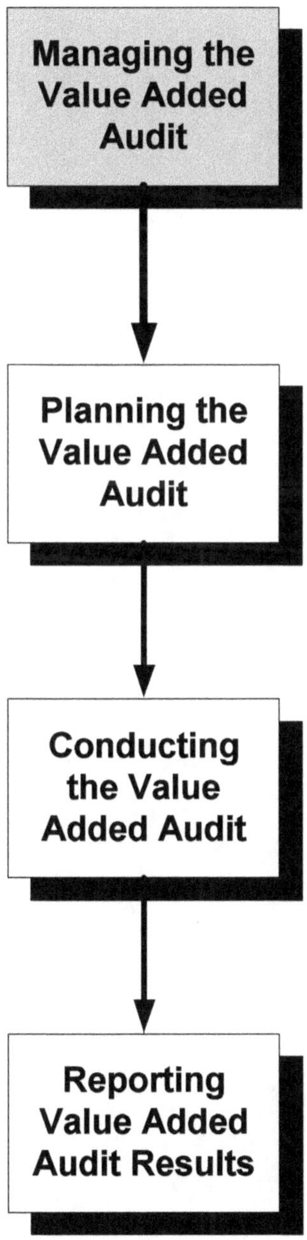

CHAPTER 6
MANAGING THE VALUE ADDED AUDIT

Senior management and board level commitment is a truism about auditing effectiveness and value adding. No commitment - no value added audits. It's pretty straightforward.

SENIOR MANAGEMENT COMMITMENT

The graveyard of business is littered with companies that failed to recognize the inevitable changes.
Anonymous

Successful value added auditing requires the full participation and cooperation of an organization's senior and board level management. If a value added audit is imposed without the full participation and acceptance of an organization's management, it will fail. The organizational governance and improvement benefits of value added audits must be understood.

WHAT MAKES VALUE ADDED AUDITS WORK?

Two words: senior management and the board of directors. Senior management must be committed in action and voice to making it work. The culture of the organization must reinforce and reward this. Sometimes, this is called 'tone at the top.' If the tone at the top of the organization reinforces the importance of value added audits, they will be adopted. It's that simple. If not, they are window dressing to appease regulators or some other party.

What makes senior management want value added audits? There are several external and internal drivers. The external drivers for change may be active shareholders, regulators, and competitors. As mentioned, Congress, New York Stock Exchange, and the Securities and Exchange Commission are requiring value added audits. If the competition is adopting new risk management and control practices, then the board of directors and CEO will change accordingly. This often occurs, when the media reports on fraud, regulators notice off-the-book transactions, or security analysts report high-risk investments.

As well, there are a number of internal drivers. Value added audits simply make good business sense. They result in more transparent, reliable and accurate business decisions. They control and mitigate risks. They decrease time: time to market, time to develop, throughput time, etc. They increase revenues.

WHAT GETS MANAGEMENT'S ATTENTION?

In general, senior management usually focuses on the following:

- **Regulatory requirements.** The cost of regulatory noncompliance can be very high in terms of bad publicity, civil risks, and criminal exposure.
- **Competitor.** A competitor has developed a new product or process. It has taken the market lead, increased market, and increased sales.
- **Customer pressures.** A customer may require process or product as a condition to be on the approved supplier's list.
- **Major product failure.** A product failure may result in a death, injury, or high litigation exposure.
- **High cost of risk and poor customer satisfaction.** Series of small risk events can result in poor customer satisfaction.
- **High rewards from value added projects.** Value added, process improvement and breakthrough projects can result in marketshare increases, lower costs, satisfied customer, or a competitive advantage.
- **Internal benefits.** Internal concerns about operational inefficiencies and ineffectiveness.

THE 'TONE AT THE TOP' – THE BOARD LEVEL

Risk management and process management flourish when senior management is actively involved in their implementation. But, what level of involvement should the board of directors have? Senior management in most companies is aware of the importance of adding value, managing risks, and improving processes.

The board of directors through its auditing committee is already aware of the importance of financial audits and is now focusing on operational and supply chain controls. We're seeing more interest in the importance of risk management, operational risk - controls, operational effectiveness, and process improvement. How is this determined? Value added audits provide this assurance.

The Sarbanes/Oxley Act and Securities and Exchange Commission suggest the board of directors assume a more active role in operational risk management. Why are boards largely passive in regards to process management, risk management, and operational improvement? Probably the major reason is because they view process improvement as an issue that should be addressed by operational management. More often, senior management and the boards of directors want this information.

The IIA standards again outline the role of value added auditors:

> "When internal auditing is accepted and acknowledged by an organization's leaders as a management activity, internal auditors can fulfill their most fundamental role – helping management and the board achieve their objectives. And, competent internal auditing professionals bring to the table objectivity, integrity,

expertise in communication, the ability to identify enterprise-wide risks, and the skill to assess the effectiveness of control put in place by management to mitigate those risks."[1]

ORGANIZATIONAL COMMITMENT

The number #1 indicator of value added audit success is the commitment of senior management, in particular the Chief Executive Officer (CEO) and the Board of Directors. Commitment can range from lip service to active, almost daily involvement. For the audit manager, senior management commitment is the benchmark for determining how successfully and rapidly the benefits of value added audits will be realized.

Senior management may be a different person depending on the scope of the value added audit. For example, if the value added auditor has to investigate fraud or high risk areas, then the board and CEO commitment are critical for a successful audit. If the value added auditor evaluates a supplier, then senior purchasing, production, and quality management support is essential.

An organization gets its messages and rewards from senior management. If employees don't believe senior management is committed, they won't be committed. Executives must be willing to attend value adding auditing courses, learn process concepts, review the status of risks, review internal control effectiveness, implement appropriate technologies, attend critical meetings, and communicate the importance of value added auditing. Senior executives should be willing to support value added audits, risk management, and process improvement principles in their own jobs so they become part of the organizational way of life.

Attitudes and actions that reflect the board of director's policy are then reinforced by strong and consistent messages from middle and first line management. Senior management can emphasize risk management, internal control, and continual process improvement by establishing and developing an improvement ethic and value added culture. If middle management supports the value added audit in policy as well as in action, employees learn that risk management, internal risk -control, and continual improvement are critical to the organization.

However, strong and consistent management signals on the importance of value added audits can be undermined by conflicting operational realities. If management only pays 'lip service' to value added audits, employees receive the reverse message that all the talk of value and improvement are an expedient measure, here today and gone tomorrow. Management may talk about adding value and exhort everyone to action through slogans. But when a conflict arises between product delivery and an audit finding and management says 'shove the product out the door,' the value added audit finding is dismissed because it has been subordinated to operational expediency.

SENIOR MANAGEMENT INFLUENCES

No commitment – no value added auditing. It's pretty straightforward. Senior management leadership and involvement influences the following:

- **Culture.** Senior management cultivates, encourages, and reinforces an organizational culture that values ethics, transparent reporting customer satisfaction, employee empowerment, and continual improvement.
- **Climate of trust.** Only leaders create an environment of trust.
- **Importance.** Senior management emphasizes the importance of improvement practices, principles, and behavior. Senior management reinforces approved behavior. Senior management establishes clear corporate mission, vision, customer and goals. As well, management integrates risk into critical business issues.
- **Communication.** Senior management has the ability to communicate performance expectations and understanding of strategic and operational plans. Management also communicates and reinforces value added audit messages.
- **Empowerment.** Management can increase the authority and responsibility of people to pursue value enhancement and empower people to think and to act.
- **Recognition.** Only management can recognize and reward individual and team accomplishment.
- **Role modeling.** Leaders role model the behavior they expect from others. They define core values and ensure behavior is consistent with the values.
- **Ethical practices.** Management sets the expectations and policies for the organization's business ethics and standards.
- **Authorize.** Senior management authorizes the development and installation of value added systems, processes, and procedures.
- **Resources.** Resources must be authorized and allocated to value added projects, systems, and processes.

How to Get Management Commitment

Management commitment is required for risk management and process improvement. The critical issue is how to obtain management commitment. It may involve a number of steps, including:

- Benchmark audit leadership in excellent companies.
- Have executive level and board level peer-to-peer discussions on the purpose and benefits of value added audits.
- Commission an internal survey on governance, risk management and internal control effectiveness.
- Conduct a competitive analysis on corporate governance.
- Conduct a risk assessment.
- Obtain case studies on value added implementation.
- Understand and focus on management hot buttons, specifically corporate governance, profitability, customer satisfaction, and regulatory compliance.
- Read the Securities and Exchange Commission, Sarbanes/Oxley Act, corporate governance, Malcolm Baldrige National Quality Award, and ISO 9001 criteria/regulations.

SENIOR MANAGEMENT CONTROL QUESTIONS

1. Is there a formal and documented disclosure process to comply with Sarbanes/Oxley Act?
2. Does the company have a disclosure committee for regulatory reporting?
3. Has the board of director's audit committee and senior management established disclosure protocols and guidelines that comply with reporting and certification regulations?
4. Does senior management have a disclosure process to monitor operational changes, both internally and externally?
5. Is there a process to evaluate how operational changes impact financial disclosure requirements?
6. Has the board of directors and senior management defined and developed a threshold or appetite for risk tolerance?
7. Has senior management developed plans for managing and monitoring the company's exposure to risk and losses?
8. Has senior management developed plans for managing upside or opportunity risk?
9. Has senior management developed plans for managing downside risk?
10. Are risk plans deployed down the organization and into the supply base?
11. Does the board of directors understand, agree, and support management's assessment of risk tolerance?
12. Are organizational risks mapped to disclosure requirements for public reporting?
13. Is there a process for designing and implementing operational internal controls?
14. Are changes (risks) in the following monitored:
    a. New acquisitions and mergers?
    b. New supplier-partners?
    c. New management systems?
    d. New IT systems?
    e. Changes in senior management?
    f. Litigation?
    g. Market/customer risks?
    h. Brand image risks?
15. Are business risk identified for:
    a. Enterprise (entity)?
    b. Primary business units?
    c. Critical suppliers?
    d. Processes?
    e. Key Activities?
16. Are internal controls of above risks based on:
    a. Control environment?
    b. Risk assessment?
    c. Control activities?
    d. Information and communication?
    e. Monitoring?
17. Is there a process for measuring, monitoring, and reporting each risk?

- Discuss issues with a well-known governance and/or value added auditing consultant.

MIDDLE MANAGEMENT COMMITMENT

Middle management is always the wild card in value added auditing. Middle management has the most to gain and the most to lose. Middle management has the most to lose because many middle managers are losing or have lost their basic control function. Remaining middle managers have the most to gain because they have a much broader span of control as well as fewer people to get the work done. If process owners can self manage, self assess, and self correct their work, the middle managers job should become easier.

The challenge is that middle management has to know how to translate corporate risk management and control into compatible systems and procedures. The following are tips for enrolling middle managers into assisting value added auditors to conduct audits and to implement audit findings, suggestions, or corrective/preventive actions:

- Senior and middle managers can establish a partnership to develop strategic risk management and improvement goals.
- Middle managers are held accountable for compliance with requirements, internal controls, and risk management.
- Middle managers are encouraged to lead their staffs in seeking breakthrough challenges in addition to pursuing continual improvement.
- Middle managers are empowered to lead risk management initiatives in their areas of responsibility.
- Middle managers are rewarded by their performance.

FIRST-LEVEL SUPERVISION SUPPORT

First-level supervision is also going through the same changes as middle management. With self-managed teams, senior management is asking what value does first level supervision add to a work environment where process teams are responsible for controlling and managing their work.

In networked computer environments, what is the role of the supervisor? Traditionally, this middle person evaluated designs, managed day-to-day operations, and communicated results to senior management. Now, process team members can evaluate designs. Teams may also self-manage design projects. And, senior management can access project files to control processes.

First level supervision must evolve from supervising to coaching, mentoring, and problem solving. Most importantly, senior management must understand and support the evolution of roles between supervisor and team members. Instead of hierarchal and perhaps adversarial control mechanisms, both groups must trust and depend on each other to get the work done. This is difficult for traditional first level supervision. Some supervisors will adapt and others will fall by the way side.

CROSS FUNCTIONAL COLLABORATION

The functional manager or key process owner is responsible for the processes and products coming out of them. The audit manager and auditor should be sensitive to these requirements. The audit manager may have different objectives than the line manager or core process owner. The line manager is under daily pressure to keep customers satisfied, keep production up, improve quality, and meet regulatory requirements. Value added audits may fall by the way side.

Operational managers and more often self-directed teams must deal on a daily basis with:

- Managing risks.
- Ensuring effectiveness of internal risk - controls.
- Dealing with conflicting customer requirements and messages.
- Satisfying demanding and irritable customers.
- Resolving personnel issues.
- Complying with regulations.
- Dealing with lack of resources.
- Fixing process breakdowns.
- Handling personnel turnover.
- Resolving grievances and other personnel challenges.

CHIEF AUDIT EXECUTIVE

Skilled leaders share traits that elevate them above the majority of managers. While most managers concern themselves with doing things right, leaders focus on doing the right thing.
Warren Bennis, educator

The top audit person, the chief audit executive, may be at a director or a vice president level. This person's major responsibility is to work with senior management integrating customer concerns into the company's business strategies, goals, and plans. Linking the organizational strategic plan and the audit function's plan may be the most crucial function of the chief audit executive.

CHIEF AUDIT EXECUTIVE

Depending on what the organization does, the chief audit executive can be the head of internal auditing, quality auditing, chief risk officer, or chief compliance officer. This person should report and communicate risk management conclusions and recommendations to senior management and the board of directors.

The chief audit executive should prepare a statement of the adequacy of internal operational controls to mitigate risks and report these to senior management. In this way, senior management has the information to strategize how they will manage risk, specifically decide to accept, mitigate, control, or reduce it.

CHIEF AUDIT EXECUTIVE STRATEGIC RESPONSIBILITIES

The chief audit executive (CAE) is also responsible for determining how, when, and to whom the audit results are communicated. The CAE in a quality or ISO 9001 (system auditing) function may be the head of quality auditing. The CAE is responsible for strategic audit issues.

The chief audit executive is generally responsible for:

- Reporting and satisfying the board of director's audit committee.
- Integrating quality, internal, customer-supplier, and other audit reports into a consolidated report for senior management and the board of directors.
- Developing the general audit charter of roles, responsibilities, and authorities.
- Obtaining senior management approval of the charter.
- Obtaining adequate resources for conducting value added audits.
- Ensuring value added audits are conducted efficiently and effectively.
- Ensuring value added audits conform to professional standards.
- Ensuring value added auditors have the required knowledge, skills, and other competencies to conduct value added audits.
- Providing direction and instructions for planning value added audits.
- Ensuring value added audits are conducted according to plans.
- Approving changes to the audit plan, implementation, and reporting results.
- Ensuring audit communications are accurate, objective, clear, concise, constructive, and timely.
- Ensuring that business and audit objectives are aligned.
- Ensuring value added audit objectives are met.
- Providing opportunities for developing auditors' knowledge, skills, and other competencies.
- Assuring audit results are acted upon in a timely and efficient manner.
- Determining value added audit work papers adequately support audit observations, conclusions, and recommendations.[2]

CHIEF AUDIT EXECUTIVE TACTICAL RESPONSIBILITIES

The chief audit executive is also responsible for overseeing tactical, value added audit responsibilities. Audit oversight and supervision depends on the proficiency and experience of the auditors and the complexity of the audit. In general the higher the dollar amount, the higher the risk, the higher the level of required auditor proficiency, and the higher the level of oversight. Experienced auditors may be used to oversee the audit work of less experienced auditors.

Audit planning is documented so audit decisions, plans, testing, results, and corrective/preventive actions form an auditable trail of documentation. Audit results are documented and communicated to senior management and the board periodically. Senior management or the management representative may approve overall audit policies and procedures. If there are changes in audit procedures, senior management should be notified in order to review and approve the changes.

The chief audit executive also deals with operational management and with external financial auditors. The chief audit executive has specific tactical level responsibilities, specifically ensuring the auditing function:

- Has auditors with professional knowledge, skills, and experience.
- Has auditors with knowledge of the organization' s industry, systems, and processes.
- Is objective and/or independent.
- Has specialized services.
- Can anticipate and be responsive to the needs of the organization.
- Has reasonable continuity of key engagement personnel.
- Can maintain appropriate working relationships.
- Can achieve contract commitments.
- Can deliver overall value to the organization and to external stakeholders.[3]

CHIEF AUDIT EXECUTIVE SETS DIRECTION

The chief audit executive is also responsible for ensuring that audits are planned, conducted, and reported according to policies and procedures. The lead value added auditor is responsible for conducting the audits according to these procedures.

Again, the goal of all value added audits is consistency. The chief audit executive ensures auditors use their best professional judgment and there are no significant adverse results. If there are significant audit deficiencies, the chief audit executive resolves differences in professional judgment over significant issues, which may involve:

- Reviewing the audit plan.
- Interpreting evidence.
- Understanding auditee's processes.
- Determining need to gather further evidence.
- Resolving ethical issues.
- Determining level of acceptable risk.
- Interpreting documentation.
- Resolving differences between audited and auditor.

MANAGING VALUE ADDED AUDITING

Small organizations usually have one auditor who budgets, plans, schedules, conducts, reports, and reviews value added audits. In larger organizations, value added auditing may be a separate group usually consisting of a small team of professionals.

Managing the Value Added Auditing function is the same as managing any function. Managing involves planning, supervising, evaluating people, problem solving, and decision making. The key element is that we are moving to risk based, problem solving and risk based, decision making. Key elements of 'Audit Supervision' can be found in the sidebar on the next page.

AUDIT SUPERVISION

GAGAS 6.53 Audit supervisors or those designated to supervise auditors must properly supervise audit staff.

6.54 Audit supervision involves providing sufficient guidance and direction to staff assigned to the audit to address the audit objectives and follow applicable requirements, while staying informed about significant problems encountered, reviewing the work performed, and providing effective on-the-job training.

6.55 The nature and extent of the supervision of staff and the review of audit work may vary depending on a number of factors, such as the size of the audit organization, the significance of the work, and the experience of the staff.

The head of value added auditing does the same things that all managers do; specifically, plan, coordinate, control, and staff the organization. The difference is that managing value added auditing is more like managing a research and development team than directing an operational area. In an established auditing team, value added auditors are professionals. They know what needs to be done, so daily supervision and controlling are not part of the value added audit manager's job.

Within a value added auditing group of highly skilled professionals, the chief audit executive budgets, schedules, and monitors the overall performance of the auditing group. It is assumed that the professionals can self-manage their assignments, planning, conducting, and reporting their results.

In a new department with inexperienced value added auditors, the chief audit executive is more directly involved. In this case, the CAE may:

- Obtain corporate resources.
- Manage the group budget.
- Receive value added audit requests from the customer and distill customer requirements into an audit brief.
- Instruct value added auditors who, what, where, and when to audit.
- Schedule audits.
- Review and approve value added audit plans.
- Oversee the conduct of value added audits.
- Review working papers, reports, findings, and corrective action.
- Review and approve value added audit reports.
- Determine customer satisfaction with value added audit reports.

This type of management would seem confining to professionals who know their craft. However, if the value added audit group is inexperienced then such oversight is necessary. With a highly skilled, professional value added auditing team, the value added auditor may perform the duties with little supervision. For example, the individual auditor would discuss the audit with the customer, develop an audit plan, conduct the audit, report results, and follow up with the customer. The individual auditor wouldn't be concerned with the overall scheduling and budgeting of the audits.[4]

STAFFING VALUE ADDED AUDITS

Leadership appears to be the art of getting others to want to do something you are convinced should be done.
Vance Packard

Before conducting the audit, the value added auditor should understand the players and environment in which they interact. The players in a value added audit are the audited, auditor and customer. Value added auditing is a major improvement tool for managing, assuring and controlling the quality of product and services.

PEOPLE AND PROCEDURES

The board of directors wants more information about operations, risks, and controls. They traditionally went to internal auditing to obtain this information. As discussed, we're seeing more integration of internal auditing and quality auditing so both groups can speak with one voice with the external auditors, senior management, and the board of directors. In this way, there is less confusion and 'this is what I think is happening' conversations. This integration starts with the heads of internal auditing and quality auditing.

Internal auditing, operational auditing, and quality auditing should work together to coordinate audits, ensure consistent audit coverage, and avoid duplication. The Institute of Internal Auditors describes it as:

> "The scope of internal auditing work encompasses a systematic, discipline approach to evaluate and improve the effectiveness of risk management, control, and governance processes."[5]

VALUE ADDED AUDIT STAFFING

Staffing an audit is always a sensitive issue. Value added auditors can be full time staff, part time reporting to a functional area or even be contractors. A number of issues should be considered when selecting a value added auditor, specifically:

- Auditor competencies.
- Auditor education.
- Training requirements.
- Business and operational process knowledge.
- Technical standards and specifications knowledge.
- Level and type of internal control knowledge.
- Financial and budgetary knowledge.

The chief audit executive has the authority and is responsible for selecting and developing auditors. The following can be considered minimum requirements for developing auditors:

- Develop job descriptions for each type of audit (risk management, system, process, audit, etc.).
- Select competent auditors.
- Provide industry, process and product specific training.
- Support continued professional development.
- Provide coaching and mentoring throughout the audit.
- Provide political and organizational cover.

VALUE ADDED AUDITOR REQUIREMENTS

The Japanese employ open discussion with generalities that leave room for movement and compromise. They have nineteen ways of saying no - suggestive of the extreme finesse with which their language navigates the shoals of conflict, avoiding it if possible.
Richard Pascal, Professor

Value added audits are conducted in different environments and in organizations with different mandates, objectives, and structures.

VALUE ADDED AUDITING SKILLS

It is assumed that value added audits are performed with due professional care and proficiency. In a large organization, this means that an audit is assigned to a person with the background, knowledge, aptitude, and skills suitable to conduct a specific audit properly. There are several schools of thought on the type of education and experience that qualifies a value added auditor. One believes that desire alone is sufficient to be a value added auditor. Another believes that value added auditors should have formal degrees, be certified, have specific training, have operational experience, and preferably have managerial experience.

Some organizations are retraining inspectors to become ISO 9001 auditors. It is believed that because these people have experience inspecting products, they would be excellent candidates to detect abnormal conditions. But it is a question whether inspectors have the ability to fix root cause problems in an environment that stresses increasingly interdisciplinary technical solutions.

On the other hand, highly degreed and experienced people may have the opposite effect. They may not have the line experience to detect abnormal operational conditions. Their solutions may be too technical and costly where a simpler solution may work better. Also, many highly educated people may see value added auditing as operational monitoring and basic problem solving beneath their abilities, and this attitude could backfire on the line.

The following requisite skills of value added auditors:

- Professional proficiency.
- Due professional care.

COMPETENCE

GAGAS 3.69 The staff assigned to perform the audit must collectively possess adequate professional competence needed to address the audit objectives and perform the work in accordance with GAGAS.

3.70 The audit organization's management should assess skill needs to consider whether its workforce has the essential skills that match those necessary to perform the particular audit. Accordingly, audit organizations should have a process for recruitment, hiring, continuous development, assignment, and evaluation of staff to maintain a competent workforce. The nature, extent, and formality of the process will depend on various factors such as the size of the audit organization, its structure, and its work.

3.71 Competence is derived from a blending of education and experience. Competencies are not necessarily measured by years of auditing experience because such a quantitative measurement may not accurately reflect the kinds of experiences gained by an auditor in any given time period. Maintaining competence through a commitment to learning and development throughout an auditor's professional life is an important element for auditors. Competence enables an auditor to make sound professional judgments.

- Sufficient value added auditing knowledge and skills.
- Objective and independent appraisal abilities.
- Project management knowledge and abilities.
- Ability to unearth relevant, accurate, and reliable information.
- Knowledge of diverse processes, project, and functions.
- Knowledge of standards, processes, and systems.
- Flexibility to adapt to different conditions.
- Collaborative problems solving skills.
- People skills.
- Communication skills.
- Proper temperament.
- Diligence.
- Analytical skills.
- Critical attitude.

These skills are described in the following subsections:

PROFESSIONAL PROFICIENCY

The success of a value added audit strongly depends on two key factors - professional proficiency and exercise of due professional care. These are a function of the value added auditor's abilities, skills, aptitude, attitude, and experience. These elements as well as the organizational environment are discussed throughout this chapter. The terms due professional care and professional proficiency have technical definitions.

Professional proficiency is the required set of skills required to conduct a value added audit properly. This means that a person or persons with the necessary skills, knowledge, attitudes, and experience can plan, conduct, and report a value added audit. A team sometimes conducts value added audits. Team members are specialists with industry-, process-, or product-specific knowledge.

DUE PROFESSIONAL CARE

Due professional care is the minimum level of professional proficiency that the value added auditor should exercise while conducting the value added audit. Due professional care does not imply perfection or infallibility. It implies that the auditor conducts the value added audit in a reasonably prudent and competent manner to provide a reasonable level of assurance to the organization that risks are managed and processes are in control.

In exercising due professional care, the value added auditor should be alert to inefficiencies, waste, costs, ineffectiveness, mistakes, flaws, nonconformances, or anything that may impede continual improvement or decrease customer satisfaction. If the value added auditor carefully follows a logical methodology, like the one expressed in this book, then it can be assumed that professionalism and due professional care have been exercised.

SUFFICIENT VALUE ADDED AUDITING KNOWLEDGE AND SKILLS

The value added auditor must have sufficient knowledge and skills to conduct a professional value added audit with due professional care, which means the value added auditor has:

- Knowledge of and proficiency in managing, planning, conducting, and reporting the result of a value added audit.
- Ability to apply professional auditing knowledge and standards in different situations and environments.
- Knowledge of risk - control, assurance, and continual improvement principles.
- Knowledge of risk management and controls.
- Experience in or knowledge of the fundamentals of such subjects as engineering, manufacturing, computers, purchasing, design, metrology, economics, or accounting.

OBJECTIVE AND INDEPENDENT APPRAISAL ABILITIES

Traditionally auditors were objective and independent. In some cases, this is changing. More often value added auditors are being asked to consult with internal organizations. This creates a situation where the auditor now must be objective but may not be necessarily independent, since the auditor has ties and affiliations with the client organization.

This may create some confusion. Regulatory compliance auditors must be independent and objective. Objectivity is one of the most important characteristics of compliance auditing. Objectivity implies that the value added auditor is not biased by external elements. The organization, the customer, wants to rely on the auditor's judgment. The organization assumes that the value added auditor is independent 'in fact' and 'in appearance'. 'In fact' means that the value added auditor has no ties or affiliations that would

> **INDEPENDENCE**
>
> **GAGAS 3.02** In all matters relating to the audit work, the audit organization and the individual auditor, whether government or public, must be independent.
>
> **3.03** Independence comprises:
> **a.** Independence of Mind. The state of mind that permits the performance of an audit without being affected by influences that compromise professional judgment, thereby allowing an individual to act with integrity and exercise objectivity and professional skepticism.
> **b.** Independence in Appearance. The absence of circumstances that would cause a reasonable and informed third party, having knowledge of the relevant information, to reasonably conclude that the integrity, objectivity, or professional skepticism of an audit organization or member of the audit team had been compromised.
>
> **3.04** Auditors and audit organizations maintain independence so that their opinions, findings, conclusions, judgments, and recommendations will be impartial and viewed as impartial by reasonable and informed third parties. Auditors should avoid situations that could lead reasonable and informed third parties to conclude that the auditors are not independent and thus are not capable of exercising objective and impartial judgment on all issues associated with conducting the audit and reporting on the work.

bias judgments. 'In appearance' means that the value added auditor has no associations that may imply bias.

There seems to be some confusion about 'objectivity' and 'independence.' A traditional arms length audit requires both objectivity and independence. There is a firewall between the audited and the auditor as shown in the 'Objectivity and Independence' figure on the next page. More often, auditors are conducting collaborative audits, which are consultative engagements. These are objective, but not independent. Threats to independence can also arise as listed in the 'Threats to Independence' sidebar on page 133.

PROJECT MANAGEMENT KNOWLEDGE AND ABILITIES

The value added auditor must use organizational resources efficiently, effectively, and economically when conducting the value added audit so the customer and organization are satisfied with audit results. The differences between efficient, effective, and economical are discussed later in the chapter. In planning and conducting the audit, the value added auditor must consider and optimize many factors including material, information, people, facilities, and monies.

ABILITY TO UNEARTH RELEVANT, ACCURATE, AND RELIABLE INFORMATION

The value added auditor must be able to unearth relevant, accurate, and reliable data, distill it, and develop information that is useful to the organization. Data should also be timely, useful, complete, objective, and understandable. Data often has to be manipulated so the reader can understand and use it, which may mean collating, graphing, comparing, or in some way massaging it.

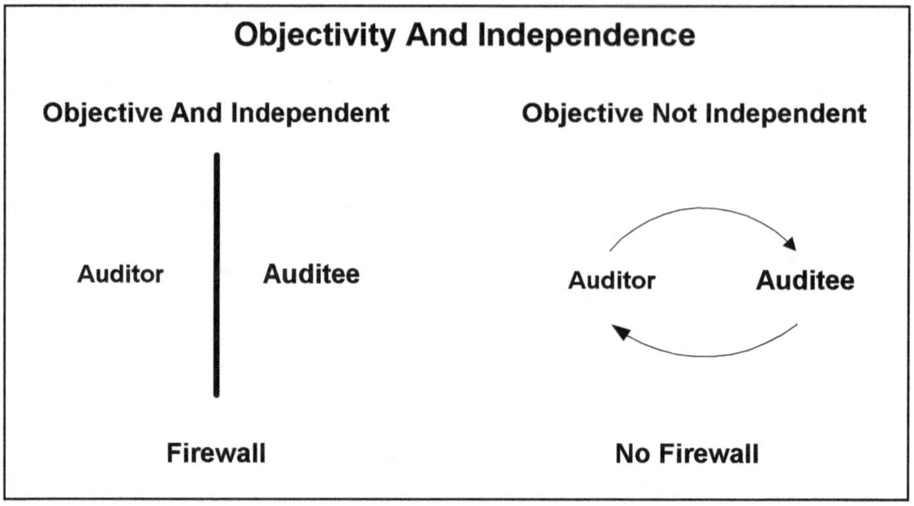

KNOWLEDGE OF DIVERSE PROCESSES, PROJECTS, AND FUNCTIONS

The customer may want an objective assessment. Objectivity implies that the value added auditor has sufficient knowledge, skills, background, and experience to evaluate diverse external and internal operations and functions. The value added auditor may be called on to assess engineering, quality, manufacturing, purchasing, accounting, customer service, and other organizational processes. Objectivity also implies that the value added auditor follows a methodology to identify problems, examine them, and if required, supply recommendations to prevent their recurrence.

This means that the value added auditor becomes familiar with individual departments and operations and, more importantly, how they interrelate. This is especially important as projects become so complex that multidisciplinary teams work together to solve problems or develop products. In this atmosphere, the value added auditor's role is as much an independent evaluator as facilitator who appraises existing process management and controls and works with process owners to improve them.

KNOWLEDGE OF STANDARDS, PROCESSES, AND SYSTEMS

The value added auditor needs operational related knowledge to analyze different technical disciplines, functional areas, and operational processes. In regulated industries, the value added auditor must know the content of specific codes, regulations, and standards, such as the Food and Drug Administration (FDA) 'Good Manufacturing Practices'.

The Department of Defense MIL-Q-9858' Quality Program Requirements' at one time was the highest-level quality standard for military contractors and aerospace companies. Many commercial companies adopted the standard, so quality auditors checked compliance to this document. Now, however, companies are adopting industry specific standards, such as AS 9100 in the aerospace industry.

THREATS TO INDEPENDENCE

GAGAS 3.13 Threats to independence are circumstances that could impair independence. Whether independence is impaired depends on the nature of the threat, whether the threat is of such significance that it would compromise an auditor's professional judgment or create the appearance that the auditor's professional judgment may be compromised, and on the specific safeguards applied to eliminate the threat or reduce it to an acceptable level. Threats are conditions to be evaluated using the conceptual framework. Threats do not necessarily impair independence.

3.14 Threats to independence may be created by a wide range of relationships and circumstances. Auditors should evaluate the following broad categories of threats to independence when threats are being identified and evaluated:
a. Self-interest threat - the threat that a financial or other interest will inappropriately influence an auditor's judgment or behavior.
b. Self-review threat - the threat that an auditor or audit organization that has provided no audit services will not appropriately evaluate the results of previous judgments made or services performed as part of the no audit services when forming a judgment significant to an audit.
c. Bias threat - the threat that an auditor will, as a result of political, ideological, social, or other convictions, take a position that is not objective.
d. Familiarity threat - the threat that aspects of a relationship with management or personnel of an audited entity, such as a close or long relationship, or that of an immediate or close family member, will lead an auditor to take a position that is not objective.
e. Undue influence threat - the threat that external influences or pressures will impact an auditor's ability to make independent and objective judgments.
f. Management participation threat - the threat that results from an auditor's taking on the role of management or otherwise performing management functions on behalf of the entity undergoing an audit.
g. Structural threat - the threat that an audit organization's placement within a government entity, in combination with the structure of the government entity being audited, will impact the audit organization's ability to perform work and report results objectively.

3.15 Circumstances that result in a threat to independence in one of the above categories may result in other threats as well. For example, a circumstance resulting in a structural threat to independence may also expose auditors to undue influence and management participation threats.

FLEXIBILITY TO ADAPT TO DIFFERENT CONDITIONS

Flexibility is the ability to adapt to different environments, people, plants, languages, standards, attitudes, customs, technologies, suppliers, cultures, products, and processes. Flexibility is also the ability to understand customer needs and auditee's operations; recognize deviations from a standard or unstable process conditions; arrive at logical, efficient solutions; and apply past experience and knowledge to different situations. In each value added audit, the auditor can thereby develop novel, understandable, and workable solutions.

COLLABORATIVE PROBLEM SOLVING SKILLS

To achieve the value added audit objectives, a participatory, multidisciplinary, problem-solving style adds more value than an adversarial, problem-finding audit. Ultimately, the value added auditor is providing a service to the organization and to the client to improve processes, products and services. The value added auditor still provides an objective appraisal, but the value added auditor may provide this to the person or area being audited to improve operations. In this case, the client also becomes the internal customer of the audit. Also, the value added auditor is more internal consultant than compliance enforcer.

PEOPLE SKILLS

It's important for the value added auditor to maintain good relationships with the customer, the audited, and the entire organization. Understanding and being sensitive to people's and stakeholder's needs helps in obtaining information. The value added auditor is a service provider to the entire organization, which is the internal customer base, and after all, today's audited may become tomorrow's customer.

Also, the term auditor raises a red flag to many people who have a natural suspicion and resistance to those who monitor their processes and controls. The value added auditor needs to work with these people to improve systems, processes, and products; value added auditor cannot antagonize or alienate them if the auditor intends to be effective.

People skills are the critical success criteria of value added auditors. Often technical value added auditing skills can be learned, whereas people skills have to be developed. Technical audit skills are relatively easy to learn, but it is more difficult to encourage and motivate people to take ownership of their processes, to notice unusual conditions, to eliminate the root cause, and to improve continually. That 'so what,' 'it's not my job,' and 'what's in it for # 1' attitudes have to be eliminated. People need to be motivated, trained, and induced to identify process and system problems and take the initiative to solve them.

COMMUNICATION SKILLS

Results of the value added audit must be communicated in an understandable and usable fashion to the customer, the person or group authorizing the audit. This is a critical value attribute of all assessments. Too many times, audit results are incomprehensible. The audit lists pages of data or checklists that have not been distilled or presented in a form that provides the reader with a clear understanding of the audit's objectives. Results should also be usable by the customer; otherwise the information will not be used to improve processes or systems.

The value added auditor must be able to hear, listen, coach, propose, and maybe assist in the implementation of new ideas. The value added auditor may even act as cheerleader for the process changes necessary for improvement. The word cheerleader may sound facetious, but initially value added auditor may have to recommend and even ad-

TECHNICAL KNOWLEDGE

GAGAS 3.72 The staff assigned to conduct an audit in accordance with GAGAS should collectively possess the technical knowledge, skills, and experience necessary to be competent for the type of work being performed before beginning work on that audit. The staff assigned to a GAGAS audit should collectively possess:

a. Knowledge of GAGAS applicable to the type of work they are assigned and the education, skills, and experience to apply this knowledge to the work being performed.
b. General knowledge of the environment in which the audited entity operates and the subject matter.
c. Skills to communicate clearly and effectively, both orally and in writing.
d. Skills appropriate for the work being performed; for example, skills including:

- Statistical or nonstatistical sampling if the work involves use of sampling.
- Information technology if the work involves review of information systems.
- Engineering if the work involves review of complex engineering data.
- Specialized audit methodologies or analytical techniques, such as the use of complex survey instruments, actuarial-based estimates, or statistical analysis tests, as applicable.
- Specialized knowledge in subject matters, such as scientific, medical, environmental, educational, or any other specialized subject matter, if the work calls for such expertise. Statistical or nonstatistical sampling if the work involves use of sampling.

vocate change. Any change is difficult and threatening, and the manner in which the auditor communicates the necessity for change can affect its subsequent success.

The value added auditor must be able to converse with process owners as well as business unit heads and senior management. Communication skills include listening, hearing, speaking, and writing abilities.

Communication implies both receiving and sending information throughout the value added audit. The value added auditor must be able to listen actively and maybe intuit information from the customer and the audited. If the value added auditor can't convey ideas while conducting the audit, trust and credibility won't develop with audited. Finally, if the auditor cannot distill and present the information in understandable and workable recommendations, then value added audit effectiveness is lowered.

Many people attracted to value added auditing have high technical skills and knowledge but poor communication skills. This can hinder the ability of the value adding auditing group to establish its value. The value added audit group may generate excellent reports that are technically accurate, but if they are perceived poorly, audit results and recommendations are not followed.

PROPER TEMPERAMENT

The lead value added auditor also needs the proper temperament. Successful value added auditors should have a shirtsleeve, can-do attitude shown in a way that does not irritate people. The value added auditor should be honest, tactful, decisive, and cooperative. There may be other necessary personal characteristics. The important point is that the value added auditor needs the right attitude to gain the respect and assistance of people while conducting the audit.

DILIGENCE

An important characteristic is diligence, in planning, conducting, and reporting value added audit results without extensive supervision. Diligent value added audit performance is the ability to:

- Understand organizational and stakeholder audit requirements.
- Assess the requirements for each value added audit assignment.
- Observe value added auditing standards and other directives.
- Carefully plan the value added audit and prepare working papers.
- Carefully review the existence of adequate controls to manage risks.
- Examine every major and relevant aspect of the value added audit.
- Remain within the value added audit scope and exercise objectivity.
- Compile and analyze evidence that is sufficient for the value added audit.
- Communicate conclusions to the audited.
- Be alert for any indication of evidence that can impact the value added audit results and possibly warrant more intensive or extensive auditing.
- Report critical deficiencies and risks immediately.
- Treat privileged information with discretion.
- Maintain ethical conduct at all times during the value added audit.
- Report on the value added audit results clearly and conclusively.

ANALYTICAL SKILLS

As stressed throughout this book, value added audits are analytical audits. Analytical and investigative skills result from classroom learning and from value added audit experience. The value added auditor can learn how to conduct a value added audit following the methodologies explained in this book. The auditor can also take college classes or attend seminars: To fully develop these skills requires experience gained by conducting different types of value added audits in different environments. In other words, the value added auditor can obtain information and minimum skills in the classroom, but the auditor must learn the craft on the job. Critical technical skills are listed in the sidebar on page 135.

CRITICAL ATTITUDE

Monitoring and assessing for wrongdoing and fraud are important elements of value added auditing. The value added auditor must be alert to the possibility of intentional errors, omissions, conflicts of interest, fraud, and irregularities.

PROFESSIONAL JUDGMENT

GAGAS 3.60 Auditors must use professional judgment in planning and performing audits and in reporting the results.

3.61 Professional judgment includes exercising reasonable care and professional skepticism. Reasonable care includes acting diligently in accordance with applicable professional standards and ethical principles. Professional skepticism is an attitude that includes a questioning mind and a critical assessment of evidence. Professional skepticism includes a mindset in which auditors assume neither that management is dishonest nor of unquestioned honesty.

3.**62** Using the auditors' professional knowledge, skills, and experience to diligently perform, in good faith and with integrity, the gathering of information and the objective evaluation of the sufficiency and appropriateness of evidence is a critical component of audits. Professional judgment and competence are interrelated because judgments made are dependent upon the auditors' competence.

3.63 Professional judgment represents the application of the collective knowledge, skills, and experiences of all the personnel involved with an audit, as well as the professional judgment of individual auditors. In addition to personnel directly involved in the audit, professional judgment may involve collaboration with other stakeholders, external specialists, and management in the audit organization.

3.64 Using professional judgment is important to auditors in carrying out all aspects of their professional responsibilities, including following the independence standards and related conceptual framework; maintaining objectivity and credibility; assigning competent staff to the audit; defining the scope of work; evaluating, documenting, and reporting the results of the work; and maintaining appropriate quality control over the audit process.

3.65 Using professional judgment is important to auditors in applying the conceptual framework to determine independence in a given situation. This includes the consideration of any threats to the auditor's independence and related safeguards, which may mitigate the identified threats. Auditors use professional judgment in identifying and evaluating any threats to independence, including threats to the appearance of independence.

3.66 Using professional judgment is important to auditors in determining the required level of understanding of the audit subject matter and related circumstances. This includes consideration about whether the audit team's collective experience, training, knowledge, skills, abilities, and overall understanding are sufficient to assess the risks that the subject matter of the audit may contain a significant inaccuracy or could be misinterpreted.

3.67 An auditor's consideration of the risk level of each audit, including the risk of arriving at improper conclusions, is also important. Within the context of audit risk, exercising professional judgment in determining the sufficiency and appropriateness of evidence to be used to support the findings and conclusions based on the audit objectives and any recommendations reported is an integral part of the audit process.

3.68 While this standard places responsibility on each auditor and audit organization to exercise professional judgment in planning and performing an audit, it does not imply unlimited responsibility, nor does it imply infallibility on the part of either the individual auditor or the audit organization. Absolute assurance is not attainable due to factors such as the nature of evidence and characteristics of fraud. Professional judgment does not mean eliminating all possible limitations or weaknesses associated with a specific audit, but rather identifying, assessing, mitigating, and explaining them.

The value added auditor must still have a critical attitude because as the issues of improvement and competitiveness have increased in importance so has the risk that an audited may not be totally candid. For example, many companies are seeking single-source, world-class suppliers to provide high quality products and services. Competition is fierce to demonstrate high quality, low cost, and technical proficiency. Those that become single-source suppliers to major companies receive long-term contracts with guaranteed margins.

The same risk of abuse can exist internally as well. Internal divisions, plants, and even departments must often provide products or services that compete with external suppliers. If internal suppliers are not competitive in terms of overall value, quality, service, and cost with external suppliers, it is common to source the work to external contractors. The plants and people who previously performed the work have a set time to improve and if performance is as high or higher than outside contractors, plant and personnel are retained. If not, the plant is sold, and the people are transferred or laid off. So even internally, the value added auditor should be aware of the possibility for abuse.

ORGANIZING THE VALUE ADDED AUDIT FUNCTION

In the simplest terms, a leader is one who knows where she or he wants to go and gets up and goes.
John Erskine

Value added auditing has more power and visibility than traditional quality or operational auditing since it deals with the survival of the organization. As part of this ascent, value added auditing has become accepted very quickly. Value added auditing reports to the chief audit executive or president of a business unit.

As value added auditing gains more credibility and acceptance, the caliber of the people attracted to value added auditing is increasing. Ten years ago, quality auditing and other types of auditing were sometimes line people who had been promoted into these positions because they knew how a product was produced. As customer satisfaction, continual improvement, safety, risk, quality of worklife, and competitiveness have increased in importance, the educational, skill, and knowledge levels of value added auditors have also increased.

IS THERE A NEED FOR A VALUE ADDED AUDITING FUNCTION?

If an organization does not have a value added auditing group, how does one get started? An organization's internal auditing group may be focused on internal financial controls and not have operational knowledge or process specific background.

The question should first be asked: 'Is there a need for a value added auditing group?' This need will trigger the process for creating the function. The driver for establishing the need can be internal or external to the organization. External needs, especially from customers and regulators, drive the process much faster, and justification is easier. For example, the Sarbanes/Oxley Act requires an internal audit function. Regulatory authorities and commercial customers commonly audit contractors and suppliers for quality, service, cost, delivery, and technology. These groups also expect suppliers to audit their suppliers and so on down the production chain. These customers also require suppliers to audit internal operations. Internal needs include the ability to measure improvements, assess quality systems, evaluate risks, evaluate internal controls, or evaluate product failures.

Once a strong case has been made for a value added auditing group, the next objective is to develop a formal proposal. A team is organized with the charge to develop a proposal and the team drafts a discussion piece focusing on the need and the value of such a group. To counter internal politics, a potential refusal, or a deadlock among competing internal groups, the team should involve, solicit, and distill the views of all internal stakeholders. The end proposal presents the consensus or the highest denominator of views of the organization's internal groups. The proposal is then submitted to senior management or the board of directors.

A purist would argue that every person, supervisor, team, or department is responsible for monitoring and controlling their processes. So, manufacturing is responsible for production improvement, engineering is responsible for design improvement, and everyone is responsible for customer and stakeholder satisfaction. Each group is responsible for self-monitoring and self-regulating their performance. To follow this logic, supplier auditing resides in purchasing. Manufacturing auditing resides with the people responsible-the line operator, process engineer, and line supervisor.

CHOOSING A LEAD AUDITOR

If a team conducts a value added audit, someone should be designated lead auditor. The lead auditor is usually familiar with the auditee's operational processes and systems. Also, he or she has demonstrated insights and skills that were developed through on-the-job experience especially through participating or leading similar value added audits. The lead auditor should also have demonstrated leadership in a team environment. He or she is probably dealing with highly trained professionals who know what has to be done. In such an environment, the lead auditor:

- Sets the direction of the entire value added audit.
- Works with other professionals to obtain data.
- Analyzes information and evidence.
- Makes meaningful value added audit determinations that satisfy the customer and organization.

ESTABLISH A REPORTING SYSTEM FOR THE VALUE ADDED AUDITING GROUP

To whom should value added auditing report? This is determined by how value added auditing in general is perceived and used within the organization. As mentioned, this group more often to the board of directors. Some maintain that value added auditing should be a high-level staff function and others maintain that it should also be integrated into line functions. If value added auditing reports to a high level officer, it would probably be the chief audit executive, chief quality officer, or chief compliance officer. If the reporting relationship were to the line organization, then value added auditing would be integrated into the area that uses the information, which is its customer base. For example, if an operating group were conducting supplier audits, then the auditor may report dotted line to purchasing and solid line to the compliance officer. If a group were analyzing the improvement performance of engineering projects, then the value added auditor would report to engineering.

Both proponents offer convincing arguments. As a high-level staff function, value added auditing has the autonomy, independence, objectivity, perspective, and ability to conduct company and supplier wide audits. Professional value added auditors reporting to high level of management or to the board of directors should be extensively trained to conduct independent analyses and examinations. Top management could rely on these reports for independence and accuracy. Political pressure to skew report results should be minimized. The audit reports would be sent to the chief financial officer or to the chief executive officer for review and subsequent action. This reinforces the image of value added auditing being independent from the area being audited.

Integrated into a line function, value added auditing would work with and report to the people or area responsible for quality, risk management, or business process practices. In this scheme, value added auditors have a better knowledge of specific operations and a close relationship to process owners. However, there is a tradeoff. This closeness could impair objectivity and independence. The auditor may lose the larger perspective, the overview of the entire organization's improvement thrust, as well as the instant ability to communicate with corporate decision makers.

VALUE ADDED AUDITING REPORTING PART STAFF AND PART LINE

Some organizations believe that the optimum arrangement is to have a staff of professionally trained value added auditors as well as a more informal group that can be called upon as needed. Professional value added auditors serve as staff to the entire organization. In other words, the entire organization is the customer base of these auditors. The informal group consists of people whose jobs is something other than auditing. This informal group is trained to conduct an audit and to recognize and correct system and process deficiencies.

SUMMARY

Senior management and the board are ultimately responsible for the organizational actions, conduct, performance, and ethics of value added auditing. In general, senior management is responsible for:

- Complete, relevant, reliable, and credible financial and operating information.
- Economic, effective and efficient use of the organization's resources.
- Safeguarding of the organization's assets.
- Compliance with laws, regulations, ethics, and contracts.

- Identification, management, and control of risk exposures.
- Established objectives and goals for operations, processes, projects, and programs.[6]

A value added audit works best in companies where it has become part of the value or corporate culture of the company. Once a value audit is perceived as a necessary element to the pursuit of profits, then continual improvement, risk management, value added auditing, and the pursuit of excellence become part of a company's philosophy for success. The pursuit of value in all of its meanings then provides a sense of common direction for all employees to guide their daily behavior. This provides the value added auditor a level of assurance that policies, procedures, and work instructions are followed consistently.

If senior management mandates or imposes ethics, risk management, or process improvement on an organization and the people within, employees may perceive this as another mandate from above and just another fad. And, management fads, such as MBO, reengineering, and quality circles, have come and gone. Seldom are fads around long enough to be internalized into an organization. There is little time for the initiative to be evaluated before something else takes its place. This too can happen with the value added auditing unless there is a company and supplier wide culture that supports the pursuit of value.

The CAE is the senior organizational audit executive. This person has the authority, resources, and responsibility to manage the value added audit function and meet business objectives. More often, the CAE reports to the board of directors. This person assists the organization to develop critical audit objectives.

The chief audit executive also establishes policies and procedures for value added auditing. This standardizes organizing, conducting and reporting of the value added audit. Standardization is key to audit quality and ultimately to audit stakeholder satisfaction.

PART III: PLANNING THE VALUE ADDED

VALUE ADDED AUDITING LIFECYCLE(TM)

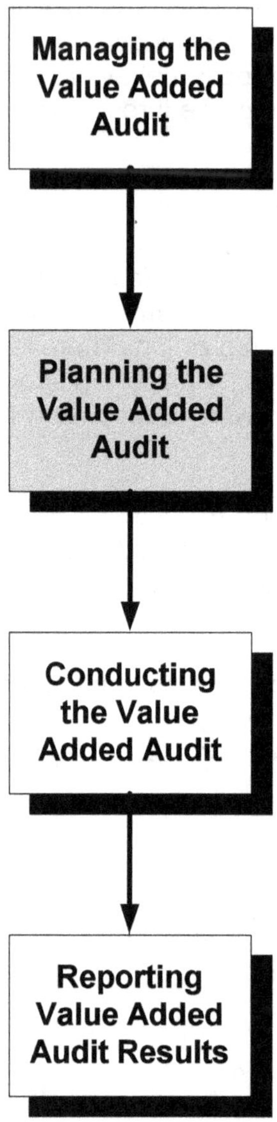

PLANNING THE VALUE ADDED AUDIT

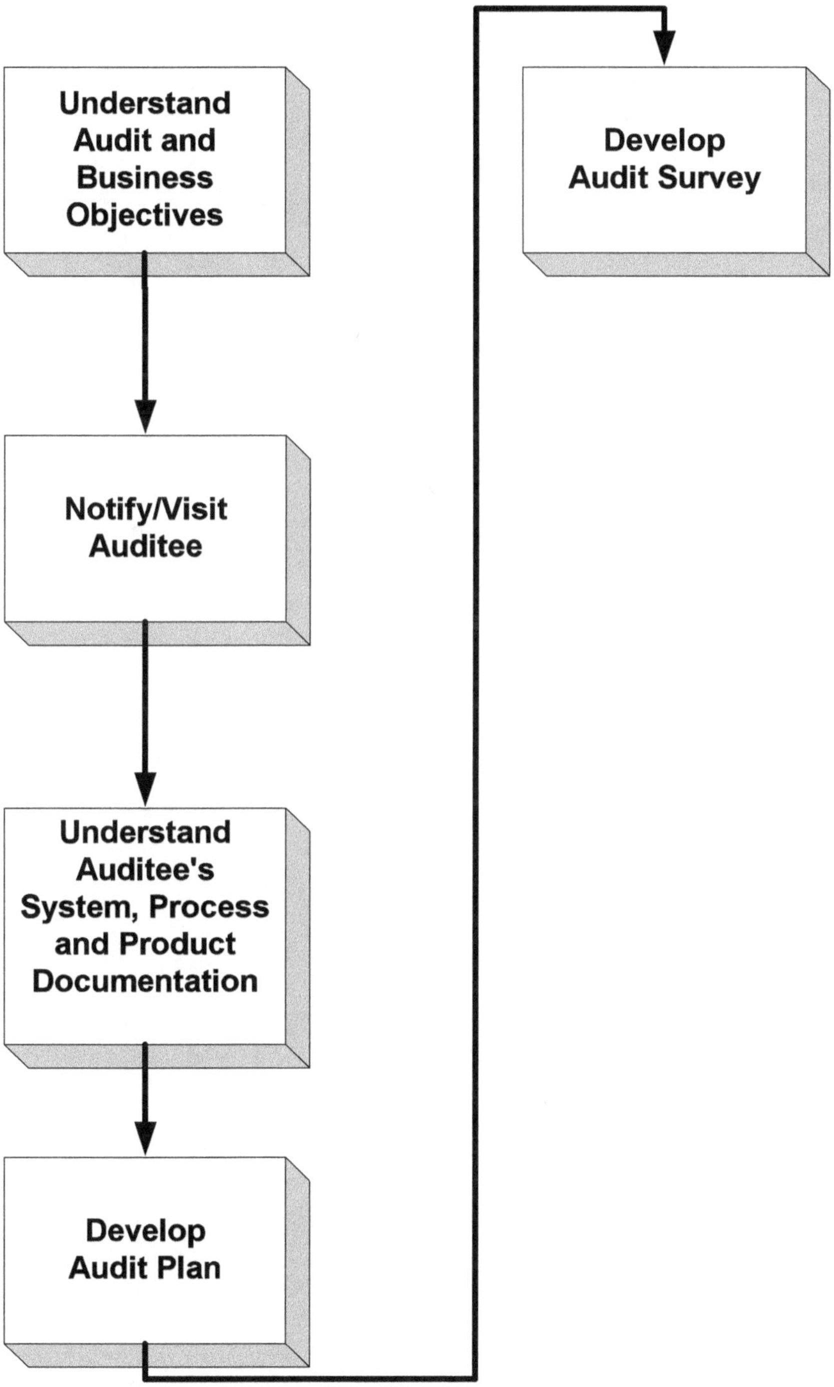

CHAPTER 7
STEP 1 – UNDERSTAND AUDIT AND BUSINESS OBJECTIVES

As a rule, the value added auditor should gather, sift, and analyze as much information as possible before actually conducting the value added audit. Thus, by the time the auditor is ready to carry out the audit, the value added auditor is familiar with the auditee's operations. The value added auditor has spent time at headquarters reviewing past audits, interviewing the audited, and planning the audit.

VALUE ADDED AUDIT PREPARATION

Businesses aren't run by geniuses. It is a matter of putting one foot after another in a logical fashion. The trick is in knowing what direction you want to go.
James R. Barker

The customer or stakeholder for a scheduled or routine value added audit is the internal organization, which requires some form of periodic assurance, assessment or monitoring that remains the same from year to year.

Previous value added audits also provide clues as to what was done, by whom, when, and why. So, planning and managing routine value added audits become straightforward.

PLANNING SEPARATES NOVICE FROM THE PROFESSIONAL

This critical planning often separates the novice from the experienced value added auditor. Once on-site, the novice auditor frequently wanders aimlessly, trying to decide where to start and what to do. The novice auditor relies on a general compliance, yes/no checklist.

The value added auditor may develop a tailored checklist, so the auditor will not waste time and everyone's patience reviewing paperwork and requesting information that could have been done at the auditor's facility more efficiently and effectively. For example, without a sense of direction the novice auditor may send a boilerplate, preliminary survey to the audited and during the audit perform a cursory analysis of unimportant matters. There is no focus to the audit. At worst, this value added audit reveals no insights on risk and critical conditions are not uncovered.

So on arrival at the auditee's site, the value added auditor should have developed a detailed plan of where to go, what to do, when to do it, who to see, how to test it, and when to

PLANNING STEP 1 – UNDERSTAND AUDIT AND BUSINESS OBJECTIVES

PLANNING THE AUDIT – GENERAL

GAGAS 6.06 Auditors must adequately plan and document the planning of the work necessary to address the audit objectives.

6.07 Auditors must plan the audit to reduce audit risk to an appropriate level for the auditors to obtain reasonable assurance that the evidence is sufficient and appropriate to support the auditors' findings and conclusions. This determination is a matter of professional judgment. In planning the audit, auditors should assess significance and audit risk and apply these assessments in defining the audit objectives and the scope and methodology to address those objectives. Planning is a continuous process throughout the audit. Therefore, auditors may need to adjust the audit objectives, scope, and methodology as work is being completed. In situations where the audit objectives are established by statute or legislative oversight, auditors may not have latitude to define or adjust the audit objectives or scope.

observe it. As much as possible, this can be done at the auditor's office. This preparatory work leads to a more efficient, economic, and effective audit. General planning tips are listed in the above sidebar. Proper planning enhances assurance and reduces the following types of risks:

- Asking basic or stupid questions.
- Chasing false leads.
- Missing significant risk.
- Not understanding controls.
- Chasing insignificant items.
- Spending too much time in one area.
- Going in circles.
- Pursuing unimportant searches.
- Coming to a dead end.
- Neglecting major deficiencies.
- Missing critical conditions.
- Overemphasizing minor deficiencies.
- Misunderstanding the customer/audited.

The result is wasted time and resources. The auditor, and audited become frustrated.

WHY IS VALUE ADDED AUDITING PLANNING SO IMPORTANT?

The essence of value added audit planning is captured in GAGAS 6.07, specifically:

> "Auditors must plan the audit to reduce audit risk to an appropriate level for the auditors to obtain reasonable assurance that the evidence is sufficient and appropriate to support the auditors' findings and conclusions."

Value added planning is important at the strategic and tactical levels because it:

- **Provides direction.** Value added audit planning provides direction for the organization to implement the value added system. Once an overall strategic direction is established, then more detailed tactical plans can be formulated for divisions, tablished, then more detailed tactical plans can be formulated for divisions, departments, and processes.

- **Provides a structured framework.** Value added audit planning provides a structured framework for decisions related to reaching correct and valid audit conclusions. Without a framework or plan, the value added audit is rudderless and cannot achieve its goals. Each department may interpret risk or improvement requirements differently, which affects implementation. A structured framework unifies different interpretations, goals, and tactics into a common effort that focuses on stakeholder satisfaction.
- **Reveals opportunities.** Value added audit planning reveals opportunities to improve operational efficiency, reduce costs, improve profits, reduce risks, increase marketshare, or please customers.
- **Facilitates process management, assurance, and control.** Contingency planning anticipates potential problems. Once areas of potential problems are discovered, plans can be developed to prevent their recurrence.[1]

VALUE ADDED AUDITING PLANNING STEPS

Each new value added audit in a continual improvement or risk-control environment will be a little different than previous audits. However, basic steps are common to all value added audits, which are discussed in the following chapters:

- Planning Step 1 - Understand audit and business objectives.
- Planning Step 2 - Notify/visit audited.
- Planning Step 3 - Understand auditee's system, process, and product documentation.
- Planning Step 4 - Develop audit plan.
- Planning Step 5 - Develop audit survey.

The value added auditor does need to complete all the value added audit planning steps if the organization is the customer of a routine, scheduled audit. For example, the value added auditor already knows the organization's requirements and the audited processes.

The above list of steps is mainly useful for preparing and organizing a value added audit that is a new audit. Some value added assessments (such as a walkthrough, attestation, product test, or reaudit) would not entail all of the above planning steps. On the other hand, an internal evaluation of core process risks and controls supplier's processes and systems would probably include most of the steps.

AUDIT STAKEHOLDERS

Be everywhere, do everything, and never fail to astonish the customer.
Margaret Getchell

The value added audit customer may be external or internal to the auditor's organization. A typical customer may include top management. Top management may request a system audit to assess compliance to specified internal procedures. This assessment may result from a desire to apply for the Malcolm Baldrige National Quality Award or to comply with customer requirements.

There are a number of possible audit stakeholders, including:

- **Organization**. In many cases, the customer is someone in the auditor's own organization. Many value added audits are routine and periodic. They are scheduled yearly and involve checking existing systems and processes for internal control effectiveness, risk management, ability to meet specifications, or rate of continual improvement.
- **Board of Directors.** The Sarbanes/Oxley Act has redefined the reporting and responsibilities of the board of directors and senior management. Senior management now wants to know the effectiveness of risks and controls throughout the organization and into the supply chain.
- **Standing committee.** An organization's standing committee, such as a Material Review Board (MRB) or Engineering Control Board (ECB), may request a focused assessment of a quality system, corrective action, product failure, or supplier nonconformance.
- **Customer/manufacturer.** A potential or existing customer may require that a supplier's quality systems and processes conform to the customer's requirements. The assessment would check compliance against a supplier standard, international standard, process specification, or engineering print.
- **Regulatory agency.** A regulatory agency may request a supplier to contract with a third party who will audit the supplier's quality management system or processes against a recognized national or international standard. If approved, the supplier is certified and placed on a register.

UNDERSTANDING AUDIT CUSTOMER'S AUTHORITY AND RESPONSIBILITIES

In general, the value added audit customer (party requesting the audit) should have the following authority and responsibilities:

- **Ensures the audit adds value.** The customer is responsible for ensuring there is a business need for the audit. Need can be determined by establishing a direct link and alignment between audit objectives and organizational objectives.
- **Authorizes value added audit.** Authorizing is more formal than simply requesting a value added audit and is based on formal approval by the organization or senior management.
- **Defines audit requirements, scope, and timelines.** The customer alone or with the auditor defines requirements, which are incorporated into an audit scope of work or audit brief. These define levels of assurance and risk. The customer may also specify the level of assurance and risk he or she wants to assume. This determines how much validation and verification the auditor will try to achieve.
- **Authorizes corrective and preventive action**. If there are deficiencies or noncomplying conditions, these are noted on the audit report. The customer, the auditor, or even the audited initiates a corrective action request (CAR) and decides on follow-

up. If a corrective action request has been initiated, then the customer or another party may request a follow-up audit to determine if the problem was corrected so that it did not recur.

Although the customer is responsible for these actions, the value added auditor often carries them out by default. The value added auditor should at least raise these points with the customer, and ask if the customer wants the auditor to handle them. The decision is noted in the audit scope of work or audit brief. The brief is a distillation of what the customer wants, needs, and expects. How to develop an audit brief is discussed later in the book.

ACQUIRING AUTHORIZATION TO CONDUCT VALUE ADDED AUDIT

A problem that arises in traditional auditing environments where audits are used to monitor internal operations is authorization by the customer. In this environment, traditional auditing still has vestiges of inspecting and policing and there is fear that value added audits may be misused or abused. Resources, including time, people, facilities, and equipment may be misused and the auditee's operations may be disrupted.

Let's look at the possible downtime due to the audit. Downtime due to an audit is expensive and can affect suppliers and other upstream and downstream operations. Also in highly politicized organizations, audits can be used as a weapon, as a means to pressure or to punish.

The customer should have the authority to request a value added audit. For internal audits, authority inevitably must come from senior management and be incorporated into an internal policy. For external audits, authorization is found in external contracts, agreements, or mandates. For external supplier audits, auditing is a condition of a contract and the authorization is found in a purchase order or contract.

Regulatory or government audits are mandated by statute or administrative order. Examples of federal agencies that mandate audits are: the Environmental Protection Agency (EPA); the Food and Drug Administration (FDA); and the Department of Energy (DOE). These agencies have extensive staffs that perform compliance audits to ensure that contract requirements are being met and internal procedures are adequate. In many cases, this authorization does not mean that an audit will automatically be conducted, just that the customer reserves the right to conduct the audit.

IDENTIFYING THE KEY PLAYERS IN THE VALUE ADDED AUDIT

In a value added audit, there are three key players: the audited, auditor, and customer. These players sometimes go by different names. The customer is the person or group that requests the audit and also may be called the client or sponsor. The audited is the person or area being audited. The auditor is the person or team conducting the value added audit. The auditor is sometimes called the assessor or even internal consultant.

The different terms generally reflect the nature of the organization's risk and improvement efforts and the role of the auditor in them. More traditional auditing efforts stress formal

PLANNING STEP 1 – UNDERSTAND AUDIT AND BUSINESS OBJECTIVES

MANAGEMENT'S ROLE

GAGAS A1.08 Managers have fundamental responsibilities for carrying out government functions. Management of the audited entity is responsible for:

a. Using its financial, physical, and informational resources legally, effectively, efficiently, economically, ethically, and equitably to achieve the purposes for which the resources were furnished or the program was established.
b. Complying with applicable laws and regulations (including identifying the requirements with which the entity and the official are responsible for compliance).
c. Implementing systems designed to achieve compliance with applicable laws and regulations.
d. Establishing and maintaining effective internal control to help ensure that appropriate goals and objectives are met; following laws and regulations; and ensuring that management and financial information is reliable and properly reported.
e. Providing appropriate reports to those who oversee their actions and to the public in order to demonstrate accountability for the resources and authority used to carry out government programs and the results of these programs.
f. Addressing the findings and recommendations of auditors, and for establishing and maintaining a process to track the status of such findings and recommendations.
g. Following sound procurement practices when contracting for audits, including ensuring procedures are in place for monitoring contract performance.
h. Taking timely and appropriate steps to remedy fraud, noncompliance with provisions of laws, regulations, contracts or grant agreements, or abuse that auditors report.

terms such as client, audited, and auditor. The audit is specifically used to determine conformance to a standard or specification. It emphasizes the status quo in terms of validating conformance to existing standards, policies, and procedures rather than risk management, continual improvement or customer satisfaction.

Each player in a value added audit has a different perspective on and responsibility for the value added audit. If each perspective is understood and appreciated, value added audits can be made more productive and effective for encouraging continual improvement. The authority and responsibilities of each party are not cut and dry. They depend on the organizational environment, personalities and abilities of the audit staff, timing, resources, and other factors. Depending on the nature of the value added audit, the auditor may be responsible for a part of the audit that in different circumstances would be handled by the customer. The above sidebar lists 'Management's Role.'

For example, the customer wants to know why it is taking so long to get updated engineering drawings. This is an open-ended assignment. It takes 6 weeks for the latest revision of an engineering drawing to be distributed and some blue lines on the engineering print are illegible. The customer has given the auditor a performance type of assignment. 'This is what is happening. Find out why and what can be done to fix it.' In this situation, the value added auditor would develop the audit scope and objectives to determine whether engineering's internal control system for distributing documentation is working properly.

PLANNING STEP 1 – UNDERSTAND AUDIT AND BUSINESS OBJECTIVES

UNDERSTANDING THE AUDITED MANAGEMENT'S RESPONSIBILITIES

The audited has the following authorities and responsibilities:

- **Selects person to assist in value added audit.** The audited selects a person or persons to help the auditor in conducting an efficient, effective, and economical audit. This person may arrange interviews, tests, or inspection. This person also can be the auditor's full-time staff person during the value added audit, serving as part facilitator, arranger, mediator, scheduler, procurer, and ombudsperson.
- **Allows access to facilities and records.** At a minimum, the audited provides access to facilities, records, and people. This responsibility means that he or she complies with the value added audit authorization, which may mean much less than being fully cooperative.
- **Cooperates and assists the auditor.** In the best case, the audited fully cooperates and assists the auditor. In assisting the value added auditor, the audited steers the auditor into the correct areas to gather the required information.
- **Provides adequate working space.** The auditor needs space to spread out reports, test results, questionnaires, and other information. The auditor may also need space to conduct interviews and to meet with the team.
- **Reviews value added audit findings.** The audited assists and reviews value added audit findings. If the audited has sampled a non-representative population, gathered insufficient evidence, pursued false leads, or made an incorrect judgment, the audited should note these to the auditor during the value added audit.
- **Rebuts findings if necessary.** Some value added audits allow the audited to rebut audit report findings. These comments are then appended to the report and transmitted to the customer.
- **Implements corrective action if required by the customer.** If the customer requires corrective action, then the audited as process owner, is responsible for the work, is also responsible for correcting the symptom and the root cause of the problem.

ADDITIONAL AUDIT STAKEHOLDERS

Successful collaborative negotiation lies in finding out what the other side really wants and showing them a way to get it, while you get what you want.
Herb Cohen

A value added audit is essentially a project. It has a defined beginning and a defined end. It follows a series of steps. At the beginning of the value added audit, it's critical that customer wants, needs and expectations are clearly identified. Also at the beginning of the value added audit, organizational and stakeholder mission-critical needs are identified so the audit team can meet them.

The value added audit should always be aware of the overall business mission and try to align the audit objectives to the business mission or objective. A value added audit may have different stakeholder who may have differing objectives. For example if the value added auditors are acting as internal consultants, the team may work with process owners to redesign a process, which may disrupt process flow and product delivery.

Sometimes value added audit teams are formed from process owners to improve systems. This works as long as the team understands the process, can identify key process variables, and wants to improve continually. This type of improvement team usually rotates from the work area so over a two year period or so, everyone has a chance to participate. New blood, new perspectives and new ideas are continually brought in to provide fresh ideas.

STAKEHOLDER MANAGEMENT

Business and project stakeholder groups overlap in terms of having similar constituency groups, including competitors, intervener groups and others. The audit team leader must be able to identify each stakeholder group, work with them, and satisfy their requirements.

Stakeholder requirements may also conflict. For example, the client may want a more extensive evaluation, resulting in audit creep. Accounting or some other cost group wants to minimize costs. Senior management wants the audit done faster because of competitive pressures. Intervener groups such as regulatory agencies may impede the value added audit project with new requirements, conditions and stipulations. Project team members are tired of the long hours. In order to resolve the conflicting stakeholder priorities and achieve a balance, the team leader must negotiate with stakeholders to optimize audit objectives against conflicting politics, resources and timelines. The process isn't easy and usually spells the difference between a successful and a less than successful audit team leader.

CRITICAL AUDIT STAKEHOLDERS

Four critical internal stakeholder groups also influence the direction and implementation of the value added audit:

- Board of Directors.
- Chief Executive Officer.
- General management.
- First level supervision.

Board of Directors

The board of directors and senior management authorize value added audits, provide their general direction, and allocate resources for their successful implementation. The chief audit executive or similar person ensures the value added audit operates as intended and achieves its goals.

The board of directors audit committee consists of independent representatives of the board. Traditionally this committee focused on safeguarding corporate assets and ensuring the effectiveness of from all departments and processes that can affect or be affected by the value added audit.

PLANNING STEP 1 – UNDERSTAND AUDIT AND BUSINESS OBJECTIVES

QUESTIONS TO ASK THE CUSTOMER IN DEVELOPING THE AUDIT AGREEMENT

1. To whom should the audit report be addressed?
2. Who is the audited?
3. What plants, operations, systems, and processes are to be audited?
4. Is the customer authorized to request the audit?
5. Who pays for the audit?
6. Is there an upper cost involved with the audit?
7. What are your requirements?
   a. Expectations?
   b. Wants?
   c. Needs?
8. What type and level of audit do you want conducted?
9. What type and levels of controls do you want?
10. What's your sensitivity to risk?
11. Are you aware of any special organizational or audit risks?
12. What level of assurance do you want?
13. What are the critical elements of audit risk from your point of view:
    a. Detection risks?
    b. Inherent risks?
    c. Control risks?
14. Do you have an example of what you want in terms of an audit format?
15. Are there special audit requirements, such as:
    a. Tests?
    b. Interviews?
    c. Analysis?
16. When do you expect the audit to be conducted and completed?
17. What are the specific objectives of the audit?
18. Are there specific international, national, industry standards, or documentation requiring conformance or compliance?
19. Have previous audits been conducted?
20. If so, what were the findings and recommendations of the report?
21. How critical is the audit?
22. What are other important issues or questions you want addressed?

Since the Sarbanes/Oxley Act, the corporate or board level audit committee now scans the competitive environment to develop organization's audit vision, enterprise risk appetite, and strategy. It manages the external financial auditors and internal auditors. The audit committee also identifies and deploys corporate resources to achieve and implement the organization's value added audit strategy of risk management and controls.

Specifically, the board of director's audit committee provides:

1. Visible leadership.
2. Focuses attention on value added auditing.
3. Provides process management and control leadership.
4. Provides ongoing audit direction.
5. Defines value added audit responsibilities.
6. Establishes corporate level value added audit objectives.

7. Monitors state of organization's risk management and controls.
8. Overcomes organizational obstacles to success.

Chief Executive Officer

The board of directors and the Chief Executive Officer (CEO) defines and authorizes the value added audit function. He or she then provides personal, continual, inspirational, visible leadership, and commitment to the assessment. Since the CEO is the most recognizable leader in an organization, only the CEO can credibly communicate the purpose and urgency of risk monitoring and assessing core processes. And, only the CEO can lead and inspire others to pursue risk management and operational continual improvement. The CEO leads by communicating a deep personal commitment to values he or she wants the organization to have. Some CEOs are known to have the spirit of an evangelist when they communicate the importance of core processes and the value added audit.

General Management

For process improvement, risk management, and internal controls to succeed an organization's general management must support, participate, and reinforce value added auditing. General management directly establishes standards of process improvement performance, assigns jobs, establishes operating procedures, and rewards performance.

Once the value added auditing is established, general management reinforces behavior and activity that lead to maintaining controls to manage risks. On a regular basis, general management supplies messages about the importance of process improvement and encourages activities that support value added auditing.

First Line Supervision

On a daily basis, first-level supervision directs teams or workers performing work, whether it is polishing a car or cutting threads on bar stock. This is called process ownership. First-level supervision is directly responsible for facilitating and overseeing daily work. First-level supervision directly imparts to the team and employee the organization's expectations of performance and process control. First-level supervision or teams communicate technical information by interpreting engineering specifications and procedures.

EVALUATE THE 'TONE AT THE TOP'

We've always felt we had an obligation to not only bring home the bacon to our stockholders, and to do that aggressively as well, but also to be a good national citizen.
William Kieschnick, business person

Well-run organizations, regardless of their size, promote a distinct culture. The value-based culture defines how work gets done and defines what adds value. It is the explicit and implicit rules that people follow to transact business. The value culture adopted by many American companies is excellence through some form of process and project management. The culture and messages of the board and senior management are called 'tone at the top.'

PLANNING STEP 1 – UNDERSTAND AUDIT AND BUSINESS OBJECTIVES

USE OF TERMINOLOGY TO DEFINE GAGAS REQUIREMENTS

GAGAS 2.15 GAGAS uses two categories of requirements, identified by specific terms, to describe the degree of responsibility they impose on auditors and audit organizations, as follows:
a. Unconditional requirements: Auditors and audit organizations must comply with an unconditional requirement in all cases where such requirement is relevant. GAGAS uses the word must to indicate an unconditional requirement.
b. Presumptively mandatory requirements: Auditors and audit organizations must comply with a presumptively mandatory requirement in all cases where such a requirement is relevant except in rare circumstances. GAGAS uses the word should to indicate a presumptively mandatory requirement.

2.16 In rare circumstances, auditors and audit organizations may determine it necessary to depart from a relevant presumptively mandatory requirement. In such rare circumstances, auditors should perform alternative procedures to achieve the intent of that requirement. The need for the auditors to depart from a relevant presumptively mandatory requirement is expected to arise only when the requirement is for a specific procedure to be performed and, in the specific circumstances of the audit, that procedure would be ineffective in achieving the intent of the requirement. If, in rare circumstances, auditors judge it necessary to depart from a relevant presumptively mandatory requirement, they must document their justification for the departure and how the alternative procedures performed in the circumstances were sufficient to achieve the intent of that requirement.

CORPORATE GOVERNANCE AND CULTURE

Organizational culture is critical in determining how a business and organization operates. If the organization has a culture that emphasizes business ethics, there are lower risks and fewer nonconformances. This is one of the most critical issues in American business, organizational governance. The IIA defines this as:

> "An organization's governance practices reflect a unique and ever-changing culture that affects roles, specifies behavior, sets goals and strategies, measures performance, and defines the terms of accountability. That culture impacts the values, roles, and behavior that will be articulated and tolerated by the organization and determines how sensitive — thoughtful or indifferent — the enterprise is in meeting its responsibilities to society. Thus, how effective the overall governance process is in performing its expected function largely depends on the organization's culture."[2]

A well-defined, organizational culture offers the following value added auditing benefits:

- Ensures regulatory compliance.
- Ensures ethical behavior.
- Ensures the right things are done at the right time.
- Ensures what is said is done.
- Reports truthfully and accurately to stakeholders of required information.
- Ensures accountability for managing risk, corrective action, preventive action, and performance improvement.

PLANNING STEP 1 – UNDERSTAND AUDIT AND BUSINESS OBJECTIVES

THE VALUE ADDED CULTURE

The culture of an organization consists of the beliefs, values, and actions shared among most of its members. A culture implies that most of the organization shares consistent beliefs and core values. Value added culture is a matter of attitude, the commitment and belief by employees of the importance of listening to the customer, creating effective internal controls, designing reliable products, purchasing the best value added products, manufacturing products according to specifications, and delivering cheerful service. Much of this sounds like advertising copy but companies that pursue these goals of excellence see the result in higher earnings.

Before embarking on a compliance audit or a value added audit, senior audit executives must determine whether the organizational culture is appropriate or conducive to conduct the value added audit. Who should do this? Sometimes, the human resources, training, internal audit, or quality audit organization may conduct an environmental assessment or scan. How is the culture assessment conducted? The cultural assessment is a gap analysis or survey of 'what is' and 'what is required.' The 'what is' is the current culture. The 'what is required' is the core values and culture change senior management and other stakeholders want of the organization.

ETHICS

Ethics is basically how people act and work. Ethics can be thought as being able to differentiate between right and wrong. Ethics is fundamental to all value added auditing. One of the first things the value added auditor should evaluate is the organizational culture, specifically focusing on ethics. Ethics is a critical soft organizational control.

Ethics are defined by:

- Accepted/approved conduct.
- Regulations.
- Codes of conduct.
- Policies.

Once introduced, ethics is internalized by reinforcing and rewarding individual and group efforts that embody the value added culture. A corporate value added culture creates a sense of identity for those who work in the organization and makes employees feel special. Everyone pulling together in the same direction to achieve a common goal engenders corporate unity, strength, and good behavior.

SOFT CONTROLS

Organizational culture and environment are also critical to effective process management. There is much talk about this, which is often the most critical and the least understood internal control. These are often the critical 'soft' controls that assure that things get done as expected. A properly articulated and deployed value based culture may involve some of the following 'warm and fuzzy' elements:

- Word and action congruence.
- Fairness.
- Core values.
- Vision and mission focus.
- Walk the talk.
- Shared risk/shared gains.
- Satisfy and surpass stakeholder desires.
- Walk the talk.
- Lead by example.
- High reputation and ethics.
- 100 best places to work.
- Follow the rules.
- Equal opportunity.

CLARIFY CUSTOMER'S NEEDS AND EXPECTATIONS

Be everywhere, do everything, and never fail to astonish the customer.
Margaret Getchell

In a small organization, the audit request may come directly from the customer. In some situations, a stakeholder communicates the assignment to the chief audit executive who assigns it to someone. In larger organizations, multinationals, for example, the audit assignment may go through several parties until it lands on the value added auditor's desk.

UNDERSTANDING CUSTOMER'S AUDIT REQUIREMENTS

A value added audit can be routine or can be requested by someone within the organization. Routine audits monitor compliance, conformance, or improvement to audit standards. Routine audits are conducted periodically, e.g., annually or biannually.

The value added audit customer is usually a stakeholder in the organization because it requires a periodic assessment of a business unit. The value added auditor would check effectiveness of internal controls or the process improvement progress. The audit reports are routed to operating department heads, business unit heads, top-level management, audited, and other interested parties.

TYPICAL AUDIT REQUEST

Audit requests given over the phone can present a problem. In an exaggerated example, a senior manager may ask: 'Jane, please look into why project X is delayed. I want to be sure it's OK.' The request is not explicit. Jane may not even be the best person to conduct the value added audit. She may not be familiar with the project. 'OK' can mean: 'Is it on time, on budget, or are controls working properly?' It can mean different things to different people. The request does not specify audit objectives, timelines, requirements, assurance, and the other factors that are important to the auditor.

PLANNING STEP 1 – UNDERSTAND AUDIT AND BUSINESS OBJECTIVES

Unless the audit request and audit authorization are well thought out and in writing, the customer's requirements can become diluted or misunderstood as they are communicated through the organization. Under those circumstances it is almost impossible to conduct a value added audit that satisfies customer's requirements.

One solution to this problem is to invest the time and effort needed to understand customer expectations and requirements. The value added auditor can then develop an audit agreement that distills the customer's needs into doable audit objectives that will be delivered on time and in the format the customer wants. The agreement essentially says: 'This is what I heard that you want and expect. Do you agree?' The agreement is developed by the auditor and returned to the customer for review and confirmation. Only in this way can customer satisfaction be assured.

DEVELOPING AN AUDIT AGREEMENT

Often the customer has not written or developed specific audit objectives, instead relying on the value added auditor to do this. Therefore, the auditor should develop an audit agreement stating what was heard and then ask the customer to review and sign it. This agreement can also serve as the formal authorization to conduct the audit. The value added auditor can also use the audit agreement to discuss and understand the scope and resources required for the audit. Sometimes the customer may not know what he or she wants only that there is a problem to be fixed. With the pressures of modern management, the quick fix is usually a short-term solution and a problem waiting to reoccur. This may indicate that the customer does not appreciate the long-term significance of the problem or what it takes to determine a root cause solution.

Even if the customer did prepare a detailed audit request, the auditor should distill the audit request into an audit agreement and send it back to the customer for review and approval. An audit agreement can be viewed as an assurance and control mechanism for the value added auditor to ensure that customer requirements are properly understood and satisfied.

Once the value added auditor has distilled customer's requirements, reviewed the permanent file, and interviewed the customer, then the auditor can develop detailed audit objectives. The value added auditor should develop objectives based on specific customer requirements. An audit objective is a workable, measurable statement that satisfies the customer's requirements. Experienced auditors can develop value added audit objectives quickly based on their experience.

Often, the auditor can develop preliminary objectives after reviewing the permanent file and interviewing the customer. However, the value added auditor may have to amend them after interviewing the audited, because the auditor will have more information that may change the basic direction of the audit. For example, what the customer thinks she requires is a simple verification of information, but an initial discussion with the audited reveals that a more extensive analysis is required, which changes the scope of the initial assessment.

PLANNING STEP 1 – UNDERSTAND AUDIT AND BUSINESS OBJECTIVES

SAMPLE VALUE ADDED AUDIT AGREEMENT

To: Customer
From: Lead Auditor
Subject: Value Added Audit Agreement
Date: Today's

Per our discussion on August 9, 20XX, the following was agreed:

Supplier X's processes that produce the following part numbers will be audited:

100, 123, 187, 188, 198, 200, 201, 223.

Supplier X produces parts per our prints. Audit will require 1 person day and cost $X. Report will be completed by August 16.

Sincerely,

Lead Auditor
cc: Audit Manager
Purchasing Manager
Permanent File

There are a number of things that can be done to understand and distill the audit customer's requirements into an audit agreement, such as:

- Identify business goals and objectives.
- Identify and assess process and operational risks.

TIPS FOR DEVELOPING THE AUDIT AGREEMENT

The auditor should be an active listener and establish rapport with the customer during the interview. The auditor may begin by asking overview and general types of questions about expectations and then ask about specific requirements. The value added auditor may have to lead the customer's thinking so that he or she can define specific requirements.

The value added auditor may find the process of developing and audit agreement to be iterative. As the customer understands and articulates his or her requirements, it may take several iterations until the agreement is satisfactorily completed. There is no one set format for developing an audit agreement. Also, there is no one set of questions that the value added auditor should pose to the customer while developing the agreement. However like a journalist, the value added auditor should ask the fundamental, when, who, why, where, and how questions.

The manner in which the questions are asked and answered is as important as the type of questions. Questioning is the science and art of obtaining information. It is science because the value added auditor can learn how to formulate and pose questions. It is art because the auditor must use all of his or her senses to intuit what the customer really wants and

SAMPLE CUSTOMER INTERVIEW

Auditor: "What end result do you want from the supplier audit?"
Customer: "I want to know that supplier X is providing good products. We recently discovered one of their parts was the cause of a field failure. I want to know whether this is isolated or if a product recall may ensue."

Auditor: "Following a little investigation, I found that the supplier provides the following part numbers: 100,123,187,188, 198, 200, 201, and 223 per our prints."
Customer: "Do we inspect or require control charts to confirm process controls?"
Auditor: "Up to 6 months ago, we inspected the parts to an AQL = 1.0% and now require control charts and capability of 1.33."

Customer: "Check the processes that produce the parts we purchase for control and capability."
Auditor: "OK. Shouldn't we also review the entire internal control structure, including materials, people, environment, methods, and machines? This looks like a high risk, critical situation."

Customer: "Yes, you're right."
Auditor: "To determine process control and capability for the above processes, I estimate it will take 1 day, but to determine the full extent of their internal controls will take about 4 person days. The approximate cost is $X. Does this sound OK?"

Customer: "OK. This is critical."
Auditor: "When do you want the audit completed?"

Customer: "As soon as possible, hopefully by August 16."
Auditor: "Great, I'll have it done and if you don't mind I'll write what we've agreed upon into a memo of understanding." (You may never meet customer's requirements unless they are written into the audit agreement and this is updated as required).

expects. What is left unsaid is sometimes as valuable as what is said. A sample interview is shown in the 'Sample Customer Interview' sidebar. Once the value added auditor has distilled customer expectations, the auditor can define audit scope, develop timelines, set objectives, analyze resource requirements, and develop a preliminary questionnaire.

There are a number of things that can be done to understand and distill the audit customer's requirements into an audit agreement, such as:

- Identify business goals and objectives.
- Identify risk appetite, risk tolerance, and assurance levels.
- Identify and assess process and operational risks.
- Collect regulations and specifications.
- Gather standards, specifications, policies, procedures, and other relevant documentation.
- Review previous value added audit plans, working papers, results, recommendation, findings, corrective actions, and preventive actions.
- Review contracts and other business documents.
- Obtain organizational chart including job descriptions.

PLANNING STEP 1 – UNDERSTAND AUDIT AND BUSINESS OBJECTIVES

- Obtain process flowcharts and other similar documentation indicating work, money, resource and other flows.
- Obtain operational metrics, business results, and other operational business data.[3]
- Collect regulations and specifications.
- Gather standards, specifications, policies, procedures, and other relevant documentation.
- Review previous value added audit plans, working papers, results, recommendation, findings, corrective actions, and preventive actions.
- Review contracts and other business documents.
- Obtain organizational chart including job descriptions.

LINK AUDIT OBJECTIVE WITH BUSINESS OBJECTIVE

Success is that old ABC - ability, breaks, and courage
Charles Luckman, architect

A value added auditing objective only works if the audit's objective is linked with the company's governance strategies and business objectives. Otherwise, the audit won't be perceived as adding value. An organization's strategic plan basically focuses on where the organization intends to go. A company may want to enter new markets, develop new products, or offer new services.

ORGANIZATIONAL METRICS

Audit objectives are often linked with organizational metrics. For example, the value added auditor is asked to evaluate business unit or process performance against established metrics. Process measurement and business metrics may involve customer satisfaction, time-to-market, profitability, cost reduction, or operational effectiveness. If non-critical or non-essential activities are audited, the rationale of the value added audit may ultimately be questioned.

It has become conventional management wisdom that excellent companies measure critical processes, system, and products. What should be measured? Theoretically, thousands of factors, elements, attributes, features, and characteristics throughout an organization can be measured. But, what critical data should be measured, collated, and analyzed by the value added auditor? This ability is what separates a professional value added auditor from a rookie. The professional value added auditor should have the operational experience and know-how to determine which process and product characteristics should be analyzed during the audit. Again, it's critical the value added auditor has the capability to separate the 'critical few' measures from the 'trivial many.'

METRICS AND GOALS THAT DRIVE ORGANIZATIONAL PERFORMANCE

More often, value added auditing assesses organizational performance and effectiveness. Strategic objectives start at the top of the organization. Corporate wide goals are first established and then are communicated to all employees. Goals and objectives are then established throughout the organization. Daily, weekly, monthly, quarterly, and yearly process improvement progress is measured towards meeting goals. At the lowest organization level,

PLANNING THE AUDIT – AUDIT OBJECTIVES

GAGAS 6.08 The objectives are what the audit is intended to accomplish. They identify the audit subject matter and performance aspects to be included, and may also include the potential findings and reporting elements that the auditors expect to develop. Audit objectives can be thought of as questions about the program that the auditors seek to answer based on evidence obtained and assessed against criteria. The term "program" is used in GAGAS to include government entities, organizations, programs, activities, and functions.

the assembly line or the receptionist, progress may be measured on a monthly, weekly, or daily basis.

Process or product objectives are then placed in the performance objectives of everyone in the organization. In the last several years, process objectives were only incorporated in the compensation and performance plans of senior management. This was especially critical in the early stage of the value added audit where senior management was actively involved in the initiative. Performance objectives have become part of everyone's performance plans. Often, the value added auditor is evaluating the effectiveness of these.

TYPICAL AUDIT OBJECTIVES

Compliance to a standard was and is probably the most typical type of audit objective. But compliance is necessary but often not sufficient in terms of adequately evaluating system and process effectiveness as well as risks and internal control effectiveness. The compliance decision is a go or no-go decision. The value added audit decision is more analytical. It may involve an analysis of an organization's risk - control structure, which may involve many related factors, some more important or stronger than others. For example, a critical audit question to be answered is whether there are sufficient strong controls to warrant a decision that risks are effectively managed.

Sample audit objectives can be found in the "Performance Objectives' sidebar on the next page. In addition to compliance, the value added auditor may have to evaluate the following objectives:

- **Improvement.** Improvement has been described in terms of limiting variation at a target, increasing production, mitigating risks, increasing effectiveness, and decreasing defects.
- **Efficiency.** Efficiency was discussed at length in Chapter 1 and is a major tenet of operational auditing. Efficiency is the process of minimizing losses, efforts, and waste. Or expressed another way, efficient performance accomplishes business objectives and goals in an accurate, timely, and economical fashion. [4]
- **Risk.** Risk is the probability of an unwanted event and is becoming a very critical value added audit objective.
- **Effectiveness.** Effectiveness is the ability to obtain or to achieve desired business results or objectives.
- **Economy.** Total costs, including maintenance, operations, product life cycle, front-end, and development costs, are optimized. The operative word is 'optimized' be-

PERFORMANCE OBJECTIVES

GAGAS A2.02 Examples of program effectiveness and results audit objectives include:

a. Assessing the extent to which legislative, regulatory, or organizational goals and objectives are being achieved.
b. Assessing the relative ability of alternative approaches to yield better program performance or eliminate factors that inhibit program effectiveness.
c. Analyzing the relative cost-effectiveness of a program or activity, focusing on combining cost information or other inputs with information about outputs or the benefit provided or with outcomes or the results achieved.
d. Determining whether a program produced intended results or produced results that were not consistent with the program's objectives.
e. Determining the current status or condition of program operations or progress in implementing legislative requirements;
f. Determining whether a program provides equitable access to or distribution of public resources within the context of statutory parameters.
g. Assessing the extent to which programs duplicate, overlap, or conflict with other related programs.
h. Evaluating whether the entity is following sound procurement practices.
i. Assessing the reliability, validity, or relevance of performance measures concerning program effectiveness and results, or economy and efficiency.
j. Assessing the reliability, validity, or relevance of financial information related to the performance of a program.
k. Determining whether government resources (inputs) are obtained at reasonable costs while meeting timeliness and quality considerations.
l. Determining whether appropriate value was obtained based on the cost or amount paid or based on the amount of revenue received.
m. Determining whether government services and benefits are accessible to those individuals who have a right to access those services and benefits.
n. Determining whether fees assessed cover costs.
o. Determining whether and how the program's unit costs can be decreased or its productivity increased.
p. Assessing the reliability, validity, or relevance of budget proposals or budget requests to assist legislatures in the budget process.

cause lowest front-end price may entail sacrificing another element that may induce higher future costs or higher risks. Or expressed another way, economical performance accomplishes business objectives and goals with minimal use of resources (i.e., cost) commensurate with the risk exposure. [5]

- **Existence.** Existence implies that documents or test results exist and are readily available.
- **Occurrence.** Occurrence verifies that what is stated or believed to have occurred actually happened.
- **Completeness.** This has several meanings. The activities or documents form a logical, accurate, timely, and systematic trail from initiation to closure. This may mean documents have all the requisite information, or assertions that a recall of

> **PLANNING THE AUDIT – BOUNDARY CONDITIONS**
>
> **GAGAS 6.09** Scope is the boundary of the audit and is directly tied to the audit objectives. The scope defines the subject matter that the auditors will assess and report on, such as a particular program or aspect of a program, the necessary documents or records, the period of time reviewed, and the locations that will be included.

products did not omit some items, or an audit recommending corrective action resulted in eliminating the root cause of the problem.

- **Accuracy.** Accuracy has several definitions. Measurement equipment has been calibrated. Results are on the specification target.
- **Precision.** Precise measurement equipment can replicate measurements. Test results are repeatable, meaning that there is little variation in individual measurements.
- **Valuation.** Valuation deals with price, cost, or value, which can be misstated or misrepresented.
- **Disclosure.** Important information may be omitted or misrepresented, which can change or distort the meaning or significance of a report.
- **Allocation.** Monies or resources are allocated to the right areas at the right times. These may be allocated in incorrect amounts or in wrong areas. This may be intentional or unintentional.
- **Reasonableness.** An auditor may test an entire population or pull a sample so sufficient information is gathered to provide reasonable assurance for an audit decision.
- **Sufficiency.** This deals with the quantity of evidence to collect. The desired level of assurance tries to balance costs and risks.
- **Simplification.** This deals with avoiding redundancies or replication.
- **Timing.** When to collect evidence or information depends on the type of evidence and the best time to collect it.
- **Validity.** This deals with whether the right things are done right at the right time with the right resources with the right consequences during the audit to ensure that the customer is satisfied.
- **Ownership.** This has multiple meanings. Does the audited physically own the property or equipment? Do people take responsibility for their actions?
- **Classification.** This deals with whether costs, property, plant, or equipment were properly allocated and reported.

VALUE ADDED AUDIT PROCESS METRICS

Profits are part of the mechanism by which society decides what it wants to see produced
Henry Wallich

Measurement or metrics should be tied to what matters to the organization, which may be customer satisfaction, time-to-market, profitability, cost reduction, or operational effectiveness. If non-critical or non-essential activities are measured, the rationale of the improvement project and ultimately the value added audit initiative might be questioned. Over-

PLANNING THE AUDIT – AUDIT METHODOLOGY
GAGAS 6.10 The methodology describes the nature and extent of audit procedures for gathering and analyzing evidence to address the audit objectives. Audit procedures are the specific steps and tests auditors perform to address the audit objectives. Auditors should design the methodology to obtain reasonable assurance that the evidence is sufficient and appropriate to support the auditors' findings and conclusions in relation to the audit objectives and to reduce audit risk to an acceptable level.

measurement may be as critical as under-measurement. Some organizations measure everything in sight.

VALUE ADDED AUDITING METRICS

As well, metrics are the basis of determining the effectiveness of all auditing activities. Value added auditing metrics may include:

- Number audits conducted.
- Number of corrective/preventive actions.
- Auditor report cycle time (time to conduct an audit).
- Corrective action cycle time (time to correct finding or nonconformance).
- Stakeholder satisfaction.
- Audited satisfaction.
- Number of customer requirements for audits.
- Improvement metrics.
- Risk mitigation.

AUDIT BOUNDARY CONDITIONS, METHODOOGY, AND RESULTS

It has become conventional management wisdom that excellent companies measure critical processes, system, and products. What should be measured? Theoretically, thousands of factors, elements, attributes, features, and characteristics throughout an organization can be measured. So, what critical data should the value added auditor collect and analyze? A few important questions a value added auditor may ask are:

- Are there specific mission critical business objectives?
- Are there safety, health, and/or environmental requirements?
- Are these objectives deployed down the organization into processes and activities?
- Are there specific customer process or product requirements?
- Is the particular measurement attribute or characteristic related to customer satisfaction or to value?
- Are attributes prioritized in terms of critical, major, and minor criteria?
- Is the data or information important to other organizational measures, such as productivity, costs, or employee satisfaction?
- Is the data universally understood to be important?
- How will the data be used to improve performance, efficiency, quality, or effectiveness?

PLANNING STEP 1 – UNDERSTAND AUDIT AND BUSINESS OBJECTIVES

Soft and hard attributes or activities can be measured. Soft measures include internal customer satisfaction, team effectiveness, opportunity cost, and quality of work life. Hard measures include capability indices, mean-time-to-failure, performance, maintainability, six sigma, and others. Some attributes are easier to measure than others depending on organizational context.

More companies are pursuing 'world-class' measures. 'World class' to many companies means comparing systems and processes with the best in their industry, in the U.S., or in the world. Common operations measures include equipment set-up time, area and machine productivity, cycle time by product, inventory turns by product, and process improvement.

PROCESS MEASURES

Management usually establishes measurable process standards and specifications. Process objectives should be tied to financial measures, which are the common language of business. Common process measures include:

- Risk levels.
- Customer satisfaction.
- Project savings.
- Defect levels.
- Cost of quality (COQ).

Risk Levels

The Chinese character for risk has two components: 1. Upside risk and 2. Down side risk as shown in the 'Two Sides of Risk' sidebar on the next page. Upside risk can be thought of quality risk. Upside risk is the ability to find and develop opportunities; satisfy customers; and meet business objectives. Downside risk is what we often think of risk, as something to be avoided or something negative. Downside risk is the ability to identify system, process, or product weaknesses or constraints, then develop internal controls to mitigate and ultimately manage these risks.

Customer Satisfaction

Customer satisfaction is measured by surveys, customer complaints, and warranty claims. Manufacturers and service organizations can directly or indirectly solicit customer feedback.

Project Savings

Cost savings are the most common and effective yardstick for measuring process improvement or risk mitigation. Once costs are identified, areas with the highest costs are singled out for analysis and for improvement projects. Any value added audit has to be cost-effective.[6]

Defect Levels

Defect or nonconformance levels are a direct measure of value. As discussed, a product can have a critical, major, or minor defect depending on the criticality of the product attribute. Defect levels are measured on a production line, in a service activity, or in the field. De-

Two Sides of Risk

Upside Risk:

- **Optimizing and improving systems, processes, and products**
- **Meeting business objecives**
- **Seeking new opportunities**
- **Satisfying customer reuqiarmeents**
- **Exceeding expectations**

Downside Risk:

- **Identifying risk**
- **Developing controls**
- **Developing strategies**

fect rates on the production line are determined through visual examination or measurement of a product. Defect rates from the field are indicated by product returns, field failure rates, reliability rates, and warranty claims.

Cost of Quality

Quality costs are important because the value of any process improvement effort is based on its contribution to company profits, whether through reduced costs or increased sales. The cost of quality (COQ) is a system that measures the costs of various components of quality.

The Cost Of Quality (COQ) includes costs associated with obtaining acceptable quality and as a result of not achieving this quality. Quality costs include scrap, rework, dissatisfied customers, loss of sales, etc. Costs can be incurred by any cost center or work unit and these costs can be high. Chrysler estimates that it wastes close to $4 billion a year or 20 - 25% of its sales revenue because of defects in its operations.[7]

Philip Crosby in his book **Quality is Free** maintained that process and product quality achievement pays for itself. This is counter intuitive, because most people think that quality costs more than it recovers. Philip Crosby believed the cost of poor quality could amount to 25% of sales in manufacturing companies and up to 40% of operating costs in service companies.[8] Other experts as well provide similar numbers. If these numbers are accurate, then the COQ is larger than pretax earnings of many companies.

VALUE ATTRIBUTES/OBJECTIVES

Always define your terms.
Eric Partridge

David Garvin, a professor at Harvard Business School, identified eight dimensions or categories of quality or value which auditors often assess. The list has been revised into the following nine value categories:

- Performance.
- Reliability.
- Conformance.
- Usability.
- Maintainability.
- Customer service.
- Aesthetics.
- Environmental.
- Software quality.[9]

Value added auditor should be familiar with the above attributes because they may become an audit objective. For example, the value added auditor may be asked to evaluate product performance, which is a product attribute. Or, the value added auditor may be asked to evaluate service performance, which is a process attribute. Most of the above quality attributes are measurable and precise. For example, even a value attribute, like aesthetics, can be measurable and auditable.

PERFORMANCE

Performance is an important value product characteristic. It is the ability of a product to operate up to the expectations of the user. Performance is measured differently in different products. In an automobile, performance includes acceleration, braking, and handling ability and in a pair of running shoes, it includes comfort, style, durability, and weight. Performance is a user-based definition of value, which means that an automobile can satisfy different expectations of different users. For example, a racecar driver expects a different level of performance from an Indy Ford compared to a Ford Escort.

Suppliers also are expected to improve a product's performance, packaging, reliability, and maintainability. Improvement may vary based on the type of product, such as whether it is an industrial or consumer product. Let us look at a computer. Performance of a personal computer may be enhanced by making it operate faster and transmit more information. Packaging can be improved by making it more aesthetic, easier to operate (user friendly), lightweight, and more compact.

COMPLIANCE OBJECTIVES

GAGAS A2.04 Compliance objectives include determining whether:
a. The purpose of the program, the manner in which it is to be conducted, the services delivered, the outcomes, or the population it serves is in compliance with provisions of laws, regulations, contracts or grant agreements, or other requirements;
b. Government services and benefits are distributed or delivered to citizens based on the individual's eligibility to obtain those services and benefits;
c. Incurred or proposed costs are in compliance with applicable laws, regulations, contracts, or grant agreements; and
d. Revenues received are in compliance with applicable laws, regulations, contracts or grant agreements.

A2.05 Examples of objectives pertaining to prospective analysis include providing conclusions based on:
a. Current and projected trends and future potential impact on government programs and services.
b. Program or policy alternatives, including forecasting program outcomes under various assumptions.
c. Policy or legislative proposals, including advantages, disadvantages, and analysis of stakeholder views.
d. Prospective information prepared by management.
e. Budgets and forecasts that are based on (1) assumptions about expected future events and (2) management's expected reaction to those future events.
f. Management's assumptions on which prospective information is based.

RISK/RELIABILITY

Reliability is long-term quality and is the probability of a product failing after a specified period of time under certain operating conditions. In other words, reliability measures the likelihood that a product or assembly will work over time. Reliability is both a user-based and a conformance-based value attribute. For example, if a customer buys a lamp with a one-year warranty, the customer expects it to last at least one year. Otherwise, the customer will return it for repair, refund, or replacement. The lamp manufacturer knows this and does not want to spend much money replacing defective lamps. Therefore, the lamp is designed and manufactured to last at least one year.

Reliability is measured in terms of mean-time-between-failure (MTBF) or mean-time-to-first failure (MTFF). MTBF is the average time it takes a product to fail between two successive failures. MTFF is the average time it takes a product to fail based on the first time it is put into service. For example, automobile manufacturers know what parts of an automobile will likely fail, so they suggest checking and replacing parts based on a preventive maintenance schedule.

CONFORMANCE

Conformance is the ability of a product to comply with a specification, standard, or design. Conformance is also the ability of a product to comply with policies or procedures. There

PERFORMANCE AUDIT OBJECTIVES

GAGAS 2.11 Performance audit objectives vary widely and include assessments of program effectiveness, economy, and efficiency; internal control; compliance; and prospective analyses. These overall objectives are not mutually exclusive. Thus, a performance audit may have more than one overall objective. For example, a performance audit with an objective of determining or evaluating program effectiveness may also involve an additional objective of evaluating internal controls to determine the reasons for a program's lack of effectiveness or how effectiveness can be improved.

a. Program effectiveness and results audit objectives are frequently interrelated with economy and efficiency objectives. Audit objectives that focus on program effectiveness and results typically measure the extent to which a program is achieving its goals and objectives. Audit objectives that focus on economy and efficiency address the costs and resources used to achieve program results.
b. Internal control audit objectives relate to an assessment of one or more components of an organization's system of internal control that is designed to provide reasonable assurance of achieving effective and efficient operations, reliable financial and performance reporting, or compliance with applicable laws and regulations. Internal control objectives also may be relevant when determining the cause of unsatisfactory program performance. Internal control comprises the plans, policies, methods, and procedures used to meet the organization's mission, goals, and objectives. Internal control includes the processes and procedures for planning, organizing, directing, and controlling program operations, and management's system for measuring, reporting, and monitoring program performance.
c. Compliance audit objectives relate to an assessment of compliance with criteria established by provisions of laws, regulations, contracts, or grant agreements, or other requirements that could affect the acquisition, protection, use, and disposition of the entity's resources and the quantity, quality, timeliness, and cost of services the entity produces and delivers. Compliance requirements can be either financial or nonfinancial.
d. Prospective analysis audit objectives provide analysis or conclusions about information that is based on assumptions about events that may occur in the future, along with possible actions that the entity may take in response to the future events.

are thousands of standards that specify how a product should be designed, built, tested, installed, stored, maintained, and repaired. For example, a screw is a simple fastener that holds one part onto another. A screw specification may designate height, width, thread taper, threads per inch, material, and material strength.

Conformance is an engineering or manufacturing definition of quality, because it is related to a feature of a product that can be measured. Conformance is measured in terms of defects rates or levels. For example, a specification for a screw may state that a certain type of screw should have 11 threads per inch and should be made of soft steel. If a screw randomly selected and inspected from a lot has 12 threads per inch and is made of stainless steel, then the screw has two nonconformances. These are defects, even though the screw has more threads per inch and is made of a stronger material than required by the specification.

PLANNING STEP 1 – UNDERSTAND AUDIT AND BUSINESS OBJECTIVES

USABILITY

Usability is a key product value attribute. A product may have all the technical bells and whistles, but it should be user friendly. All of us have been frustrated in trying to use software that simply doesn't work. The reasons may be technical incompatibility, poor documentation, user inability, or a number other reasons. One company that has tested hundreds of software products uncovered many quality blunders, including excessive technical jargon, overuse of icons, over reliance on training to overcome poor software design, cryptic error messages, and poorly conceived help systems.[10]

MAINTAINABILITY

Maintainability is the same as serviceability. It is the ability of a defective product to be repaired easily, quickly, and economically. For example, if an automotive water pump fails, it has to be replaced. But first, the automotive equipment in the way of the water pump has to be removed. Since the compact car was designed to be light, the area under the hood is tight and everything is squeezed into a small space. This decreases automobile weight and also decreases serviceability because the automobile is harder to work on. A number of key processes ensure maintainability, including the design and service processes.

This value attribute is both a conformance- and user- based definition. In an engineering sense, maintainability can be measured. A common measurement is mean-time-to-repair (MTTR), which is the average time it takes to repair a defective water pump or service a customer. In a user-based sense, maintainability reflects how the customer feels about replacing the defective unit. Was it easy to replace? Did the person get bruised knuckles? Could it be done quickly? Or, if a serviceman at a repair station performed the work, was the service courteous, fast, and economical?

CUSTOMER SERVICE

Customer service is an important value attribute used to encourage repeat sales. Therefore, companies design products that can be serviced easily and safely. And companies are providing faster turnaround service to repair defective products. For example, Caterpillar Tractor promises it will deliver repair parts anywhere in the world within 48 hours and Mercedes guarantees 24-hour service in California and Arizona.[11]

AESTHETICS

Aesthetics consist of the intangible attributes of quality: how a product looks, feels, sounds, tastes, or smells. Obviously, aesthetics are subjective, but they are a very important user-based value attribute.

Many people buy a car not based on measurable, technical features of conformance or reliability, but on their perceptions of quality, fit, and finish of the vehicle. People are often more concerned about the 'fit and finish' of an automobile than the technology inside, such as a new turbocharger. Or, they buy based on current and faddish styling standards. The same can be said for a number of other major consumer purchases, such as in apparel and fashion industries.

PLANNING STEP 1 – UNDERSTAND AUDIT AND BUSINESS OBJECTIVES

BUSINESS OBJECTIVE CONTROL QUESTIONS

1. Are business objectives, requirements, specifications, and metrics developed and appropriate for:
   a. Enterprise level?
   b. System level?
   c. Process level?
   d. Activity level?
   e. Product?
2. Within each organizational and product level, is there sufficient metric clarity and granularity?
3. Are the objectives consistent with the overall business and strategic objectives?
4. Is management (process prime and process owners) involved in setting doable objectives?
5. Do process owners have sufficient resources to achieve the stated objectives?
6. Has the process prime and/or owner identified critical risks from external and internal sources that can impact the ability to achieve system, process, and product objectives?
7. Has the process prime and/or owner:
   a. Prioritized the risk
   b. Likelihood of occurrence?
   c. Magnitude of the impact?
8. Has the process prime and/or owner evaluated contingency plans for reducing risks to an acceptable level?
9. Do the process prime and/or owner continuously anticipate, identify and correct unusual system and process events?

ENVIRONMENTAL

Environmental groups and regulatory authorities are important stakeholders. Many perceive that resources are being used and abused. The world's environment is changing radically as indicated through the ozone depletion, unbreathable air in American cities, toxic waste sites, and the list goes on.

Businesses are now listening as a result of the Valdez oil spill, Three Mile Island, and other environmental disasters. Environmental concerns go well beyond being better corporate citizens, companies want to improve their public-customer perceptions as well as limit their legal exposure. Compliance to regulations is moving toward a more proactive approach to prevent environmental degradation. How is this done? Companies are designing highly reliable and maintainable products. Packaging uses recycled materials. Products can be disposed cleanly and easily.

Environmental quality is also becoming more critical in determining overall project or product costs. Companies are beginning to implement processes to improve environmental conditions. In many companies environmental issues were once the responsibility of environmental compliance specialists who were often inspection and correction oriented. Now, prevention planning, periodic auditing, and people involvement are being emphasized.

More importantly a philosophy, the quality of global life, is emerging. Articulate and fervent believers maintain the world is a finite resource, which is constrained by real limits to

> **SIGNIFICANCE IN A PERFORMANCE AUDIT**
>
> **GAGAS 6.04** The concept of significance assists auditors throughout a performance audit, including when deciding the type and extent of audit work to perform, when evaluating results of audit work, and when developing the report and related findings and conclusions. Significance is defined as the relative importance of a matter within the context in which it is being considered, including quantitative and qualitative factors. Such factors include the magnitude of the matter in relation to the subject matter of the audit, the nature and effect of the matter, the relevance of the matter, the needs and interests of an objective third party with knowledge of the relevant information, and the impact of the matter to the audited program or activity. Professional judgment assists auditors when evaluating the significance of matters within the context of the audit objectives. In the performance audit requirements, the term "significant" is comparable to the term "material" as used in the context of financial statement engagements.

growth. Another new philosophy is sustainable development. Assertive governments are developing mandates to improve business environmental performance. ISO has developed series of environmental management standards, called the ISO 14000 series.

SOFTWARE QUALITY

Software specifications are an integral part of all management information systems. Software specifications are different than other specifications. But software quality, just like any tangible product, must satisfy user's requirements, comply with relevant standards, be cost-effective, and be measurable. For example, software quality must also be interoperable, intraoperable and exhibit a number of other product characteristics.

ADDING VALUE

Customer focus is the key to all value added auditing success. Someone once said that perceptions are reality. Well, this applies to auditing value. If management and critical stakeholders think the audit adds value then process stakeholders and management will use audit results to manage risk and to improve decision-making. If management sees the audit as a means to maintain the status quo, appease a customer, then audits will lose real and perceived value over the long term.

There are a number of ways to indicate value. One way is to change the way a person manages, from inspection to prevention, from management by anecdote to management by facts, from risk mitigation to risk management, or from rework to prevention. Other ways to add value include managing processes, minimizing waste, minimizing process variation, improving operational effectiveness, minimizing costs, or managing by exception.

Let's look at one example of adding value; to manage by exception. Management by exception implies if the process owner sees a spike in a process metric then this indicates something unusual has occurred. For a process owner or functional manager to manage by exception, processes should be defined, flowcharted, and stabilized. If there is an unusual occurrence, unusual condition, or an exception, then this is noted on a control chart. The process owner can then correct the problem and then reestablish the process within the process control limits.

PLANNING STEP 1 – UNDERSTAND AUDIT AND BUSINESS OBJECTIVES

SUMMARY

There has been much soul searching about the effectiveness and value-add of audits to improve operational performance and profitability. Too often, quality or internal audits are treated as separate activities or initiatives apart from the core business and planning processes. As stand alone activities, internal or quality audits are given less priority than daily operational concerns such as meeting daily production targets, responding to a customer request, or fixing a down machine. As well, internal audits seem to be given less priority than long-term concerns, such as customer satisfaction, profitability, and cost reductions, which are directly viewed as key business performance and success indicators. There is now much more concern to integrate and align value added improvement into a company's long-term business objectives and plans as well as daily operations.

Reliable and accurate performance and risk management information are now much more critical because they indicate if a particular business objective, purpose, task or function was achieved.

CHAPTER 8
STEP 2 – NOTIFY/VISIT AUDITEE

The value added auditor should inform the audited of the impending value added audit as early as possible. If the audited is the last to know of the audit, then he or she may resent the intrusion. As a rule, there should be few or no surprises in a value added audit.

The method by which the audited is notified depends on the type of audit. If it is an internal value added audit, notification can be sent by email with an attached note. If it is a critical external audit, then a formal letter of transmittal over the chief audit executive's signature should accompany the notification letter.

NOTIFY THE AUDITEE

To me, the most important element in management is the human being. So, the first essential is to treat people with consideration.
Yoshiki Yamasaki

The value added auditor may draft a letter, and the head of value added auditing would review it and then send it to the appropriate party for signature. The tone of this letter can be informal or formal depending on who is being audited and the nature and purpose of the audit. Internal audit letters usually are informal while external are more formal.

PURPOSE OF THE AUDIT LETTER

The audit letter serves several purposes:

- Announces the impending value added audit.
- Identifies its purpose and objectives.
- Proposes an audit date.
- Defines the type of audit as well as its parameters (audit scope).
- Expresses authorization of the audit.
- Introduces the lead value added auditor and other audit personnel.
- Sets the beginning and ending dates of the audit.
- Requests assistance and cooperation.
- Requests a contact person be identified.
- Mentions the distribution of audit results.
- Identifies audit lead and team members.
- Identifies areas or issues of special interest.
- Lists special requirements and resources.
- Lists the distribution of the report.

PLANNING STEP 2 – NOTIFY/VISIT AUDITEE

SAMPLE LETTER INFORMING SUPPLIER OF VALUE ADDED AUDIT

To: Supplier's Top Manager
From: Vice-President, Quality
Subject: Customer-Supplier audit

The value added-auditing group will conduct an audit of plant ABC on from Date X to Date Y. The purpose of the audit is to examine Quality Management System and processes as specified by the quality standard and purchase order 101. Section 3.13 of our quality standard specifies that we notify you 4 months prior to the audit as well as 2 months prior to the preliminary survey.

Jim Brown, the lead auditor, and Carol Smith, his assistant, will conduct the audit. They will contact you shortly to discuss audit specifics. They will also request a preliminary meeting be scheduled at a mutually beneficial time to discuss audit objectives and to answer your questions. They can be reached at (123) 456.7890 if you've any follow-up questions.

The enclosed survey questions were developed to inform us of the status of your continual process improvement program. We would appreciate it if you answer the questions fully and attach any relevant information to the survey. Please return the completed survey within 6 weeks of the audit.

After the questions have been reviewed, please appoint a contact person to arrange for scheduling facilities, equipment, interviews, and resources. We want to minimize disruptions in your work area and appreciate your assistance in this audit. Again, thank you for your attention to this very important matter.

Sincerely,
Chief Audit Executive

The letter is usually addressed to the process manager or process prime of the area being audited. If the function or area being audited is internal and a team is accountable for the area's work, then the team in general or the team leader may be addressed. The important point is to request the assistance of the audited and to have the audited assign someone to be responsible for assisting the auditor. See above 'Sample Letter Informing Supplier of Value Added Audit' sidebar.

The more time the value added auditor has to prepare and plan, the less time it will take to conduct the audit. 'No surprises' is a key element of all value added audits. Unexpected audit results can surprise and can result in higher unpredictability.

SENDING THE AUDITEE AN AGENDA

Send the audited an agenda of the meeting, preliminary audit schedule, and list of audit objectives as soon as possible. The audited should know the audit schedule, what to expect from the audit, information and resources required, timelines, and post audit meeting time.

Prior knowledge of the auditee's operations and systems enhance the value added auditor's effectiveness. The auditor cannot conduct an efficient and effective value added audit if he or she doesn't know the auditee's organization, systems, processes, and products. The time to learn about the auditee's industry, organization, systems, processes, and products is not during the audit.

An audit may disrupt the auditee's operations. An audit or audit team that's tripping over the workers and requiring elaborate explanations over basics hinders the audit effort more than helps it.

VISIT THE AUDITEE

A small leak will sink a great ship.
Thomas Fuller

A preliminary plant or on-site visit may help the value added auditor obtain impressions of the facilities to be audited. Again, the visit may be no more than a plant walkthrough. It provides an overview of policies, standards, procedures, risks, and process flows. People are met. Processes and systems may be reviewed. Process risks and controls may be identified. Questions are answered. Again, its purpose is familiarization.

A plant walkthrough can also be used to obtain general impressions. These impressions may prompt further examination during the audit.

MAKING SURE THERE ARE NO SURPRISES

'No surprises' is more than simple courtesy. For example, if the process survey is sent sufficiently early, then the audited can review the questions and get back to the value added auditor for additional explanation. The value added audit will proceed faster. The audited can review internal processes and prepare for the audit. This doesn't mean putting on the best face, but having the required paperwork, people, and equipment on hand so that the auditee's operations are not disrupted and the audit is conducted effectively.

In this visit, the value added auditor may want to gather initial impressions as well as track material, paper, work, or other flows through the processes. In terms of material process flows, the value added auditor would track the raw material as it comes in the shipping dock, check processing steps, check incoming components and subassemblies, monitor the assembly of parts, and then check testing and process testing through the operation. Also, this assessment may also provide the level of required assurance if risks are low, controls are strong, or time is limited.

CONDUCTING PREAUDIT MEETING

If possible, the auditor should interview the audited before conducting the value added audit. This can be done on the phone or in a sit-down meeting. The latter is more effective but may be impossible if the audited is far away. Interviewing techniques are discussed in greater detail in the next chapter.

A major reason for the interview is to introduce the value added auditor. The auditor can explain the purpose, timelines, methodology, and scope of the value added audit. The auditor can ask for the auditee's help in making the value added audit as quick, efficient, and effective as possible. The audited can assist the auditor in any number of ways, including providing space, scheduling interviews, acquiring special equipment, or testing samples.

Preliminary contact with the audited can be by telephone followed by a formal letter, email, or fax. In some cases, a personal visit especially with an internal group may be required. During the initial visit, the auditor should establish rapport with the audited and try to understand the auditee's operations while the audited tries to understand the value added audit's objectives and the auditor's requirements for conducting the audit.

The preliminary meeting may follow up on the initial survey or can be separate from the initial survey. Often the audited is sent the survey, then there is a meeting to discuss and follow up on the survey responses.

In a formal, extensive value added audit, the interview is called a preaudit meeting. The purpose of the preaudit meeting is to kick off the formal value added audit. Each party exchanges information. The value added auditor can learn about the audited and the audited is assured of the purposes of the audit and fears are allayed.

VIDEO CONFERENCING - THE NEXT BEST THING TO BEING THERE

If the audited is nearby, then a meeting can be easily arranged. If the audited is distant, then a videoconference or skype session can be conducted. These are expensive but are becoming more popular as time, travel, and availability constraints increase.

The videoconference is similar to a sit-down conference except two or more parties may be thousands of miles distant. Each party goes to a studio with a board, cameras, and table. Each site then transmits in real time the participants' voices and images to the other sites. The conferees ask questions, discuss issues, and see responses on a screen. This type of conference is useful for conducting the preliminary audit meeting, post audit meeting, and interviews. In the near future as costs decrease, videoconferences will become more prevalent.

INTERVIEWING THE AUDITEE

Interviewing is the process of obtaining information through social interaction. It requires the ability to listen actively, to understand complex situations, and to articulate thoughts. This skill is required in planning, conducting, and reporting audit results.

Of the various techniques discussed in this chapter for obtaining information, interviewing for many auditors is the most difficult. Many people attracted to value added auditing are highly skilled technical professionals who have worked by themselves for much of their lives. They designed or purchased products by themselves. Teamwork was not relevant. Often, these professionals hid behind their knowledge and had trouble interacting with others.

PLANNING STEP 2 – NOTIFY/VISIT AUDITEE

RECORD OF IMPRESSIONS[1]

Management Yes/No
I. Do supervisors appear concerned about the value added audit?
2. Is concern demonstrated by action?
3. Are supervisors compensated for implementing value added results?
4. Do supervisors seem to be respected by employees?
5. Does there appear to be a continual improvement ethic?
6. Do supervisors seem to know their jobs?
7. Are process management and controls well understood and established?
8. Is supervision sympathetic to employees?
9. Do process engineers assist supervision?

Employees
I. Do employees appear concerned about value added audits?
2. Is concern demonstrated by action?
3. Are employees compensated for implementing value added results?
4. Do employees seem to know their jobs?
5. Are process control concepts well understood?
6. Are employees relatively satisfied?
7. Is there a continual training program?
8. Are employees cross-trained?
9. Are people trained in the following:
a. Statistical process control?
b. Problem solving?
c. Team building?
d. Six sigma?

Environment
I. Is the work environment:
a. Clean?
b. Safe?
c. Well lighted?
d. Orderly?
2. Is there adequate test equipment?
3. Is equipment properly maintained and repaired?

Process controls
I. Is there sufficient measuring equipment?
2. Is it calibrated?
3. Do personnel know how to use them?
4. Are they properly handled, stored, and secured?
5. Is there a comprehensive process control structure in place?
6. Is corrective action and preventive action effective?
7. Who is responsible for corrective action and preventive action?
8. Is continual improvement pursued?

A new skill set is now required of value added auditors. They now need profound business practice and process knowledge so they can do risk assessments. The professionals may have to serve as facilitators or even advocates of the continual improvement process. The auditor in the interview process has to establish rapport with the audited. This rapport reinforces a participatory, problem-solving environment, which is required to improve processes and manage risks.

Types of Questions to Ask
The value added auditor can obtain the same information by asking or phrasing questions in a different manner. A question that solicits resentment or resistance should be reformulated. The same question can be asked in different ways.

- **Closed.** Did you perform the inspection?
- **Forced.** Did Jane, Tom, or you perform the inspection?
- **Direct.** What are your responsibilities?
- **Open.** Tell me about your process improvement program.
- **Probe.** In what ways was corrective action completed?
- **Encouraging.** Your process controls work well; do you have any other suggestions for improvement?
- **Linkage.** You mentioned that process improvement in Plant X can be implemented easily, can you elaborate on that?[2]

Prepare for an upcoming interview
If it's an extensive interview, the value added auditor may develop an agenda, send it to the auditee, and stick to it. There is a tendency to go into the interview as an interrogator. Questions can be rapid and hostile, creating a suspicious and fearful atmosphere, resulting in the auditee becoming reactive and hostile. This does not encourage or reinforce a climate of cooperative problem solving. Setting the proper tone in any interview is important. The value added auditor should be open, trusting, and candid, not authoritarian, policing, suspicious, or accusatory. This is especially critical in the entrance and exit meetings. If a cooperative and open tone is established in the first meeting, the reservoir of goodwill will continue through the value added audit. This helps to establish the necessary goodwill for subsequent meetings.

The auditee's cooperation and participation in the audit is necessary so that the auditee will accept and use the audit results to manage risk and improve the process. In a participatory problem-solving approach, the auditee is candid and the auditor does not fix blame in the audit report. Cursory interviews are often conducted as part of a plant tour or walkthrough of the auditee's facilities. The walkthrough was discussed earlier in the book.

The 'Record of Impressions' figure on the previous page lists questions to consider when touring the plant. If this is the first visit to the facility, the value added auditor should ask broad, open-ended questions focusing on overall systems, processes, and controls. The purpose is to understand the auditee's operations. If the value added auditor is reauditing a facility, then the value added auditor is already familiar with the operations so the value added auditor can focus questions on changes that occurred since the last audit.

PLANNING STEP 2 – NOTIFY/VISIT AUDITEE

INITIAL AUDITEE QUESTIONS

- Who is the auditee?
- What are the auditee's known risks?
- What are the peculiarities of the auditee's industry?
- What products are produced or services are delivered?
- What are the auditee's core processes?
- What are process risks?
- What specific risk events can occur?
- What is the likelihood of the event?
- What is the potential magnitude of the event?
- What are internal controls?
- Has an audit been conducted recently and what were the results?
- Were corrective action and/or preventive actions effective?
- Did the problem recur?
- Were people trained in problem solving?

SETTING A FRIENDLY TONE FOR THE MEETING

Good relations with the auditee are also very important, especially if the auditee is in the same organization. The value added auditor should make every effort to solicit the auditee's opinions, suggestions, and ideas. A value added audit can cause considerable disruption of the auditee's operations. Auditee personnel often must assist the audit team by obtaining documents and performing tests. Early conferences establish the rapport that is the foundation of a good working relationship. In general, this meeting sets the tone, agenda, schedule, and tests of the value added audit, so be friendly with the auditee. The tone of the meeting and subsequent audit should be participatory and candid.

The value added auditor shouldn't present himself or herself as a policeman or inquisitor. In a collaborative environment, the value added auditor and the auditee should see each other as partners in the value added audit and any subsequent improvement efforts. The value added auditor is a service provider to the organization. The auditee may also be a supplier of services, products, or information to the customer, who has retained the auditor to gather information.

A personal visit is more than a friendly touch, it may be politically smart. The person being audited may later be the customer requesting an audit and maintaining harmonious relationships in an organization aids the continual improvement effort.

PREPARE LIST OF ITEMS NEEDED FROM AUDITEE

Quality, not quantity, is my measure.
Douglas Jerrold

The next step is to obtain information about the auditee, an easy task if the auditee is internal to the organization. If external and if it is a first audit, then value added auditor may have to conduct a more thorough examination. The process of understanding the auditee is covered extensively in the 'Planning the Value Added Audit' chapter.

SENDING AUDITEE SURVEY

The value added auditor should send the preliminary survey at least 6 weeks before the value added audit. The auditee should take no more than 2 weeks to review the audit information and then request a meeting. The auditor should address and clarify issues that were brought up in the survey. The auditee may also have questions.

In this meeting, the value added auditor should prepare a list of items that may be needed to conduct the audit. The list of information depends on the nature of the value added audit, but the following list addresses important areas, including:

- Management value added policies, standards, and procedures.
- Financial condition.
- Legal obligations.
- Industry position and comparisons.
- Organizational structure.
- Internal operations.
- Internal risk - controls.
- Value added culture, philosophies, and attitudes.
- Engineering prints.
- Technology standards.
- Process flowcharts.
- Specifications.
- Purchase orders.
- Tests.
- Statistical control charts (SPC).
- Policies, procedures, and instructions.
- Information material and other flows of resources.
- Cost information, including quality costs and improvement costs.
- Benchmark and other information.
- Test measurements.
- Problem areas and effectiveness of corrective actions.
- Continual improvement information.
- Value added planning.

OBTAINING AUDITEE PROCESS INFORMATION

Obtaining flowcharts may be the most critical element of value added audit planning. A flowchart illustrates the movement of resources, including information, money, raw material, people, or paperwork. The major advantage of flowcharting is that it shows what is occurring as a product flows through a sequence of process steps and what value is added in each step. By using a flowchart, the value added auditor gains a quick understanding of the workings of the system to be investigated. A flowchart can sometimes be obtained by checking the permanent file. If the flowchart is over a year old, the auditor should request a new one from the auditee. Templates are available for document, information, and material flows. The template has standard symbols representing the most common activities.

SUMMARY

If the value added auditor has previously conducted an audit of the auditee, understanding the auditee may simply mean reviewing the current processes and updating information. Or, it may mean obtaining basic information on the following:

The nature, level, and extent of the data gathering are related to the value added audit. If the audit entails little risk and is a cursory examination then the value added auditor doesn't have to obtain much information. If the audit involves high risk, then a higher level of assurance is required, and the audit is more extensive and intensive.

If previous value added audits have been conducted, the value added auditor should review those results to develop an audit plan and to determine areas of focus. Reviews of past audits shouldn't stop the auditor from reviewing areas that were satisfactory in the past.

PLANNING STEP 2 – NOTIFY/VISIT AUDITEE

CHAPTER 9
STEP 3 – UNDERSTAND AUDITEE'S SYSTEM, PROCESS, AND PRODUCT DOCUMENTATION

In this audit step, the value added auditor starts developing a good understanding of the auditee by evaluating system, process, and product documentation. Just as customer requirements must be understood, the value added auditor should understand the auditee's system and process documentation.

REVIEW AUDITEE DOCUMENTATION

Structure is not organization.
Robert H. Waterman, writer and consultant

By consulting with people who conducted or were involved with a previous audit, the value added auditor can obtain valuable, first hand information that can make the current value added audit more effective, efficient and economic. If a team conducted the previous audit, it may make sense to add one of these people to the new value added audit team. In some cases, a member of the initial audit team becomes the lead auditor for the new audit.

REVIEWING PERMANENT AUDIT FILE

The value added auditor should review the permanent file to determine what was done in previous audits. Previous audit results as well as general background information should be in the permanent file. The information should include all the critical information from each of the three audit stages. The file may contain:

- Audit plans.
- Interview results.
- Preliminary survey.
- Quality documents.
- Engineering prints.
- Internal audit memos.
- Risk-control analysis
- Meeting minutes.
- Audit reports.

PLANNING STEP 3 – UNDERSTAND AUDITEE'S SYSTEM, PROCESS, AND PRODUCT DOCUMENTATION

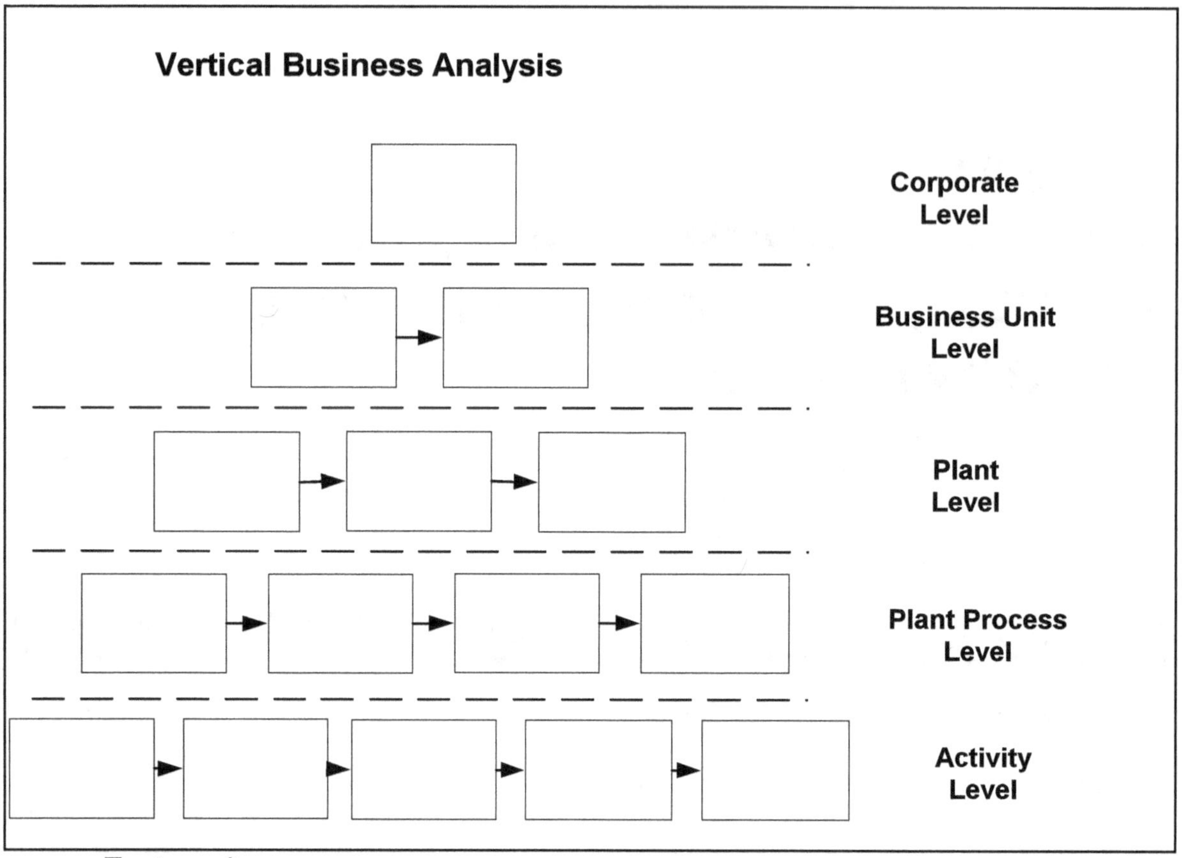

- Test results.
- Corrective and preventive action requests.
- Inspection results.
- Internal control reports.
- Calculations.
- Risk assessments.
- Deficiency findings.

REVIEWING STANDARDS AND OTHER RELATED DOCUMENTS

Often, value added audits evaluate compliance or adherence to standards, procedures, or other documents. The permanent file should contain accurate, complete, and understandable, audit documents that apply to the auditee. If the last audit was more than a year ago, then it is fairly certain the value added auditor may have to request current documents from the auditee. The reason why they are so important is because these documents specify with what the auditee must comply and conform.

The following is a partial list of related valued added audit documents, which may have to be reviewed prior to the audit:

- Contracts.

- Purchase orders.
- Benchmark tests.
- Quality manual.
- Supplier quality standards.
- Customer feedback.
- Regulations/mandates.
- Standards.
- Industry standards.
- National standards.
- Service manuals.
- Policies.
- Procedures.
- Shipping, storage, and packaging information.
- Housekeeping standards.

VERTICAL BUSINESS ANALYSIS

The value added auditor should first assess the business environment, context, industry, and overall systems of the auditee. The overall analysis of the auditee provides the value added auditor with a macro perspective.

In this analysis, the auditor looks at the auditee in terms of global competition, national competition, and industry position. This provides background information in terms of the auditee's relative strengths and weaknesses and is especially useful for analyzing suppliers. For example, many companies just want to work with world-class suppliers and this type of analysis can reveal a supplier's competitive benchmarks and core processes.

This vertical analysis may start at the corporate level and go down to the business unit, plant, plant process, and activity levels. One of the things I've learned through many assessments is that risks are found in business, process, and product interfaces or interactions. Each level has organizational, system, process, product, and work level documentation as shown in the 'Vertical Business Analysis' figure on the next page. By the time the value added auditor has worked through several levels, the auditor should have a good overview of the auditee. Depending on the scope, risk, controls, and assurance required, the value added auditor must assess the level of detail that he or she must needs to review. If the auditor needs more information, then this can be requested of the auditee and can be reviewed and assessed during the conduct of the audit. This analysis can proceed from top down as shown by analyzing:

- Global competition and benchmarks.
- National competition and background information.
- Industry position and margins.
- Quality Management System.
- Supplier core processes.

PLANNING STEP 3 – UNDERSTAND AUDITEE'S SYSTEM, PROCESS, AND PRODUCT DOCUMENTATION

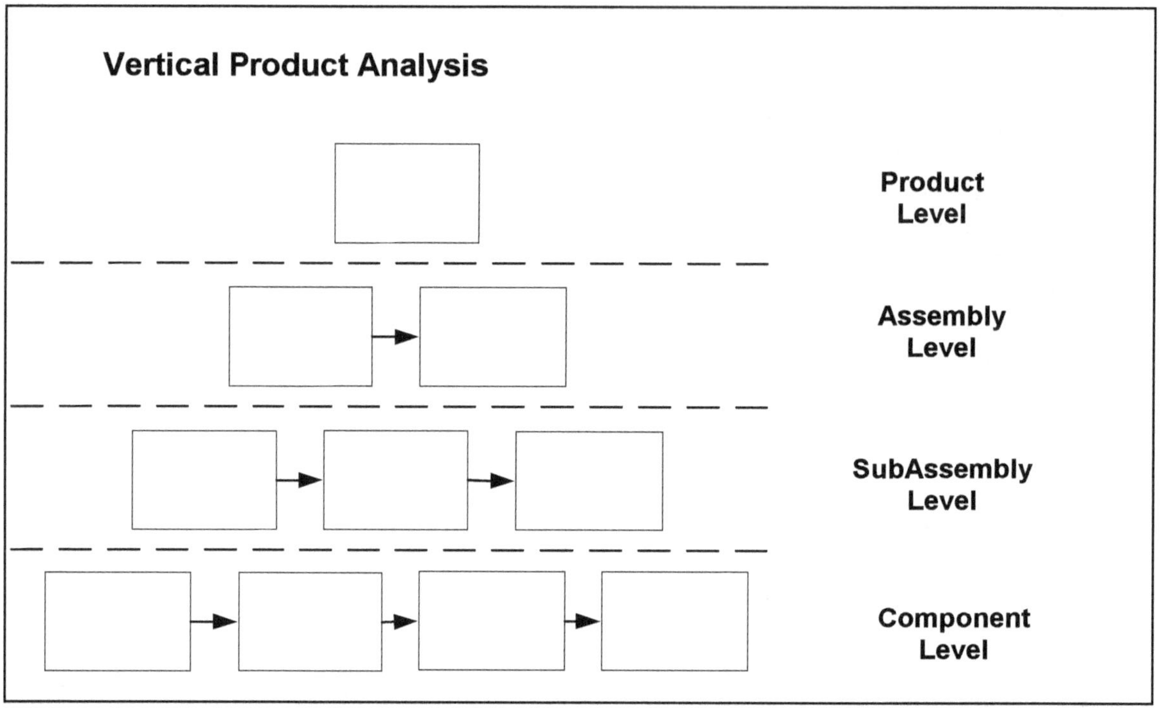

VERTICAL PRODUCT ANALYSIS

Organizations are systems of interrelated parts. However, to assess a part of the system properly may require that the whole system is understood as well as the interfaces or interaction of component parts. For example, a product may have many parts, subassemblies, and assemblies. A product assessment determines if the product performs up to system requirements; but a root cause analysis may require investigating a part of a subassembly that is provided by a third-tier supplier. And, the supplier may be located 10,000 miles away.

So, the value added auditor should assess a company's product in much the same way. The auditor can assess the overall product, then assembly, subassembly, and finally component level documentation as indicated by the above 'Vertical Product Analysis' figure. The documentation may take the form of drawings, bills of material, standards, procedures, flowcharts, or work instructions. Each level of documentation reveals more information on the product's value characteristics, such as those previously discussed, such as in product performance, reliability, usability, or maintainability.

INTERFACE PROCESS ANALYSIS

The value added auditor can also assess process interactions and interfaces during this step. This analysis gives the auditor an interrelationship or 'white space' perspective of the auditee. In this case, the value added auditor may look at the inputs, the process, and the outputs. The auditor may look at how one process feeds into another or how a supplier provides products to a customer's process. At the lowest level, this is relatively easy. For example, the inputs for a machine process may be raw material that is being fabricated. The outputs would be products that meet specifications. The interface analysis may reveal important information involving time, money, specifications, resources, material, and training risks.

Looking at a simple machine process, the value added auditor can determine if specifications are being met, time the operation, determine what value was added at what cost and how much material was consumed, examine operator training, and so on. This same process analysis can be done in a professional, administrative, and service environment. At the lowest level, the input, process, and output analysis is relatively simple. As one goes up the corporation into company wide and supplier wide systems, the levels of complexity increase quickly. This is discussed in more detail in 'Conducting the Value Added Audit' section of this book.

EVALUATE COMPLIANCE WITH STANDARDS AND DOCUMENTS

It is a paradox that the greater the decentralization, the greater the need for both leadership and explicit policies from the top management.
Bruce Henderson, consultant

Compliance audits particularly add value to comply with regulatory or statutory requirements. They are also a baseline assessment in the conduct of a more analytical assessment. In a value added audit, there may be audit criteria or standards against which to objectively determine compliance, fulfillment, or adherence. Often, this is a critical objective of an audit. ISO 19011 defines audit criteria as a: "set of policies, procedures, or other requirements against which collected audit evidence is compared."[1]

STANDARDS AND MEASURES

Where are operational specifications or contract requirements of the process, product or service spelled out? While an audit customer may have specific requirements, these are more often expressed in terms of organizational requirements stated in a performance standard, drawing, business objective, or value added documentation. This documentation may be internal or external documentation that the organization adopted or must comply with. Examples of the latter include international, national, and industry standards that companies adopt in order to standardize processes, procedures, and products. For example, much of the world has adopted the International Organization For Standardization (ISO) 9000 system of quality standards. To a large extent, ISO 9001 has become the global quality standard.

Where are standards and measures found? A partial list of documents that specify operational requirements follows:

- Quality standards.
- Policy statements.

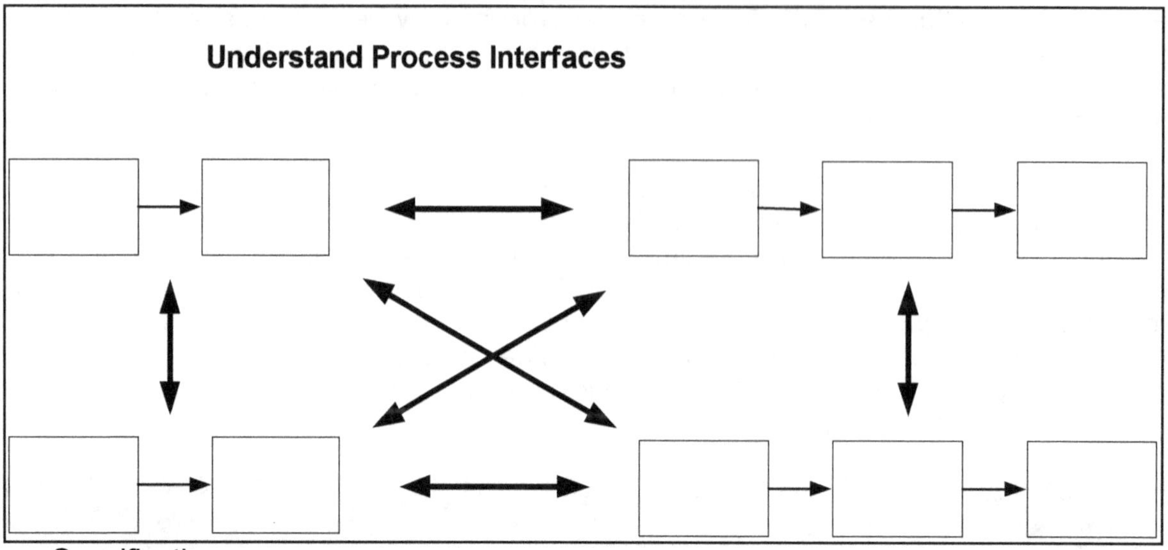

- Specifications.
- Industry standards.
- International standards.
- Trade practices.
- Service standards.
- Workmanship instructions.
- Work instructions.
- Purchase orders.
- Contracts.
- Benchmarks.
- Company directives.
- Policies.
- Procedures.
- Workmanship practices.
- Recommended practices.

There are many other performance-related documents that have customer-related information. The value added auditor's challenge is to identify the appropriate standard, business objective, or metric against which to evaluate operations, processes, products, conditions, situations, and functions.

GENERAL COMPLIANCE AUDITING

Compliance auditing is still required to meet regulatory or statutory requirements. Procedures must be in place, they must be followed consistently, and paperwork must verify they are consistently followed.

Compliance audits are an entry-level assessment. These audits assume certain factors. The organization has a defined system and process to be assessed. There are standards, which the organization must meet. The auditor evaluates conformance or compli-

PLANNING STEP 3 – UNDERSTAND AUDITEE'S SYSTEM, PROCESS, AND PRODUCT DOCUMENTATION

STANDARDS AND DOCUMENTATION BENEFITS

- Communicate value added requirements.
- Control and document key value added practices.
- Train employees in process and product requirements.
- Indicate compliance to requirements.
- Illustrate how the company handles personnel training, engineering design changes, and other important operational information.
- Establish a reference manual of approved, standardized, and accepted procedures and practices.
- Maintain uniform and proceduralized practices.
- Indicate how to maintain capable and stabilized processes.
- Serve as a model for continual improvement.
- Define methods and requirements for training.
- Indicate continued compliance with standards.

ance against these standards. The standards can be specifications, laws, drawings, policies, procedures, flowcharts or instructions. The important thing is that standards spell out what things need to be addressed and how they will be done.

In much the same way, organizations have standards that they follow so audits are consistent from audit to audit. For example, process standards may define control, capability, and improvement requirements. The goal of process standards is to ensure that a process is stable, meaning it is consistent in its operation so process results, products, and services are consistent.

Also, an organization may have audit standards. Why? While the processes and activities being audited may be different, the process by which auditors conduct audits is consistent. Again, consistency is the hallmark of compliance auditing.

PROCEDURES

A standard or guideline procedure can detail how the audit is managed, conducted, and reported. Each audit consistently follows the procedure.

Procedures can be 'shall' or 'should' documents. 'Shall' procedures prescribe and specify in detail how audits are conducted. 'Should' procedures are guidelines as to how audits are conducted. What are better? 'Shall' procedures work well when there is new staff, large turnover, high-risk areas, big dollar amounts, regulatory requirements or when there is a high need for standardization. 'Should' procedures work well when there is relatively low risk and people understand and follow work protocols? These people have intimate knowledge of the process and system. There are few nonconformances or risk events.

PLANNING STEP 3 – UNDERSTAND AUDITEE'S SYSTEM, PROCESS, AND PRODUCT DOCUMENTATION

DOCUMENTATION AND STANDARDS CHALLENGES

Doing little things well is the way towards doing big things better.
Anonymous

Lack of, conflicting, or misunderstood standards, policies, procedures, and work instructions can result in a product recall, excessive variation, contradictory practices and poor decisions. All of these make a value added audit difficult.

BUSINESS OBJECTIVES IN PLACE OF AUDIT STANDARDS

So what does the auditor do if there are no standards, drawings, policies, or procedures against which to compare audit evidence? This will be the case if the value added auditor is evaluating risks, internal control effectiveness, or consulting with process owners to prevent nonconformances. The auditor may have to identify overall business objectives or process metrics to use as the basis of the value added audit.

In a value added audit, the auditor would verify, evaluate, and report compliance or adherence in terms of performance standards. The value added auditor may assess processes and products against these documents. Processes and products are compared against business objectives and standards. Processes are compared against procedures, process flowcharts, and work standards. Products are evaluated, tested and inspected to determine conformance to engineering specifications or standards.

Some organizations follow the credo that employees are capable of determining what is right and doing it. However, problems can arise. The person or team guided by instinct, good will, skills, and knowledge may do what it deems right or correct at the moment. But, trouble may develop if the person or team does not anticipate developments, understand upstream/downstream process requirements, or have insufficient background knowledge. Written, consensual documentation should serve as the roadmap and milestone for value added activities.

DOCUMENTATION CHALLENGES

If organizational standards, policies or documents exist, they should be used and adhered to. However, this may not be the case for several reasons:

- Policies may not have been distributed to all managers or operating groups.
- Teams may not be using them because they have not been updated.
- New employees have not been introduced to policies, procedures, or work instructions.
- In a diverse workplace, new employees may not be able to read or understand procedures and work instructions.
- Policies and procedures have not been updated.
- Employees are not aware of the existence of work instructions.
- Policies may be contradictory.

- Departmental or workouts follow different practices than those spelled out in the documentation.

WHAT TO DO IF THERE IS NO DOCUMENTATION

Most organizations have established operational standards, metrics and expectations, which are operationalized through procedures and work instructions. If these exist, then the value added auditor can use these as one metric in the audit analysis. The auditor determines whether the business objectives, standards, metrics, processes and work instructions are acceptable to meet audit objectives and then determine if they are being met.

What happens if there are no technical, procedures, policies, specifications, standards, or other types of documents? This may raise a major red flag. The lack of defined, accepted, and established process documentation could be among the most glaring deficiencies the auditor could find. Why? Because the value added documentation details customer and organizational requirements. The value added auditor would have nothing against which to check systems or processes to establish what is acceptable and unacceptable. In terms of improvement, standards and specification documents state what is minimally acceptable, what are the benchmarks, and what is the rate of improvement.

If no documentation, standards, or metrics exist to form the basis of an audit opinion, the value added auditor has several options:

- May proceed no further and inform the auditee and the customer of your decision.
- May conduct a cursory evaluation of undocumented and uncontrolled systems and report results in the audit.
- May recommend to the auditee what steps to take to obtain the appropriate specification documentation and develop process controls.
- May work with the auditee to develop realistic standards and documentation.
- May report opportunities for improvement (OFIs).

Any of the above requires a rescoring of the audit, which should be reflected in the updated audit brief. The decision of what option to follow rests with the auditor's customer or supervisor and depends on time constraints, importance of the audit, and other risk factors.

SUMMARY

As well, the auditor should be sensitive to the auditee's needs. This may sound strange because it implies that the auditee is also the auditor's customer. In a value added management perspective, this is true. The internal value added audit is conducted ultimately to improve the organization. The auditee is part of the organization. The results of the audit are primarily intended for the person authorizing the audit, but the results are also intended to improve the auditee's processes. So in this sense, the auditor, customer, and auditee are all partners, or mutual customers, in the organization's improvement process.

The decision threshold for a value added audit is 'reasonable assurance.' If the auditor can develop this level of assurance, then he or she can stop at this planning step. This would usually occur when the auditor is familiar with the auditee's processes, risk is low, and controls are working properly.

As discussed, ISO 9001: 2015 and other standards are focusing on ' process effectiveness', 'risk - control', and 'efficiency of operations.' There is less emphasis on formal documentation. Any process documentation, specifically procedures, should be developed to make operations more efficient and effective. Documentation that is developed should make operations, processes, and systems more consistent by ensuring that:

- Customer requirements are understood and defined.
- Customer requirements are internalized into the company.
- Critical product attributes are identified.
- Critical process parameters are identified.
- Critical system, process, and product risks are identified.
- Processes are in control, capable, and improving.
- Operational methods are consistent and documented.
- Operators are trained in these methods.
- Operations and systems are monitored and audited.
- Deviations are symptom and root cause corrected.

CHAPTER 10
STEP 4 - DEVELOP AUDIT PLAN

The value added auditor should develop a formal audit plan before conducting the actual audit. The plan broadly specifies audit objectives, risks, controls, and tests for each segment of the audit. The audit plan, as a guide for conducting the audit, can be compared to a road map. It details which road to take, what are the milestones, how long it will take, what are the risks, and who is responsible for what, by which date. If the audit team or members do not have a plan, they might get lost in terms of evaluating unimportant areas or neglecting important areas.

AUDIT PLANNING

Any business must always plan ahead, either to capitalize on success or to reverse the trend if not successful.
Anonymous

By this time, the value added auditor has laid the groundwork of the value added audit plan. The audit plan should be aligned and reinforce organizational mission requirements. This should be reflected in the purpose of the audit. For example, if the audit objective is to determine compliance with a standard, then the audit plan ensures audit standards or requirements are clearly stated and understood.

ELEMENTS OF THE AUDIT PLAN

If the audit objective is to determine process effectiveness then the auditor would ensure that objective data is available to verify and validate effectiveness. If the audit scope requires a risk assessment then the project audit plan would identify organizational risks and exposures then detail how these are mitigated, controlled, or managed.

Regardless of the type of audit, audit objectives should be aligned with strategic goals and requirements. Strategic objectives detail what the organization wants to achieve. Think of this as the organizational destination. Along this journey, there are a number of steps. Each step has a number of potholes, obstacles, dead ends, switchbacks, and forks in the road. Each of these represents an operational or audit risk that should be controlled.

The project plan at a minimum should address the following:

- Identify value added audit customer.

PLANNING STEP 4 – DEVELOP AUDIT PLAN

PREPARING A WRITTEN AUDIT PLAN

GAGAS 6.51 Auditors must prepare a written audit plan for each audit. The form and content of the written audit plan may vary among audits and may include an audit strategy, audit program, project plan, audit planning paper, or other appropriate documentation of key decisions about the audit objectives, scope, and methodology and the auditors' basis for those decisions. Auditors should update the plan, as necessary, to reflect any significant changes to the plan made during the audit.

6.52 A written audit plan provides an opportunity for audit organization management to supervise audit planning and to determine whether:

a. The proposed audit objectives are likely to result in a useful report.
b. The audit plan adequately addresses relevant risks.
c. The proposed audit scope and methodology are adequate to address the audit objectives.
d. Available evidence is likely to be sufficient and appropriate for purposes of the audit.
e. Sufficient staff, supervisors, and specialists with adequate collective professional competence and other resources are available to perform the audit and to meet expected time frames for completing the work.

- Identify the auditee.
- Describe the deliverable (s).
- Establish audit objective or goals.
- Define a work breakdown structure (simple PERT or other chart identifying the critical project steps and audit methodology.
- Define the scope of the value added audit.
- Identify budget and timeline.
- Identify standards or other documentation to determine adherence.

Additional issues to consider in developing the audit plan could include:

- Overall business objectives.
- Peripheral value added auditing objectives.
- Sponsor of the value added audit.
- Dates and results of the last audit engagement.
- Updated assessments of risks and effectiveness of risk management and control processes.
- Requests by senior management, audit committee, and governing body.
- Current issues related to organizational governance.
- Major changes in organization's business, operations, programs, systems, and controls.
- Opportunities for improvement.
- Changes to and capabilities of the audit staff.
- Areas impacted by the audit plan.
- Flexibility to cover unanticipated demands on the internal audit activity.[1]

PLANNING STEP 4 – DEVELOP AUDIT PLAN

AUDIT PLANNING – RISK

GAGAS 6.11 Auditors should assess audit risk and significance within the context of the audit objectives by gaining an understanding of the following:

a. The nature and profile of the programs and the needs of potential users of the audit report.
b. Internal control as it relates to the specific objectives and scope of the audit.
c. information systems controls for purposes of assessing audit risk and planning the audit within the context of the audit objectives.
d. Provisions of laws, regulations, contracts, and grant agreements, and potential fraud, and abuse that are significant within the context of the audit objectives.
e. Ongoing investigations or legal proceedings within the context of the audit objectives; and
f. The results of previous audits and attestation engagements that directly relate to the current audit objectives.

SEGMENTING THE AUDIT AND MAKING IT EASIER TO PREPARE

The audit plan is also a management tool for the lead auditor to track different segments of the value added audit. A large audit consists of a series of small audits conducted by several auditors. This is called chunking or segmenting the audit. A lead auditor periodically assesses the progress of the segments and in the end pulls all the segments together into a comprehensible and comprehensive report.

Each segment may have its own mini audit plan. These subplans address the same issues as the larger plan but at a more specific level. For example in a company value added assessment, the audit team may focus on improving customer satisfaction based on indicators from focus groups, surveys, warranty costs, or complaints. At a lower level, the value added auditor would focus on proxy elements of customer satisfaction such as defect rates, cost of quality, safety, and aesthetic design. At the end of the audit, the chunks are pulled together to form the complete value added audit. If the whole audit is to evaluate a company or operation accurately, the constituent audits should be accurate, which largely depends on developing integrated monoplanes.

For example in an audit of a major supplier, a value added auditor would be assigned to conduct the audit of each segment. In a large audit, the team will usually develop the entire plan jointly. Individual team members then construct plans for their assigned audit areas. Then team members come together, comment, and review the subplans as they relate to overall audit objectives.

The value added auditor could also use questionnaires to organize the value added audit and to help answer the above questions. A positive response to a question implies that the specific control system exists and is operating properly. A negative response implies the opposite. Anytime the value added auditor encounters a negative response, the auditor can either issue a corrective action request (CAR), preventive action request (PAR), or investigate the matter further by testing the system. Today, many value added

AUDIT PLANNING – ADDITIONAL CONCERNS

GAGAS 6.12 During planning, auditors should also:

a. Identify the potential criteria needed to evaluate matters subject to audit.
b. Identify sources of audit evidence and determine the amount and type of evidence needed given audit risk and significance.
c. Evaluate whether to use the work of other auditors and specialists to address some of the audit objectives.
d. Assign sufficient staff and specialists with adequate collective professional competence and identify other resources needed to perform the audit.
e. Communicate about planning and performance of the audit to management officials, those charged with governance, and others as applicable.
f. Prepare a written audit plan.

auditors use open-ended questionnaires. This is because they are quick, effective and focus on customer's requirements.

IDENTIFYING AUDIT RISKS AND CONSTRAINTS

Every audit has certain risks and constraints. This is probably one reason why an activity, product, or process is being audited. So, it is important that the value added auditor attempt to identify, understand, eliminate, or minimize possible audit constraints and risks before the conduct of the audit. What are possible constraints or risks?

1. **Scattered information.** Scattered information makes an audit more difficult to conduct. Email, fax and instantaneous communications facilitate the transfer of documents but do not facilitate analysis and verification, which require a person onsite. In some cases, a third party may be contracted to perform tests, observe tests, confirm analysis, or conduct other tasks.
2. **Internal opposition.** Value added audits may be threatening and internal opposition may arise. This occurs in organizations where value added auditors are perceived as enforcers, inspectors, or informers. This won't occur if each employee recognizes that his or her goal is customer satisfaction, control of risky processes, and continual improvement.
3. **No authority to request audit.** If the customer does not have the authority to request an audit and this is required of the auditee, it may result in resentment. This sometimes occurs when customers start a process improvement program with suppliers who don't want to participate and the purchase order doesn't stipulate an audit.
4. **Process not capable of meeting business objectives and other requirements.** If a process is not capable of meeting requirements then it is difficult to conduct an effective audit. Capability is specified on the engineering print, specification, standard, or similar documentation. For example fundamental to all engineering design are realistic, manufacturable specifications. It's critical that design engineers develop specifications based on the ability of the manufacturing equipment. Specifications shouldn't be too loose or too tight.

VALUE ADDED AUDIT PLANNING CHECKLIST

- Is there a comprehensive audit plan?
- Are the audit scope and objectives clearly identified?
- Has the permanent file been reviewed?
- Were prior audit reports reviewed?
- Were prior value added auditors or specialists interviewed?
- Is a summary of previous deficiencies prepared and reviewed?
- Are value added audit risks and controls identified?
- Was the auditee interviewed?
- Was sufficient background information gathered?
- Was a preliminary survey conducted?
- Are flowcharts prepared and reviewed?
- Are constraints for conducting the audit identified?
- Are audit dates and timing set?
- Are value added audit objectives clearly defined and documented?
- Are value added audit milestones identified?
- Is a methodology for conducting the audit developed?
- Are sufficient resources and staffing requirements identified?
- Is the audit assigned to one person or team?
- Does each team value added auditor understand his or her responsibilities?
- Does each team specialist understand his or her responsibilities?
- Has the auditee identified a contact person?
- Are the auditee's operations, systems, processes, or products to be assessed identified?
- For each area being assessed, are risks and internal controls identified?
- Is there a list of tests to be conducted?
- Is required equipment identified?
- Are staff specialists identified and contacted?
- Is auditee assistance in data collection and testing requested?
- Is there a procedure or checklist for determining that all elements of the plan are completed?
- Is a preliminary survey completed?
- During the preliminary interview were the auditee's ideas and comments solicited?
- Is the preliminary survey discussed with value added auditing management, the customer, and the organization?
- Has the customer identified how, when, and to whom audit results are communicated?

5. **Process not in control.** A process not in control implies that special cause variation exists and is causing abnormal variation. This variation can be isolated, identified, and eliminated.
6. **Unavailable information.** One of the more vexing problems in a value added audit is unavailable information. The information has not been intentionally hidden. The problem is that data have not been collected, collated, and presented in understandable form. This is a widespread and common problem.

AUDIT FOCUS

Planning is the process by which profitable growth is sought and attained in a changing and uncertain world.
Anonymous

Risk assessments are a fundamental element of today's business. If value added auditing can address these directly and consult with senior management, then real and perceived value is being added. Risk issues today keep senior management awake at night. They may be a scan of changing marketing conditions, product development slippages, lack of operational controls, or poor quality products.

A risk assessment approach to value added auditing allows auditors to focus on areas with the highest level of control risk. The auditors and management can use this tool to identify areas of highest risk as well as identify the consequence of the company if the corporate process or activity is not controlled properly.

EXECUTIVE AUDIT PLANNING

Strategic planning involves identifying customer wants, needs, and expectations regarding quality, risk, products, delivery, technology, and service. Additional critical elements involve selling products and services at a price that attracts customers and is produced at a sufficiently low cost so profits can be generated.

Many organizations still focus on tactical and compliance auditing emphasizing compliance control, product testing, and inspection. While these are definitely important, the strategic issues dealing with customer satisfaction, corporate governance, supplier capabilities, organizational culture, and business systems sometimes are much more important. This can be demonstrated by the large number of companies that are implementing ISO 9001 whose focus is on documenting quality systems to demonstrate compliance to the standard as opposed to implementing business processes that improve performance. Strategic objectives that get management attention include continual process improvement, innovative products and services, risk management, customer satisfaction improvement, and cost reduction.

And too many companies are still managed by priorities and politics that don't matter to the customer or that generate little income. Also, too many audit objectives tend to focus on criteria that aren't important to the customer. It's critical that all value adding auditors understand and emphasize customer priorities and to satisfy them in every possible way. Value added auditing then becomes a means for continually evaluating and ultimately improving critical elements of the organization as well as the supplier base to enhance customer satisfaction and generate revenue.

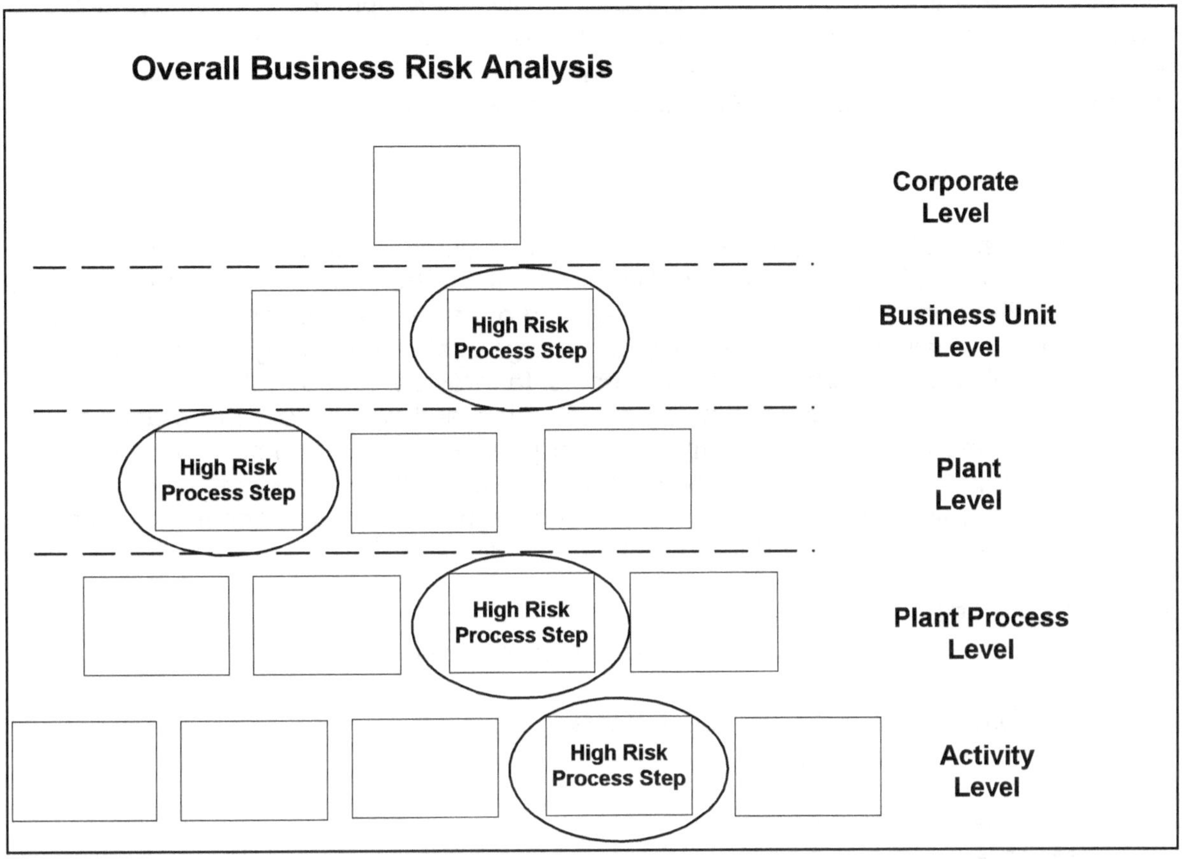

Value added auditors and audit management should know the organization's strategic direction and then align their audits with this direction. Value added auditors should be specifically familiar with the following governance and strategic issues:

- Organizations strategic direction.
- Organizational and business unit plans.
- Senior management direction.
- Corporate policies and procedures.
- Core processes.
- Policies and procedures.

SEGMENTING USING THE RISK APPROACH

Many novice value added auditors evaluate all activities similarly. This is a mistake. As discussed, an experienced value added auditor ranks or prioritizes areas of the highest risk and focuses time on these activities. In any value added audit, there are some factors that have higher risk quotients. For example, processes or activities that deal with customer requirements, regulatory issues, senior management concerns, or board of directors' concerns are rated higher than others. Also, chronic and systemic issues would also be rated higher than those that are isolated. Strategic issues are also rated higher than tactical issues. Regulatory, legal, product development, IT development, and marketing are also high-risk areas. Process areas or process steps with the highest risk would be identified and audited more closely than areas with no or moderate risk as indicated in the 'Overall Business Risk Analysis' figure on the previous page.

So, the value added auditor would focus on assuring there is:

> "a systematic, disciplined approach to evaluating and improving the *adequacy* and *effectiveness* of risk management, control, and governance processes and the quality of performance in carrying out assigned responsibilities. The purpose of evaluating the adequacy of the organization's existing risk management, control, and governance processes is to provide reasonable assurance that these processes are functioning as intended and will enable the organization's objectives and goals to be met, and to provide recommendations for improving the organization's operations, in terms of both efficient and effective performance. Senior management and the board might also provide general direction as to the scope of work and the activities to be audited."[2]

PRIORITIZING RISK AREAS

The value added auditor can also prioritize audits in terms of:

- Customer and regulatory requirements.
- Probability of risk events.
- Dollar materiality.
- Process or product complexity.
- Management competence.
- Asset liquidity.
- Existence and effectiveness of controls.
- Worker abilities.
- Process stability.
- Process capability.
- Process/product changes.
- Time of last audit.
- Complexity.
- Regulations.
- Areas of highest risk require more attention, evidence, and audit due diligence.

WHEN TO CONDUCT AN AUDIT

Nothing is more satisfying when timing and delivery occur in perfect sequence.
Anonymous

There is no consensus as to when a value added audit should be conducted. Some organizations have a policy that whenever a major change occurs in a design, process, system, supplier, procedure, or standard, a preliminary assessment should be made to determine if it results in abnormal process variation, which can diminish value. The assessment does not have to be a full value added audit, but perhaps is just a brief evaluation of the risk consequences, causes and correction of the process change.

The purpose of the assessment is to determine whether existing internal controls are affected by the change. The change may be in a supplier or in a specification. In the former, a change in raw material from a new supplier may radically alter a manufacturing process. The raw material may have varying chemical or physical properties. In the latter, a tightened specification may mean that the process producing the part is no longer capable of producing it to specifications. In general, if the customer believes there is possible product or service deterioration, a value added audit should be initiated.

Value added audits can be scheduled or unscheduled. Unscheduled audits are the result of some unforeseen event or major change and may require an immediate investigation or just a cursory assessment. For example, a major failure, especially if there were a potential for injury, would result in an immediate investigation. Or, if a supplier changed a process, then the customer might conduct an assessment to determine whether the process was still under control and capable of meeting specifications.

MAJOR RISK FACTORS REQUIRING INVESTIGATION

It's impossible to state categorically what factor should mandate a value added audit. However, one policy is to assess the need for an audit whenever a major system, process, or product change occurs that may result in customer dissatisfaction, legal exposure, high costs, or some other type of risk. A partial list of such changes is:

- High product failure rate.
- Customer complaints.
- Recurring quality problems.
- Frequent shortages of materials.
- Late delivery of materials.
- Defective materials.
- Lack of product review.
- High quality costs.
- Process change.
- New contract.
- Special/new process.
- Special/new product.
- Product modification.
- Process modification.
- Equipment modification.
- Procedural change.
- Environmental change.
- Testing change.
- Specification change.
- New or modified regulatory requirement.

- Packaging, shelf life labeling, and storage change.
- New raw material.

A change in any of these can have a major impact on the end product and may result in additional risks or abnormal variation that cannot be controlled internally. So when a major change in an auditee's process occurs, the first question the value added auditor should ask is: 'What is the risk and the probability of enterprise system, process, or product failure?'

The extensiveness and intensiveness of the assessment is up to the customer and the level of risk/assurance he or she wants to assume. The relationship between risk and control is discussed in the following chapters. Obviously, high assurance, high risks, and few controls would entail a more costly, effective, time consuming, and resource consuming assessment than one with lower assurance, lower risk, and documented controls.

DEVELOP THE VALUE ADDED AUDIT SCHEDULE

Routine audits can be scheduled and planned well in advance. For example if the audit customer or client knows that internal departments or suppliers are due for a periodic and routine audit, then the value added audit organization can develop an audit schedule.

In developing the audit schedule, the chief audit executive or manager knows that resources to conduct the audit are finite and an emergency or unscheduled event can shift resources from a scheduled audit to an emergency audit. An important consideration is to schedule audits when the auditee's activities can be observed, people are available, and disruptions are minimized. Specific activities to be observed can be testing, processing, or fabricating material. It is sometimes more cost-effective for the value added auditor to observe tests than to conduct tests. Inspection and testing are expensive especially if products are destroyed in the process or operations are disrupted or stopped. Activities dealing with health and safety would have a higher priority than routine activities.

The audit schedule can be developed jointly, with the auditor and the auditee negotiating the best time to conduct the value added audit. Or, the value added auditor can develop a schedule based on the customer's needs and send this to the auditee for comment and approval. Negotiating a schedule acceptable to all parties is preferred. When the schedule is constructed, it should be distributed early so that the value added auditor and the auditee can anticipate changes.

Scheduled audits are often conducted within windows that are mutually acceptable to the customer, auditor, and auditee. The window may depend on the availability of people, facilities, equipment, or space. It may be a function of when certain parts are being produced or a regulatory authority may determine it.

PLANNING STEP 4 – DEVELOP AUDIT PLAN

BALANCING AUDIT COST AND EFFECTIVENESS

Value added auditors should balance audit cost and effectiveness when budgeting audit time requirements. Audit intervals can also be determined by the size of the organization. Value added audits are subject to more analysis, interpretation and discretion, which may require more time to conduct than a compliance audit. The audit manager may develop a preliminary time budget and present this to the audit client. If the audit time estimates are too conservative, then the audit department may have to issue a report that does not accurately reflect the risks, controls, effectiveness and state of the area being assessed. On the other hand, if the time estimate is too high, the client may not approve of the audit especially if the client is paying for it.

Even determining the length of time the value added audit will take entails risks for planning the nature, extent, and time of the audit. Another method is to assign weights to the vulnerable areas or high-risk areas being audited. Higher risk areas have a higher score and these audit areas require a higher level of assurance.

FREQUENCY OF VALUE ADDED AUDITS

The frequency of value added audits may also be stipulated by a regulation, perceived risk, contractual requirement, standard, purchase order, or national standard. Audit frequency is also determined by:

- **Level of desired assurance**. The level of assurance or risk acceptable to the customer is the major factor in determining audit frequency. A higher level of assurance implies more frequent audits.
- **Criticality of the product.** Critical products, such as heart monitors, integrated circuits, and space loads present more risks and higher costs. Higher risk exposure implies higher frequency and more rigorous audits.
- **Regulatory requirements.** The customer, a regulatory authority, may require annual or biannual audits to maintain certification as a preferred supplier.
- **Contractual requirements.** Commercial contracts can also specify that audits be conducted on a set schedule.
- **Level of deficiency.** If a previous audit reported critical or major deficiencies that if went unreported and uncorrected could lead to loss of life or loss of customer confidence, then audits would be scheduled more frequently.
- **Type of deficiency.** If a deficiency would result in high costs in terms of legal exposure, loss of business, product recalls, or loss of life, audits would be scheduled more frequently.
- **Audit cost.** If an audit is expensive to conduct then it would be scheduled less frequently.
- **Auditee location.** If the auditee is a foreign supplier, then an audit would be expensive to conduct and may be contracted to a third party audit or be scheduled infrequently.

If regulation, mandate, or administrative ruling stipulates audit frequency, the audit organization works around the prescribed schedule. If the customer has control over the audit frequency, then the customer should review the above factors to determine the optimum frequency.

TYPES OF AUDIT SCHEDULES

Keep an ear to the ground and watch the handwriting on the wall.
Herbert Prochnow

Audits can also be classified in terms of time, which includes audit frequency or the order in which the audit is conducted. Time-related audits can be:

- Scheduled.
- Preaudit.
- Random.
- Cyclical.
- Post/follow-up.

SCHEDULED

A scheduled audit is planned and announced. This audit is scheduled so as not to interfere with the auditee's operations. It is the bread and butter of most value added auditing. It verifies whether different external and internal groups are controlling quality, risk, cost, delivery, and other critical process factors so that internal and external customers are satisfied. It identifies and prevents emerging problems from causing deficiencies.

A periodic audit is often scheduled yearly. The value added auditor should send the auditee a notice 1 to 2 months ahead of the audit. The value added auditor and the auditee know the requirements for this audit, who will do what, and what types of resources are required. Not all organizational areas or suppliers are audited. Management or the customer will determine who will be audited and the general objectives of the audit.

Scheduling this type of an audit is usually based on:

- Criticality of the product or service.
- Criticality of the system or process.
- Essential product or project development.
- Dollar volume.
- Risks and controls.
- Quantities.
- Changes in operations.

PREAUDIT

A preaudit is an initial, general assessment of a supplier, process, or product. A preaudit usually consists of yes/no or open-ended discussion questions. Responses to the questions are used to understand the auditee, to establish a baseline to measure future improvements, or to identify potential deficiencies. A preaudit has a general or specific focus. A general preaudit may cover the auditee's process improvement program, financial condition, location, products, functions, systems, and processes. Its purpose is to obtain an overview of the auditee's operations. A preaudit may solicit specific process or product information including process improvement training, internal controls, and process capability indices.

A preaudit is often used to evaluate the auditee. For example, if the customer has had an extensive risk management and process improvement program for years and a new supplier wants to become certified, and then the preaudit serves as a preliminary screen to determine whether an order should be placed.

A customer may preaudit a supplier by sending a survey form listing the areas that will be investigated and asking whether the specific areas have risk management and process improvement systems in place and operating. The survey is sent to the customer 2 or more months before the audit. Upon reviewing the preaudit survey, the value added auditor would then determine if special tests must be conducted or special facilities are necessary. The survey also tells the supplier what will be investigated and communicates intentions in terms of deficient areas that must be corrected by a set date. A follow-up may be conducted to ensure that corrective actions were initiated and successfully prevented a recurrence.

RANDOM AUDIT

A random audit is unscheduled and unannounced. It does not focus on one area, but may check a sample of a number of areas to obtain an overall impression. Sometimes a random audit is conducted on recommended corrective actions to determine if the auditee's actions were effective. It is similar to a post audit. A random audit is often used to assess emergency systems or contingency plans. For example, a random audit may be conducted to determine preparedness and responsiveness of internal groups for such emergencies as an evacuation, fire, or oil spill.

The following are situations where random audits may be conducted when there is a risk of:

- Illegal practices.
- Unethical procedures.
- Emergency drills.
- Fraud or negligence.
- Security breaches.
- Emergency conditions.

The positive image of value added auditing can be destroyed by a poorly conducted unannounced audit. By its nature, it can cause insecurity and resentment. It reinforces the image of auditors policing operations. No one likes to have someone who is perceived as an inspector suddenly barge into one's work area demanding information and implying negligence.

CYCLICAL AUDITS

Cyclical audits are often compliance audits. They are conducted according to a predetermined schedule, yearly or every three years depending on the perceived risks. They add value and confidence in highly regulated environments.

A cyclical audit is also a continual audit that assesses a project, process, or system on an ongoing basis. For example, contract assembly work or a construction project may require a resident auditor to observe construction, tests, production, or assembly. A resident auditor is required for products or projects involving high risk, high cost, regulatory review, or one-of-a-kind products. Certain construction projects, such as nuclear power plants or military hardware developers, have resident auditors because of a contractual requirement. It makes no sense to observe and audit an operation if a screw or other fasteners are being produced - unless of course the screws are going into a heart monitor or space vehicle.

The resident auditor represents the customer. This auditor may reside at the construction site or close to the supplier's facility. Weekly or monthly, the auditor submits progress reports on process improvement, budgets, tests, delivery, production, or other business targets. If the variance between actual progress and targets is unacceptable, the resident auditor or the customer may request an explanation or corrective action.

POST/FOLLOW-UP AUDIT

A post-or follow-up audit checks that corrective action was implemented correctly and effectively so it eliminated the symptom and the root cause of the problem. Sometimes, it is conducted much later than the original audit to confirm that the problem did not recur.

A post audit is not necessary for every problem or finding reported in the initial audit. This would be redundant and sends the wrong message that the auditee is not trusted to fix his or her problems. However, for significant deficiencies or high risks, a post audit may be justified.

DEVELOPING AN AUDIT SCHEDULE

Control circumstances and do not allow them to control you.
Thomas A. Kempis

The value added audit schedule outlines who will do what, when, where, and how. Implicitly, the schedule also addresses why the value added audit will be conducted. Resources have to be identified and acquired. Developing the audit schedule is a joint effort. The audit manager usually works with the auditee to develop a mutually acceptable audit schedule. This may be a team effort from beginning to end. Or, the audit manager may determine what are the scheduled audits, prioritize these, forecast what unscheduled audits may occur, then develop a preliminary schedule that is sent for comment, review, and approval to upper management, internal auditors, customers, and auditees.

HOW TO DEVELOP THE VALUE ADDED AUDIT SCHEDULE

Developing the audit schedule is a team planning and organizing process that brings the auditor, customer, and auditee to the table to discuss the purpose, timelines, methods, scope, and sequence of the value added audits. If internal customers, both auditees and customers, are brought together sufficiently early, then one party can't complain about not being consulted.

Once the joint schedule has been developed, it should be distributed widely to all impacted parties, internal as well as the supplier base. This is more than a courtesy. It informs the auditee of the impending audit, informs the customer the audit has been scheduled, and informs staff personnel that their expertise is required at certain times for conducting audits. It informs specialists that resources are required such as computers and measurement equipment. Also, if there are schedule conflicts, the schedule can be modified and everyone informed of the changes.

An audit schedule can be developed yearly and updated quarterly. It's often better to develop it quarterly to incorporate any unforeseen changes A typical audit schedule is a Gantt chart, showing the audit activity, start date, completion date, resources required, and audit leader.

The advantages of the Gantt chart are obvious. The project and its major requirements can be seen and understood easily. The Gantt chart shows who will conduct the audit when and where. If there are scheduling conflicts these can be easily detected. The auditor and the auditee can see the audit's beginning and ending dates. Resources can be scheduled easier. The supervisor can also compare audit project budgets and times with the actual times.

JIT AUDIT SCHEDULING

Traditionally, compliance or regulatory audits were conducted based on a fixed schedule. More value added auditing is being done on a just-in-time basis. The assessment, assurance, or consulting are done on demand or on an as-required basis. A line manager or process owner may ask the audit function for consulting assistance.

When scheduling a just-in-time audit, the lead auditor or auditing manager should:

- Identify customers and their special requirements.
- Obtain information about the auditee.
- Identify assistants, specialists, and resources.
- Develop a budget.
- Develop a schedule.
- Decide who conducts the audits.

PLANNING STEP 4 – DEVELOP AUDIT PLAN

ASSESSING THE NEED FOR A LONG-TERM AUDIT SCHEDULE

Is there a need for a long-term schedule of 2 to 3 years? The arguments for and against are illuminating. In addition to the planning, budgeting, and scheduling advantages of a long-term schedule, it also provides the following value:

- Demonstrates commitment to 'good governance' procedures.
- Promotes 'tone at the top' ethic.
- Demonstrates commitment of the organization to long-term, company wide continual improvement.
- Establishes measurement baselines and benchmarks for risk management and control.
- Notifies external regulatory authorities and stakeholders that value added auditing is part of the organizational strategic goal.
- Establishes general guideline milestones, accountabilities, objectives, and goals for the value added group.

The argument against developing a long-term schedule is that major and rapid change seems to be the norm these days. Several changes that may make a long-term schedule obsolete are:

- Life cycles are down to 2 or less years for many products.
- New legislation changes the playing field.
- New domestic and foreign competitors enter the market with newer, cheaper, and higher quality products.
- Budgets are at zero-base levels.
- Managements and philosophies change.

IDENTIFY AUDIT RESOURCES

One machine can do the work of fifty ordinary people. No machine can do the work of one extraordinary person.
Elbert Hubbard, Business person

Some value added audits require specialized equipment, facilities and personnel. Often these can be anticipated, especially if the audit has been conducted previously. For example, product testing may require certified specialists, tools, and facilities. The value added auditor should know whether these people are available to participate in the audit. Sometimes specialists will become part of an audit team for the duration of the audit.

The lead auditor is ultimately responsible for value added audit planning, conducting, and reporting. The auditor may be a single person or a team. If a team, the auditor is a senior person with appropriate operations experience and technical background.

Value added audits can be performed by the following groups:

- In-house auditors.

- External contractors.
- Retired employees.

IN HOUSE AUDITORS

The advantage of in-house value added auditors is that they bring a known level of business knowledge and professional care to the audits that may not be known to contractor auditors. These factors increase the customer's level of confidence and the perceived reliability of the audit. However, they may lose their edge and become stale over a period of time. They may not work as hard and therefore miss deficiencies, risks, or unusual events. In addition, internal auditors must have enough work to justify their existence or they become an overhead burden.

EXTERNAL CONTRACTORS

External contractors, sometimes known as third party assessors, are often used to audit suppliers. Several firms have an extensive list of auditors on retainer or on contract in many parts of the world to perform value added audits. Their major advantage is their ability to conduct an audit in different parts of the world quickly, efficiently, and cost effectively. Their major disadvantage is that the customer sometimes has only the resume of the auditor's background. Unless someone has used them before and can vouch for their abilities, may be a risk.

RETIRED EMPLOYEES

Retired employees with operations or financial backgrounds are sometimes used as contract auditors. These people are often contracted to perform audits of suppliers. Their major advantage is that they know the customer's requirements and products. Their major disadvantage is that they may have a narrow focus and may have lost their investigative edge.

METHODS OF ASSIGNING AUDITS

Assigning audits is not easy. Auditors have conflicting priorities, internal obligations, and external commitments. Internal commitments may include special company projects, conflicting audits, committee assignments, reports to be written, or internal consulting. Value added auditing often requires domestic and overseas travel; family time is sacrificed as well as professional commitments. The load on the auditor can be heavy. Assigning audits has to be handled carefully.

Assigning audits to internal auditors must be done carefully, and can be accomplished by:

- Seniority.
- Lot.
- Rotation.
- Inducement.
- Negotiation.

Senior auditors have preference in the first method. To encourage fairness, audits may be selected by lot or be rotated among the auditors. Auditors can be induced to perform difficult audits by being offered compensatory time or bonuses. Another way is to post the list of pending audits and have auditors negotiate among themselves who will conduct which audits.

STAFFING VALUE ADDED AUDITS

Staffing and allocating resources to some types of audits is pretty much cut and dry. For example, ISO 9001 and other management systems audits are allotted a specific number of person days to complete. The number of square feet in a facility or the number of workers may also determine the length of time an audit will take. These audits are usually compliance and follow a set procedure.

Value added audit staffing is more variable and can be based on a number of factors, including:

- Auditor's abilities.
- Number of auditors on team.
- Audit scope.
- Process risks.
- Time constraints.
- Level of required audit confidence and assurance level.
- Type of risk - controls.
- Level and types of process controls.
- Process capabilities.

WORK BREAKDOWN STRUCTURE (WBS)

A work breakdown structure is an audit plan. An audit plan is developed for each engagement. Depending on the intent, scope and type of audit engagement, the audit plan can be one page or multiple pages.

WBS ELEMENTS

The WBS below is a detailed work plan of an audit of a multi-billion dollar construction project. It was detailed enough that each team member understood the performance audit task, control objective, standards against which to assess, field work to evaluate the control objective, and the deliverables.

Unforeseen conditions can necessarily modify the Work Breakdown. The work break down is structured in terms of the following:

- **Task.** Is the unit and sequential activity of audit work.
- **Control Objective.** Is the goal and/or purpose of the specific task.

- **Standards.** Consists of the audit criteria to be reviewed for compliance, adherence, testing, analytics, evaluation, or control document review, which may include review of policies, procedures, work instructions, forms, and protocols.
- **Fieldwork.** Consists of interviews, reviews, analytics, observations, or corroboration that auditors will conduct. Fieldwork for the example was divided into the following phases: 1. Program/Project Planning; 2. Program/Project Design Management; 3. Program/Project Construction Management; and 4. Program/Project Contract Management. In each phase, gaps, risks, lack of controls, inadequate controls, and elements non-compliance were critically analyzed.
- **Deliverables.** Consists of the outcomes of the fieldwork that were used to support the Preliminary Findings and Recommendations (PFRs).

The WBS was written for each phase of the engagement. In a large value added audit, a WBS can be developed for each phase as described below:

SURVEY PHASE

Survey Phase consisted of the preliminary work to understand the scope of the engagement; review the technical/engineering programmatic governance and control model; conduct a preliminary risk assessment; develop a list of control and operating manuals; evaluate scope, cost, schedule, quality, and risk-controls; and develop inputs into the Audit Plan. Example of a survey phase WBS is shown below:

Task 1 Control Objective:

- Attend kick off meeting, attend client ½ day training, and develop understandings with client.

Fieldwork:

Standards: Professional Engineering Ethics, Ethics in Public Service, contract terms and conditions, etc.

- Schedule kick off meeting with client project manager.
- Clarify roles and responsibilities of client, auditor management, etc.
- Attend ½ day client training session.
- Clarify communication, resolution, and escalation protocols with client.
- Convene meeting with auditing professional staff and contractors to discuss ethics and independence.

Deliverables:

- Memo of Understanding between client and auditor regarding performance and expectations.
- Attend meetings and communicate ethics requirements.

AUDIT PLAN PHASE

Audit Plan Phase consisted of developing a detailed audit plan including preliminary risk assessment; list of documents and flowcharts to be used as audit standards; and preliminary list of programmatic/project/process controls to review/test. Example of an audit phase WBS is shown below:

Task 5 Control Objective:
- Determine preliminary documents to review and procedures to test in Fieldwork Phase.

Fieldwork:
- **Standards:** Program/Project Management (PM) Manual; Construction Management (CM) Manual; Quality Management (QM) Manual; Risk Management (RM) Manual; General Contractor/Construction Management, etc.
- Gather policies, procedures, work instructions, flowcharts, logs, forms, and risk maps.
- Prepare log of documents etc.
- Select projects to audit from or capital improvement projects.
- Identify documents to review from population based on prioritized risks, gaps, etc.
- Identify procedures to test based on critical project areas, project life cycle risks, PM Manual, CM Manual, QM Manual, etc.
- Develop preliminary sampling plans based on project life cycle, project type, types of controls, residual risks, dollar-size, etc.

Deliverables:
- List of documents to review and PM/CM/QM procedures to test in order to identify service gaps and analyze performance.
- List of sample projects to be reviewed.

FIELDWORK PHASE

Field Work Phase consisted of conducting program planning reviews; reviewing program architecture and controls; evaluating architect/engineer (A/E) reviews; reviewing availability of professional resources; assessing Construction Management, Quality Management, and Program Management controls; evaluating change orders; assessing purchasing controls; assessing contracting controls; determining compliance; assessing savings projections, assessing implementation of best practices; and determining five–year impacts on future costs. Example of a fieldwork phase WBS is shown below:

Task 7 Control Objective:
- Review governance and Program/Project Management Office (PMO) structure to determine if it provides adequate oversight of project management, A/E's, contractors, etc.

Fieldwork:
- **Standards:** PMO charter, PMO architecture, PM Manual, PMO authorities/responsibilities, internal control reporting, Project Management Institute (PMI) PMO guidelines; etc.
- Interview executives, managers, and engineering professionals regarding governance, overall programmatic management strategy, selection/management of A/E's, selection/management of contractors, etc.
- Review PMO model for appropriateness and adequacy to manage multi-billion dollar program. PMO is equivalent to program/project governance model.
- Review and determine capability and maturity of PMO.

- Review of PMO management and deployment capability for billion-dollar program.

Deliverables:

- Summary of effectiveness of executive oversight of PMO and management staff.
- Summary of overall effectiveness and efficiency of program operations, PMO architecture, controls, etc.

Task 8 Control Objective:

- Evaluate and determine if has suitable, appropriate, and implemented *documented controls* including: PMO controls (PM Manual); design (plans/specs); CM Manual; QM Manual; standards; metrics, etc.

Fieldwork:

- **Standards:** COSO ERM, Project Management Institute BOK, American Society of Civil Engineers (ASCE) risk checklists, Associated General Contractor (AGC) standards, etc.[1]
- Interview PMO management regarding Master Plan, management controls, oversight, etc.
- Review PM, QM, and CM manuals against 'best in class' practices.
- Determine 'effectiveness' and 'suitability' of controls over programmatic lifecycle.
- Review PMO controls policy for monitoring critical events and controlling adverse impacts of the event.
- Develop survey lists from CM, QM, and PM manuals for critical areas, including programmatic management, design, construction management, purchasing management, contract management, and closeout.
- Review 1. Existence of appropriate control-documents; 2. Extent of control-document usage; 3. Evidence to support control usage; and 4. Feedback mechanisms to improve operations.
- Review process for developing 'plans and specs.'
- Review change management processes and their effectiveness.

Deliverables:

- Verification of suitable and adequate documentation controls, including programmatic/project standards, policies, procedures, and work instructions.
- Identification of documentation gaps in PM and CM manuals, program charter, policies, and standards. Gaps include: no procedures where they should exist; inadequate procedures; or weak procedures for the risks identified.

DRAFT REPORT PHASE

Draft Report Phase was a continuous effort starting in the Survey Phase. The team surfaced Preliminary Findings and Recommendations (PFRs) in each phase of the audit and shared them with the client project manager. The draft report was continuously written throughout the audit. Example of a draft report phase WBS is shown below:

Task 22 Control Objective:

- Prepare draft report of audit.

Fieldwork:

- **Standards:** client Performance Audit Report Format, client policies, Work Papers, and prior deliverables.
- Conduct exit meetings with auditee.
- Distill PFRs from Task 1 to Task 21.
- Ensure criteria are addressed for each audit objective.
- Cross reference audit objectives to findings and recommendations and project potential cost savings.
- Develop doable and accountable recommendations for each finding.
- Summarize five-year projection or impact on implementing each PFR and benefits of implementing 'best practices'.
- Outline near term steps for to implement recommendations.
- Conduct QC review of all work, findings, and recommendations.
- Share and discuss first draft with client.

Deliverables:

- First draft of report.

FINAL REPORT PHASE

Final Report Phase consisted of the following: review by the auditee of the Per's; review to ensure evidence in work papers support audit reasoning; prepare final Findings and Recommendations; develop cost saving estimates; and finally incorporate client comments into the Final Report. Preliminary Findings and Recommendations were shared, discussed, and reported in bi-weekly meetings. Example of a final report phase WBS is shown below:

Task 23 Control Objective:

- Prepare final report and close out the performance audit.

Fieldwork:

- **Standards:** Client Performance Audit Report Format, client policies, Work Papers, 'Plain Talk standard, etc.
- Discuss draft with client and incorporate client comments into final draft report.
- Provide context and description of findings.
- Conduct additional cost savings and cost-benefit analyses.
- Develop Final Report.
- Prepare and deliver five hard copies and one soft copy of final report to client.
- Prepare testimony PowerPoint presentation on audit and share with client.
- Revise final presentation.

Deliverables:

- Submit work papers and PowerPoint to client.
- Deliver final report in hard and soft copies to client with work papers.
- Provide legislative testimony if required.

REVIEW AND APPROVAL OF THE PLAN

Long range planning does not deal with future decisions but with the future of present decisions.
Peter F. Drucker

Value added auditors must understand audit requirements, gather evidence, make analytical judgments, report results and ensure corrective/preventive actions are taken. The level of due diligence and analysis for conducting a value added audit is higher than conducting a compliance audit. For example, value added auditors may review operational efficiency, effectiveness, and risks. These would require more analysis and evaluation than simply a compliance audit.

REVIEWING THE PLAN

Once a preliminary plan is developed, the value added auditor should review it with the chief audit executive, audit manager, the customer, or someone familiar with the auditee. Each value added audit should also be initialed and approved by the reviewer. Audit assumptions and post technical analysis should also be reviewed. The reviewer should assess the nature, extent, intent, and results of the audit. This is critical if the audit is being conducted for the first time. This person reviews the plan and suggests changes based on her or his experience. This person may be able to suggest additional research, contacts, or tips.

In a large organization, the audit manager or chief audit executive would review and approve the audit plan. In a small organization, the lead auditor probably would evaluate the plan and work activities. The chief audit executive, audit manager, or lead auditor would then suggest changes prior to starting the audit.

The reviewer should maintain notes of issues arising during the audit review process. The reviewer assures that working papers provide adequate evidence to support audit results and questions raised during the review are properly resolved.

A risk based project plan would identify risks, risk priorities and possible exposures. The risks are prioritized because this ensures that process owners as well as the auditors are focusing their attention on what's critical. The Pareto Principle, the 80-20 rule, says 20% of the factors can cause 80% of the problems or risks. Process owners and auditors should focus their attention on these 'critical few' areas instead of the 'trivial many.'

MANAGE 'AUDIT CREEP'

The first 90% of the project takes the first 10% of the time. The last 10% of the project takes the last 90% of the time.
Anonymous

Audit creep is a factor that the value added auditor must always look out for. It is the human tendency of customer's expectations to increase over time. The customer expects more than was initially requested. The scope or objectives of the original audit request are unintentionally expanded. This may mean a more extensive audit resulting in lower risks but at a higher cost. All of these involve tradeoffs. If the customer is not informed of these tradeoffs as situations arise, there is a tendency to move goal posts while the audit is being conducted.

HOW TO ENSURE AUDIT SCOPE DOES NOT INCREASE OVER TIME

The major problem with audit creep is that the scope and objectives of the audit become a moving target and the value added auditor will never be able to satisfy the audit's objectives. Ultimately, the auditor's performance and credibility will be judged by how effectively and efficiently he or she conducts the audit. If the audit's goal posts are forever moving, then it becomes impossible to conduct an effective value adding audit.

Audit creep can be managed as long as rising expectations are incorporated into the audit agreement. Additional expectations often arise from preliminary audit reports with findings that the customer wants investigated. For example, the auditor may find a deficiency significant enough to justify further examination. If this additional work is not addressed in the updated audit agreement, customer expectations will not be met. Also, if the value added auditor conducts the expanded audit and consumes unauthorized resources, then upper management can come back to the auditor and say that the audit was conducted inefficiently.

During the audit, the value added auditor may discover major problems in terms of lack of controls, no documentation, and possible process chart editing. For example, a supplier may not have sufficient controls to prevent a nonconformance. The risk to the customer increases. The auditor should notify the customer, audit manager, and purchasing manager of the deficiencies or risks. Changes in the scope, timelines, and other changes would be incorporated in an updated audit agreement to the customer.

SUMMARY

A value added audit is basically a project. It has a well-defined beginning, defined end, milestones, and deliverables. Things change during the audit. Customers may want more critical or analytical information. There are unanticipated audit constraints.

Scheduling value added audits must be handled delicately. In mature organizations, value added audits are considered a means for controlling risks, improving continually, and enhancing competitiveness. In less mature organizations, any assessment may be resisted. Value added auditing may not even be conducted. Compliance audits are well understood while value added audits are not understood.

Value added auditing may cause resentment in some organizations. Internal obstacles to implementing value added systems include naysayers, redundant controls, internal politics, second-guessing, and not-invented-here syndrome.

Bottom Line: The value added auditor should know what processes and systems to review, where they are located, and who is responsible for them. The auditor can then ensure that the auditee has been notified of the audit, a schedule has been set, and resources have been obtained to conduct tests.

Whatever the problem, when the original understanding or scope of the audit changes, this is called audit creep. As much as possible, this should be controlled because audit creep results in additional resources which if not authorized depreciate audit value.

All audits should be planned. The following are the basic key elements for planning an audit:

- Identify the auditee.
- State the audit objectives aligned to business objectives.
- Treat the audit as a project with a defined beginning, end and deliverables.
- Identify audit stakeholders.
- Identify business objectives.
- State the scope and degree of testing required in the audit.
- Identify the audit standards.
- Identify process risk and controls.
- Identify technology elements of process and transitions to be assessed.
- Determine the level of verification, validation, or testing.
- Develop checklists/surveys.

CHAPTER 11
STEP 5 - DEVELOP AUDIT SURVEY

The value of an audit survey becomes clear when auditing complex organizations, systems, and processes. The amount of available data and information can be overwhelming and in some cases, it may even be in a foreign language. The survey establishes a foundation of basic understanding between the auditor and the auditee.

TYPES OF SURVEYS

All organizations are at least 59% waste - wasted people, wasted effort, wasted space, and wasted time.
Robert Townsend, writer

A survey may be sent to the auditee prior to the value added audit. If this is a first time audit, then the survey provides a baseline of the auditee's systems and processes. If this is part of an ongoing, cyclical, or requested assessment, then the survey benchmarks the changes since the last assessment. It reveals what has changed and what the value added auditor should look at. As well, it brings the process owners into the value added assessment. We'll look at two types of surveys:

- Process survey.
- Customer-supplier survey.

PROCESS SURVEY

The process survey can be generic or can be tailored to the auditee. The process survey can examine variation or changes in the following:

- Environment.
- Risks.
- Process management, assurance, and control.
- Process metrics.
- Variables.
- People.
- Process.
- Products.

Internal auditors, quality auditors, and operational auditors can review different activities and develop tailored surveys focusing on their needs. Internal auditors often use a survey that focuses on the control environment, risks, controls, information, and assurance.

PLANNING STEP 5 – DEVELOP AUDIT SURVEY

Quality auditors develop a survey that focuses on customer requirements, product quality requirements, process capability, control, and product testing. Operational auditors develop a survey that focuses on work efficiently, economy, and effectiveness. Value added auditors should be capable of many types of assessments and should be able to adapt a survey to the audit's objectives.

CUSTOMER-SUPPLIER SURVEY

The value of a survey can be appreciated when evaluating suppliers. The survey informs the auditee who may be a supplier of the impending audit, its purpose, and what it will entail. To assist the audit, the auditee may be asked to provide personnel, space, and equipment, and to arrange the necessary appointments. The auditor can save time when the supplier becomes a partner in improving communications and integrating operations. Progressive suppliers realize that this is an opportunity to communicate, cooperate, and coordinate requirements with the customer. If the supplier completes the survey satisfactorily, the supplier can request an audit to verify the supplier's responses to achieve the next higher-level certification. See 'Customer-Supplier Survey' sidebar on the next page for a sample of this type of questionnaire.

The value added auditor can use the completed survey to design a customer-supply program of detailed tests to confirm the 'yes' responses or to investigate the 'no' responses, which lead to corrective action request. A well-crafted question also can uncover hidden operational flaws and prescribe effective solutions. The value added auditor is basically someone who asks good questions and has the ability to unearth proper answers.

SURVEY VALUE

Nothing is more central to an organization's effectiveness than its ability to transmit accurate, relevant, and understandable information among its members. All the advances of organizations - economy of scale, financial, and technical resources, diverse talents, and contacts, are of no practical value if the organization's members are unaware of what other members require of them and why.
Saul Gellerman, writer

The preliminary survey is not a formal audit. However depending on the level of assurance required by the customer, it can form the basis of a rudimentary assessment. As its name implies, it provides a general overview. The purpose of the survey is to understand the auditee's processes, objectives, risks, and internal controls. The preliminary survey addresses the general issues that will be assessed in the audit.

SURVEY USES

Surveys can be used to:

- Analyze processes and systems.
- Evaluate breadth and depth of processes.
- Assess risks and controls.
- Organize the value added audit.

PLANNING STEP 5 – DEVELOP AUDIT SURVEY

CUSTOMER-SUPPLIER SURVEY

Please fill out the attached survey. The survey will be used to gain a better understanding of what you do and what has changed since the last assessment. Thank you for your help and collaboration. Your input and ideas are critical for the success of this value added audit. Please indicate your principal areas of concern, areas of risk and major changes that have occurred since the last assessment.

Environment
What has changed since the last audit regarding regulatory, safety, or health areas?
Is senior management concerned about these changes?
What organizational and process changes have you seen?
What overall environmental issues concern you?
How do you plan to address these?

Risks
What has changed since the last audit regarding risks?
What new risks have you identified?
How are new risks assessed and mitigated?
What risk issues concern you?
How do you plan to address these?

People
What has changed since the last audit regarding people?
How do you ensure people know process and product requirements?
What people issues concern you?
How do you plan to address these?

Process Management
What process variables have changed since the last audit?
What seems to work the best for you?
What process issues concern you?
How do you plan to address these?

Process/Product Metrics
What process/product metrics have changed since the last audit?
What seems to work best for you?
Are these metrics aligned with company objectives?
What metric issues concern you?
How do you plan to address these?

Improvement
In what areas has there been improvement or lack of since the last audit?
What seems to work best for you?
What improvement issues concern you?
How do you plan to address these?

General questions:
How can we help you achieve your business objectives?
Where are you most comfortable?
What are your most critical concerns?

- Audit performance.
- Provide consistent approach.
- Develop value added audit objectives.
- Focus attention and energy.
- Anticipate problems.
- Provide baseline and benchmark assessment data.
- Validate audit assumptions.
- Obtain overview of auditee's operations.

ANALYZE PROCESSES AND SYSTEMS

The questions on the audit form are usually in 'yes/no' or 'not applicable' format. Anytime there is a negative response, the auditor should elaborate and provide more information. The value added auditor may encounter exceptions, nonconformances, noncompliances, or discrepancies. The following questions can then be addressed:

- Is the problem critical, major, or minor?
- What are the organizational risks and consequences?
- Will the problem or nonconformance recur?
- What is the magnitude and likelihood of an occurrence?
- Does the problem involve safety or health issues?
- Is it a systemic, chronic, or people-driven problem?
- Is the system deteriorating, maintaining, or improving?
- Is the problem improving sufficiently, extensively, and intensively?

One of the more difficult questions confronting the value added auditor is determining the level of adherence with business objectives, standards, procedures and instructions. If the function is operating in a reasonably effective and efficient manner, despite minor noncompliances or variances, then the auditor can note this in the audit and recommend no corrective action.

EVALUATE BREADTH AND DEPTH OF PROCESSES

The value added audit is not black or white. The analytical audit deals with shades of gray that require process analysis and the educated decision by a trained auditor. The value added auditor must decide on issues of extensiveness and intensiveness. Let's look at the extensiveness of statistical process control (SPC). Extensiveness is the breadth of SPC. For example, a typical audit question would state: 'Does the auditee use SPC?' The answer may be yes, but to what extent:

- In all operations?
- Just in manufacturing?
- Just in administrative areas?
- Just for critical product characteristics?
- Just in a limited area of one plant?

Intensiveness is the level to which SPC is applied. Let's look at the intensiveness of statistical process control usage. In SPC usage, the critical issues would include:

- Is continual process improvement pursued?
- At what rate?
- Are people thoroughly trained?

ASSESS RISKS AND CONTROLS

As discussed, risk is the first filter by which many management decisions are made. Why? Companies want to engineer safe outcomes or at least deterministic outcomes. In other words, senior management and other stakeholders don't want surprises.

Surveys are also helpful in determining the size and direction of the gap between present management controls and acceptable risk to the organization

Detailed self-assessment surveys can be developed to facilitate control/risk assessments. Any organization that uses surveys to assess and evaluate control procedures should have a formal, documented process. The documented process may:

- Assess the probability of achieving business objectives.
- Identify risk and exposures.
- Assess control process that management risk.
- Develop action plans to reduce risk to acceptable levels.

ORGANIZE THE VALUE ADDED AUDIT

Surveys can be used for organizing value added audits and for evaluating internal controls. Anytime there is a negative answer to a significant survey question, the value added auditor can ask:

- What level of assurance is required?
- What is the risk of failure?
- What are the consequences of a failure?
- Is the internal control strong, moderate, weak, or nonexistent?
- Are there compensating internal controls?
- How effective are they?
- What are major variances from mission critical business objectives?
- Should there be heavy reliance on internal controls?
- Should tests be conducted to obtain evidence of effective controls?
- Should raw material (input) or output (finished product) tests be conducted?
- Should a corrective or preventive action request be issued?

AUDIT PERFORMANCE

Surveys can be general or tailored to a specific operation. For example, a survey may be tailored to manufacturing or to a service operation. In a manufacturing survey, many

SELF ASSESSMENT SURVEY BENEFITS

- Process owners become trained and experienced in assessing risk.
- Process owners learn how to strengthen control processes to management risk.
- Process owners focus on achieving business objectives.
- Informal, 'soft' controls are more easily identified and evaluated.
- People become motivated to take 'ownership' of control processes.
- Control of 'objectives – risk – controls – test' methodology becomes integrated into operations.
- 'Objectives – risk – controls – test' methodology becomes integrated into process control and continual improvement.
- Auditors become partners with process owners to develop and assess processes.
- Auditors evolve to become process control facilitators, coaches, and consultants.
- Quality and internal auditors become more knowledgeable about operations and process controls.
- Auditors assess risk-controls performance and assurance instead of designing processes.
- Management responsibility for risk management and control are reinforced.
- Auditors validate the evaluation process by performing audit tests instead of providing verification.
- Auditors use their professional judgment to determine the adequacy and effectiveness of the entire risk management and control processes.[1]

questions would deal with statistical process control, capability, and inspection equipment.

PROVIDE CONSISTENT APPROACH

Consistent approach is the hallmark of a value added auditor. Auditors can use surveys to develop a consistent approach that works for the auditee, audit stakeholders, and future auditors. The auditor still can add or delete questions or even tailor questions to specific processes or activities.

DEVELOP VALUE ADDED AUDIT OBJECTIVES

The survey helps the auditor develop value added audit objectives, establish audit guidelines, and develop an audit plan. Using this information, the value added auditor can become familiar with the auditee's operational risks, identify internal control points, and develop tests to determine their effectiveness.

FOCUS ATTENTION AND ENERGY

Every value added auditor is constrained by time, budget, and resources. The value added auditor must be able to plan, conduct, and report accurate results in a timely fashion. The preliminary survey provides clues on where to concentrate time and energies. For example, the auditor can develop a Pareto chart identifying major deficiencies and prioritize the value added audit to areas of highest concern. Surveys allow audit management to determine if auditors are looking at the right things correctly. Audit and senior management can coach auditors to focus their attention on the critical things that matter.

PLANNING STEP 5 – DEVELOP AUDIT SURVEY

A novice value added auditor tends to give equal treatment to all audit areas or dwells too much on the insignificant. Or, data is gathered that does not relate to information that applies to evidence that addresses the customer's requirements. The value added auditor must focus attention on important data, collate it, and then derive meaningful conclusions to satisfy audit objectives.

ANTICIPATE PROBLEMS

The survey is also an excellent device for anticipating requirements and problems. For example, if the auditor is conducting an audit of an offshore supplier, the value added auditor may discover that something was assumed or forgotten and the auditor must request assistance from the home office. This may have been anticipated through the initial survey.

PROVIDE BASELINE AND BENCHMARK ASSESSMENT DATA

The survey can provide a historical baseline or data points for future audits. The value added auditor and the auditee can see if processes are stable, capable, and improving or deteriorating. Surveys can become the basis for trending nonconformances, waste, improvement or risk. The surveys become part of the historical audit record. Multiple stakeholders can evaluate the data to ensure that processes are stable, capable, and improving.

The survey can also become benchmark data for companies with similar internal processes as well as comparing similar processes among different companies. The survey can then provide historical data that can be mined to detect patterns, trends, systemic/chronic conditions, risk conditions, new relationships and so on.

The survey can also be automated to provide a quick overview of a process or activity. Multiple users and stakeholder can access the survey data in a web accessible format for uses by different groups.

VALIDATE AUDIT ASSUMPTIONS

The value added auditor may discover after the survey that basic assumptions were incorrect and may have to be changed. The value added audit methodology, resources, timelines, scope, equipment, or something else may change. It's more efficient to change during the preliminary survey than to discover something during the audit.

As much as possible, the value added auditor should review objectives, budgets, and plans immediately after the survey has been received. For example following receipt of the survey, the auditor may discover significant unexpected changes in the auditee's systems or unreported deficiencies from previous audits. These will require additional examination and change in audit scope. The 2-day audit now becomes a 3- or more day audit requiring the assistance of specialists. The additional time and resources require that the auditor changes the audit scope, schedule, plan, methodology, and budget.

PLANNING STEP 5 – DEVELOP AUDIT SURVEY

OBTAIN OVERVIEW OF AUDITEE'S OPERATIONS

The survey provides an overview of the auditee's operations. Often audits are restricted in scope for example to analyzing a specific problem. The value added auditor may be asked to confirm a specific situation, investigate a nonconformance, or assess the results of corrective action.

HOW TO DEVELOP A PRELIMINARY SURVEY

Always define your terms.
Eric Partridge

Some of the evidence and information required to satisfy the customer may be developed from a preliminary survey. The following should be considered in developing a survey:

- Determine the type of survey.
- Determine objective or purpose of the survey.
- Determine type of questions.
- Tailor make a survey.
- Leave room for open ended discussions.

DETERMINE THE TYPE OF SURVEY

The survey can be oral or written. An oral survey is conducted in a face-to-face meeting, by phone, or by interactive video. The oral survey is sometimes called a preliminary interview. The purpose of the preliminary interview is to introduce the auditor and to obtain process information about the auditee.

In a written survey, the value added auditor would send the auditee a self-survey and request that he or she return it in sufficient time to be analyzed. The advantage of the survey is that the value added auditor has the opportunity to follow up on questions and the auditee has the opportunity to clarify issues and ask questions about the audit. Also, it can contain open-ended questions that may require longer responses than can be delivered easily through a phone conversation or meeting. For example, the value added auditor may want specific information that requires the auditee to investigate. The auditee has the opportunity and time to answer the questions.

A survey usually consists of yes-no question. They are usually generic and focus on many elements of control. Tailor made surveys also can be produced. There are a number of benefits of using these tailor made surveys:

- Tailored control surveys can be produced quickly.
- Surveys can save time and costs in gathering value added information from a team.
- Surveys can be used to train new operating personnel about risks and controls.

- Process owners use the forms as a memory aid as to what to look at and control.
- Auditors can validate that critical process elements are controlled.

DETERMINE OBJECTIVE OR PURPOSE OF THE SURVEY

Internal auditors usually conduct a presurvey as part of their audit planning. The scope and depth of the survey depend on the type of audit. A quality (systems) audit may conduct a bench audit of the quality manual and other quality documentation. A value added audit survey may be more extensive.

DETERMINE TYPE OF QUESTIONS

When developing the initial survey, the value added auditor should ask broad, open-ended questions. The auditee can freely supply answers to provide organizational, system, process and product context that is invaluable when conducting the audit. The auditee's responses may go off on a tangent, but the nature of the responses may provide clues to current troubles or future problems. This is another form of careful listening. Also, the value added auditor may not be familiar with the auditee's processes. By asking specific questions, the auditor narrows the focus of the survey to areas that the auditor may think are deficient and the focus may be too narrow. By answering open-ended questions, the auditee has the opportunity to contribute his or her expertise and identify areas of present or future concern.

The particular open-ended questions incorporated into the survey largely depend on the needs of the customer, the recipient of the audit report. The purpose of the survey is to understand the auditee's company, processes, and products. The value added auditor has to rely on knowledge of the customer to choose the appropriate questions. If the customer wants specific, detailed information, these questions can be incorporated in the survey.

TAILOR-MAKE A SURVEY

The value added auditor may find an existing survey that was developed from a previous value added audit in the permanent file. Or, the auditor can develop a new survey. A process survey is tailored to specific value added audit objectives. This saves time from having to develop a specific process survey for each auditee.

The value added auditor can develop a generic or tailor make a survey. The questions are intentionally broad. To assess a limited process, the auditor would tailor the questions to the process. To assess a supplier's processes, additional questions could be formulated.

As seen by the nature of the questions, the auditee may be asked to reveal sensitive information. This is an important element of the auditor-auditee relationship, which is a partnership built on mutual trust. For example, the auditee trusts the auditor not to disseminate confidential or proprietary information; the auditor trusts the auditee to be forthright and helpful in the conduct of the audit. Cynics would contend that this type of relationship is impossible in an ordinary audit. However, in a value added audit, both parties have more to gain in increasing customer satisfaction, developing safe and aesthetic products, and improving continually.

LEAVE ROOM FOR OPEN ENDED DISCUSSION

There should be uniformity in the preparation of the survey and standardization of planning, conducting, and reporting audits. The value added auditor can develop questions on the process management, assurance, and control, always keeping in mind the goals are to prevent defects from occurring, to improve continually, to mitigate/manage risks, and to satisfy the final customer. At the end of the survey, the auditor should leave room for open-ended discussions of various aspects of the audit survey.

USING A SURVEY

Fundamental progress has to do with the reinterpretation of basic ideas.
Alfred North Whitehead

The typical format of a survey, as shown in the 'General Survey Questions' figure on the next page, is to have a list of questions on each functional area or process being audited. Also, control surveys are shown in the 'Closed Purchasing Questions' and 'Closed Manufacturing Process Questions.'

Closed and open-ended questions serve different purposes. Closed questions are binary and specific, usually good for conducting a compliance audit. Open-ended questions are good for understanding the overall process and system. Open-ended questions also allow the auditee to amplify on a response.

Many questions are binary, yes/no questions, where the value added auditor would place a checkmark in the appropriate column. Questions are phrased so a 'yes' answer is acceptable and confirms the condition or situation described in the question. A 'no' response indicates that the condition specified by the question has not been met. A negative response may result in further investigation, disqualification of a question, or initiation of corrective action. If a boilerplate survey is being used and if a question is not relevant or unimportant, then the value added auditor disregards the question. Questions may be cross-referenced to specific sections of the auditee's process manual or to an applicable quality standard. At the end of the survey under the 'Comments' section, the auditor may add more information, such as referring to the working papers that indicate the results of specific testing.

Generic survey questions are usually used to evaluate an area's internal control structure. They deal with many overlapping process and system controls. The questions are useful for identifying areas where the auditee is deficient or requires improvement. In terms of indicating whether controls are working effectively, the auditor may have to customize the checklists in terms of specific customer needs and the auditee's systems by developing a list that highlights the strengths and weaknesses of the system under consideration.

PLANNING STEP 5 – DEVELOP AUDIT SURVEY

GENERAL SURVEY QUESTIONS

1. What is your full business name?
2. What do you do?
   a. Produce?
   b. Service?
   c. Provide?
   d. Distribute?
   e. Represent?
   f. Supply?
3. Who are your major customers?
4. What do you produce, design, or deliver to us?
5. How much do we purchase from you?
6. What percentage of total sales is this?
7. Do you have an organization chart for us?
8. Where are the following located?
   a. Headquarters?
   b. Divisions?
   c. Plants?
   d. Distribution centers?
   e. Service centers?
9. How many people do you employ?
   a. Management?
   b. Professional?
   c. Exempt?
   d. Hourly?
10. What are your financials?
    a. Profit/loss?
    b. Assets/liability?
    c. Cost of quality?
    d. Internal budgets?
11. Do you have a formal company and/or supplier wide process improvement program?
12. What are your organization's:
    a. Value added objectives?
    b. Value added goals?
    c. Accountabilities?
    d. Benchmarks?
    e. Metrics?
13. What are major risks in your:
    a. Industry?
    b. Operations?
    c. Competition?
    d. Others?
14. What feedback do you get from:
    a. Customers?
    b. Suppliers?
    c. Community?
    d. Competitors?
15. How is the feedback derived?
16. How do you measure:

17. How do you deal with your human resources, in terms of:
    a. Fairness
    b. Respect
    c. Training
18. Who is your chief quality officer or chief audit executive?
19. Who are your major suppliers?
20. What were the constraints in conducting the previous audit?
21. What were the major changes since the last audit?
22. How long did the previous audit take?
23. When is the best time to conduct the audit?
24. What are your major processes and products?
25. What are your internal management, assurance, and controls?
26. Where are the major internal controls?
27. What were the previous reported process deficiencies?
28. How were solutions implemented?
29. Why did problems recur?
30. Who implemented the solution?
31. Are people properly trained?
32. Have you requested or been offered any deviations or waivers in the last 2 years?
33. If yes, are you still working under these waivers?
34. What areas require the most attention or concern?
35. What are the major changes since the last value added audit?

TIPS ON USING SURVEYS

Each time an audit is completed, the value added auditor should update and keep the audit information current. The auditor should also save questions and collect tips from previous audits. For example as an audit is completed of a supplier, the auditor files this information in the permanent file and if the supplier consents, makes it available to the rest of the organization. A product improvement team may use the information to enhance or modify an existing product. Having information on field failures, warranty costs, and other data can improve product reliability, maintainability, and customer satisfaction. Engineering may want internal or supplier manufacturing capability information to develop realistic specifications.

The value added auditor should develop a checklist of each audit before the preliminary meeting with the auditee. The questions can be used to guide the auditor or the audit team in performing its audit. Questions may be divided into processes, areas, functions, or systems to be audited. Then each can be subdivided further into specific types of internal control questions.

Generic questions can be used for most value added audits. This is acceptable because most areas involve standard operations. The questions are designed to audit different organizations. The value added auditor may find that some questions are not applicable because they address a different industry, organization, process, or product. In this situation, the value added auditor may develop more applicable or specific questions. The problem is that developing new questions or tailoring existent questions takes time

CLOSED PURCHASING QUESTIONS

Yes No N/ A

Comments

I. Is there a formal make-or-buy decision process?
2. Are the following considered in the decision?
   a. Total cost of alternatives?
   b. Available capacity?
   c. Delivery?
   d. Service?
   e. Training?
   f. Available capacity?
   g. Return on investment?
   h. Inventory and storage costs?
   i. Quality capability?
   j. Technical content?
   k. Plant location?
3. Is the supplier responsible for assuring and controlling the quality of materials?
4. Are suppliers formally selected and evaluated in terms of:
   a. Quality?
   b. Service?
   c. Availability?
   d. Total cost?
   e. Technology?
   f. Just in Time (JIT) delivery?
5. Are suppliers monitored throughout:
   a. Product lifecycle?
   b. Product development?
6. Is continual improvement pursued?
7. Are supplier benchmarks set and achieved?
8. What measurements are used to gage improvement?
9. If problems occur, are they resolved quickly and efficiently?
10. Are there effective corrective and preventive action systems?
11. Is corrective action audited after implementation?
12. Are certifications received with required materials?
13. Do certifications include test results?
14. Does the supplier use a material review board (MRB) and engineering control board (ECB)?
15. Does the supplier have a system for handling non-conforming materials?
16. Does the customer request specific action from the supplier when a problem occurs?
17. Do suppliers respond in timely and accurate manner?

and knowledge of the auditee's business. If the auditor has sufficient experience then the auditor can review the list of questions, identify those that apply, and then rewrite them to focus on the customer's specific needs and the auditee's facilities.

If the value added auditor has to design a survey to satisfy the customer's requirements, the value added auditor should focus on what is important so there is little risk that the auditor will automatically check positive responses. The auditor may tailor the questions

CLOSED MANUFACTURING PROCESS QUESTIONS	Yes/No	N/A
I. Is there a step-by-step description of the entire process?		
2. Has a process flowchart been developed?		
3. Have instructions been developed for each process step?		
4. Are numerical targets with process parameters established for each process step?		
5. Are product quality characteristics identified on engineering prints?		
6. Are process control procedures written and used?		
7. Are processes in statistical control?		
8. Have capability indices been established for critical quality characteristics?		
9. Does engineering use process capability to develop specifications?		
10. Are there internal control systems separate from statistical process control systems?		
11. Do these systems form a closed loop?		
12. Do they provide real-time feed forward and feedback information?		
13. Is this information used to control and improve the system?		
14. Is SPC used in:		
a. Supplied material?		
b. In-process?		
c. Final?		
15. Has process capability been defined?		
16. Is there a safety checklist for critical process steps?		
17. Is there a list of raw materials or components for each process step?		
18. Is production equipment periodically maintained?		
19. For the entire production process:		
a. Are there sampling schedules and instructions?		
b. Are there in-process statistical controls? .		
c. Are actions taken based on process control conditions?		
20. Has everyone been trained in SPC?		
21. Are the following charts used to monitor quality?		
a. Variable (X-bar and R)?		
b. Cumulative sum (CUSUM)?		
c. Standard deviation (s)?		
d. Moving range (MR)?		
e. Moving average (MA)?		
f. Exponentially Weighted Moving (EWMA)?		
22. Are charts manually charted?		
23. Are charts computerized?		
24. Does corrective action solve root cause problems?		
25. Is corrective action initiated to remove chronic problems?		
26. Are control charts periodically reviewed for performance?		
27. Are the specifications and tolerances periodically reviewed?		
28. Are control limits modified as the process improves?		
29. Are charts with hourly sampling reviewed:		
a. Weekly?		
b. Monthly?		
c. Bimonthly?		
30. Are charts with daily sampling reviewed?		

to customer needs, business objectives, risks, controls, and to the area being audited. The auditor also may take an active role in solving problems, identifying areas of high risk and their internal controls instead of automatically checking yes/no answers.

USING THE AUDITEE'S DOCUMENTATION TO DEVELOP THE SURVEY

If the value added auditor doesn't have the time, knowledge or skill to develop new questions, the auditor may use the auditee's quality and operational manuals and create questions out of the procedures or workmanship standards. We'll discuss this using a quality example.

Sometimes, suppliers develop extensive quality manuals to satisfy customer's contracting requirements. A supplier would tell its customers that it has a quality program, and the quality manual attests to it. The customer believes that if the auditee developed extensive procedures and detailed everything then quality would be assured. This is often a false assumption.

Military contractors and aerospace companies that had to comply with MIL-Q9858 and now ISO 9001 developed exhaustive quality manuals. It cynically was called 'womb to tomb' documentation because the manuals detailed every element of quality from raw material to finished product. This occurred at the division and plant levels. It was thought that if the quality or standard operating manual described and detailed all the quality systems, processes and procedures then quality was being incorporated into the organization. The problem was that the quality manual, standards, procedures, and instructions did not ensure that:

- Quality was being implemented at all.
- Quality was being implemented properly.
- Defects were prevented from occurring.
- Quality systems were continually improved.
- Final customer was satisfied.

ISO 9001: 2015 does not require a quality manual. Value added auditors can still conduct audits using the auditee's quality manuals or other standard operating procedures. However, they should take the quality manual or other process documentation and spend the necessary time to craft specific questions and test systems. The goal is to ensure that the quality manual or other documentation is not cosmetic. A manual may have been developed, but the procedures in the manual have not been deployed.

ADVANTAGES OF A SURVEY

Ready, fire, aim.
Anonymous

The advantages of using a survey to conduct a value added audit are:

- **Facilitates planning.** Auditors, auditees, and the customer share the same agenda. They help ensure a common audit language. They help reprioritize areas to be audited. They ensure that audits add customer value.
- **Prioritizes areas of audit risk.** The value added auditor can identify and rate those questions that are essential to minimizing risk or achieving audit objectives.
- **Allows a disciplined approach.** The auditor can follow a systematic and disciplined approach to examine a particular industry, company, supplier, area, system, or process.
- **Can be used as a guide.** The auditor can use it as a guide to ensure that all audit areas are covered. The surveys cover the main points of the audit.
- **Promotes audit by exception.** The auditor can focus on exceptions, questions that generate a 'no' response.
- **Saves time and effort.** The auditor doesn't have to develop a specialized audit each time. A specific survey takes time to develop, complete, and conduct. A good boilerplate survey saves time and money. An audit survey can be used for similar situations.
- **Helps to understand process risks.** The auditor can use the survey to apply generic risk questions to understand overall organizational and process risks.
- **Has ability to tailor.** The auditor can use a generic survey and develop more specific questions to target the assessment to supply the customer with detailed information.
- **Helps understand previous auditor.** The auditor can use the same generic survey to follow and understand the reasoning of the previous auditor.
- **Ascertains facts.** The auditor can use the survey to ascertain facts such as when the survey is sent to a supplier to complete before the audit is conducted.
- **Helps understand auditee's operations.** The auditor can use the audit form and survey to logically follow and to understand the major elements of an auditee's processes.

DISADVANTAGES OF A SURVEY

The following are disadvantages of a survey:

- **Too limiting.** A yes/no survey is limiting because it focuses on binary solutions, go/no-go, black/white, good/bad, and conforming/nonconforming conditions. Processes are seldom binary and are usually some shade of gray.
- **Not intended for specific situations.** A survey focuses on generic conditions and is not designed for a specific company's operations or specific risk - controls. The auditor may have to tailor questions or may have to write an evaluation.
- **Follows prescribed patterns.** A survey leads the auditor into a prescribed methodology and questions. The auditor may become myopic and not see the obvious. The auditor becomes conditioned or hypnotized and follows a line of rote investigation.
- **Promotes overconfidence.** New auditors may perceive the survey as a piece of cake and fly through the questions without seeking corroboration.

- **Hypnotic.** Long surveys are hypnotic. An auditor may believe that it is a waste of time and skip over many questions.

CLOSE THE SURVEY

Upon receipt of the survey, the auditor can fine-tune the audit. The auditor can:

- Identify significant audit risk – control issues and reasons for pursuing them.
- Identify critical information developed during the survey.
- Develop detailed engagement objectives, procedures, and special approaches such as computer-assisted audit techniques.
- Identify critical process risk - control points.
- Develop preliminary estimates of time and resource requirements.
- Develop detailed dates for conducting and reporting the audit.

SUMMARY

Does a preliminary audit guarantee a successful audit? No. But a preliminary survey focuses the auditor's efforts closer to the target - an effective, efficient, and economic audit that satisfies a customer's requirements.

A survey is a basic planning and assessment tool that can be used to:

- Develop value added audit and risk – control objectives.
- Analyze processes and systems.
- Evaluate breadth and depth of processes.
- Assess risks and controls.
- Organize the value added audit.
- Audit performance.
- Provide consistent audit approach.
- Focus attention and energy of highest concern.
- Anticipate problems.
- Provide baseline and benchmark assessment data.
- Validate value added audit assumptions.
- Obtain overview of auditee's operations.

Also whether the initial contact was made through email or a phone call, the survey formally informs the auditee of the impending audit, its objectives, and what will be required to facilitate it. The auditee obtains specific information about the customer's requirements and expectations.

A survey in some ways is a method to identify the auditee's ability to meet the audit's objectives. Subsequent investigations by the auditor determine how accurately and precisely can the auditee hit the bull's-eye or meet the performance objectives.

Finally, a survey checklist is a critical tool in the value added auditor's toolbox. But, it does not replace the creativity, professionalism and value of a professional auditor.

PART IV: CONDUCTING THE VALUE ADDED AUDIT (FIELDWORK)

VALUE ADDED AUDITING LIFECYCLE(TM)

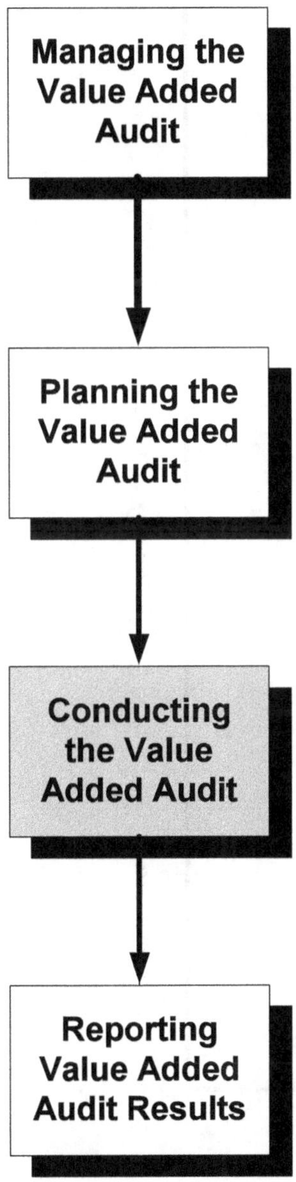

CONDUCTING THE VALUE ADDED AUDIT

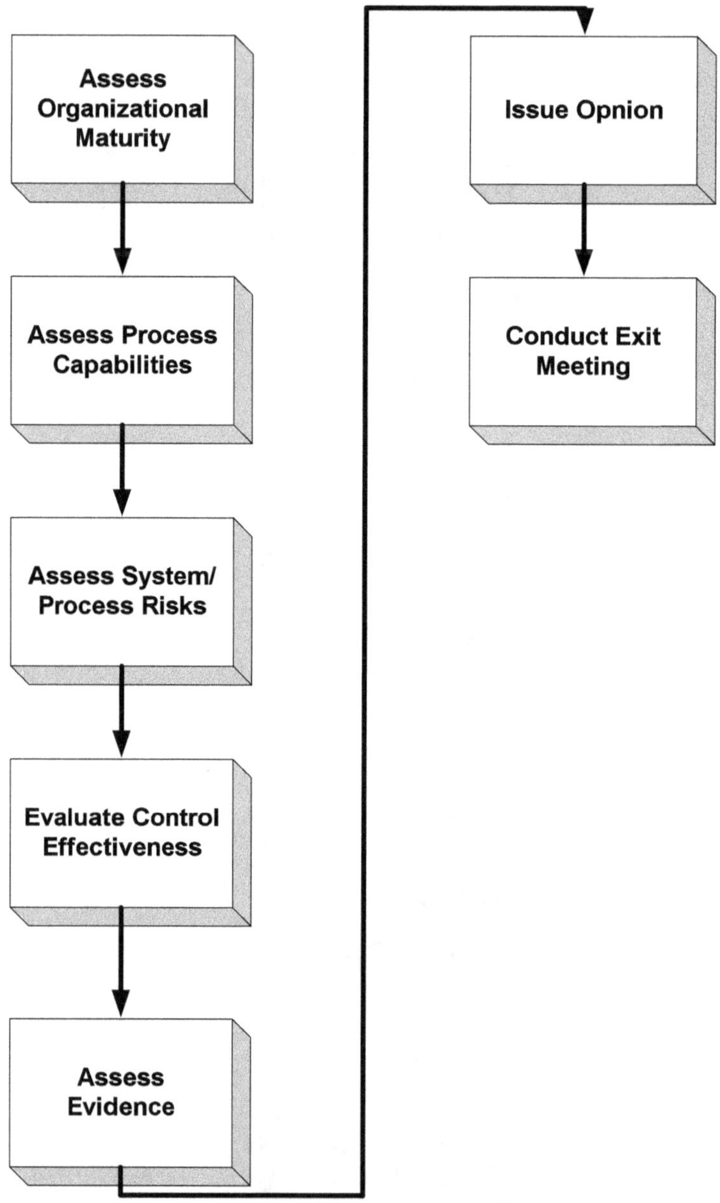

CHAPTER 12
STEP 1: ASSESS ORGANIZATIONAL MATURITY

In this step, the value added auditor develops an understanding of the auditee's overall organizational capabilities and maturity. This is a strategic assessment of the organization. In this step, we discuss two methodologies: 1. Nortel Networks process lifecycle methodology and 2. generic process assessment methodology.

Nortel Networks developed a 'lifecycle' methodology for assessing effectiveness of organizational processes. Nortel assessors followed a generic process survey to obtain evidence that process objectives are being met, planned activities are realized, results are achieved and process stakeholders are satisfied. Evidence can be in the form of records, statements of fact or other information relevant to the agreed criteria and which can be cross-checked.

Once the value added auditor understands the overall organizational capabilities, and maturity, the auditor can then evaluate specific systems and processes. This step is a tactical or operational process assessment. On a tactical level, the value added auditor may also analyze how well process owners understand their process, eliminate non value-added steps, stabilize the process, determine process capability, and finally improve the process over time. These concepts were discussed in the 'Process Management 101' chapter of this book.

ORGANIZATIONAL CAPABILITY AND MATURITY

Particulars are not to be examined till the whole has been surveyed.
Emmanuel Celler

Every organization has a distinct organizational maturity and process capabilities. This is often called the Capability Maturity Model (CMM). The Capability Maturity Model was developed in the process-quality world of the 1980's. It migrated into software and was popularized by the Software Engineering Institute (SEI) at Carnegie Mellon University. SEI developed its trademarked SEI CMM®. CMM is based on five levels of maturity as illustrated in the 'Capability and Maturity Model.' An organization 'matures' as it moves up the curve.

CONDUCTING STEP 1 – ASSESS ORGANIZATIONAL MATURITY

PROCESS CAPABILITY AND MATURITY

Several years ago, software technology was outpacing developer's ability to manage and control the process. "The ability to develop and deliver reliable, usable, software within budget and schedule commitments continues to elude most software organizations."[1] The failure rate for software projects was high. Studies indicated that up to 75% of software projects didn't meet their budgets, schedules, scope, or satisfy stakeholders. The root cause solution was to look at the software development process. As this concept matured, it migrated to assess other organizational areas.

The CMM methodology is firmly based on process management and Total Quality Management. All Cams incorporate the following TQM ideas: process management, process control, customer satisfaction, measurement (metrics), quality planning, auditing, corrective action, and preventive action.

DETERMINING ORGANIZATIONAL PROCESS MATURITIES

The Baldrige National Quality Award and other CMM appraisals are especially valuable in determining the type and ultimately the effectiveness of a value added audit. The Software Engineering Institute (SEI) audits software development firms according to its checklists and scores a company from 0 – 5. The Baldrige National Quality Award is a similar CMM assessment based on a cumulative score of 1000 points in 8 categories.

Why is it important that a value added auditor understand capability and maturity models? Organizations by their nature have different process maturity levels. For example, a company may have product inspection, process controls, enterprise-wide process management, and finally six-sigma process management. Each level from inspection to six sigma indicates a different capability and maturity level. The level of process management, capability, control, and improvement at a level 5 organization is much higher than an organization that focuses on Level 1 product inspection. So, the value added auditor shouldn't apply 'world class' audit methodologies and criteria to a relatively immature organization.

PROCESS MATURITY

The big things you can see with one eye closed. But keep both eyes wide open for the little things. Little things mark the great dividing line between success and failure.
Jacob Braude

Most process maturity and work capability models follow some type of ladder consisting of 3 to 5 steps.

SEI'S MATURITY MODEL

The Software Engineering Institute consists of these 5 levels:

- **Level 1: Initial.** Formal operational and control processes often don't exist. Work gets done through informal relationships and commonly understood protocols. Existent processes are often ad hoc or even chaotic.

- **Level 2: Repeatable.** There are few formal, documented processes. In a project or process management environment, there are few methods to track quality, schedule, and cost.
- **Level 3: Defined.** Most work follows documented, standardized, and integrated processes.
- **Level 4: Managed.** Work processes are documented detailed, quantified, controlled, and managed. A critical element of this level is the widespread use of process metrics to measure processes.
- **Level 5: Optimized.** Continual process improvement (CPI) is widely used to improve an organization's work.[2]

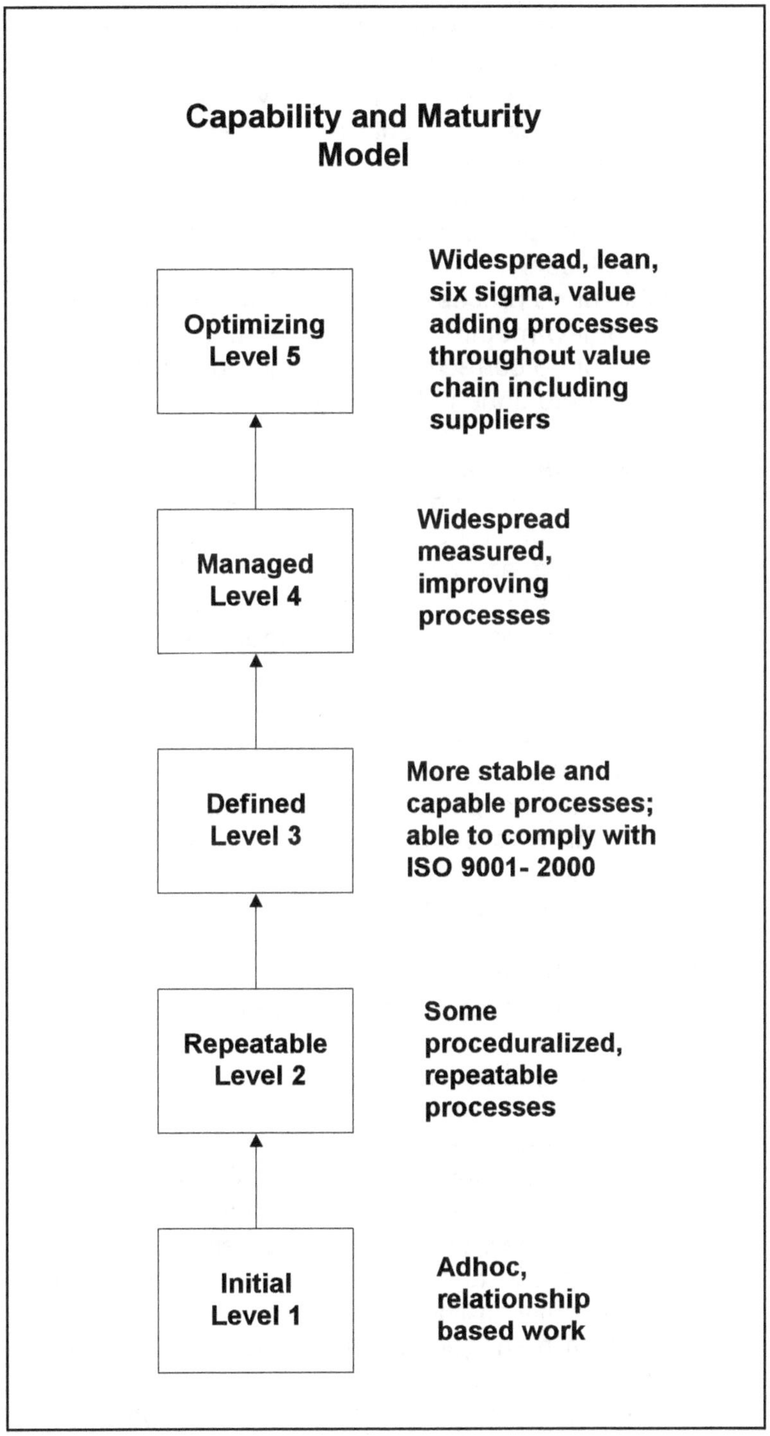

GENERIC PROCESS AUDITING CRITERIA

Many new CMM'S are being used and developed. We have developed our own version as indicated by the 'Capability and Maturity Model' figure shown on the right of page.

CONDUCTING STEP 1 – ASSESS ORGANIZATIONAL MATURITY

Level 0 –1: Initial

- Work is conducted through relationships and understood protocols.
- Processes are seldom flowcharted and understood.
- Risks are not identified.
- Stakeholders are often not known or satisfied.
- Customer requirements are sometimes understood and incorporated into processes as needed.
- No or few policies, procedures, or work/test instructions exist.
- Cost, schedule, scope, and quality measurements are either not defined or defined loosely.
- Little or no process management and control exist.
- No or few metrics and goals exist.
- Little or no corrective action and no prevention action processes are in use.
- Work gets done individually.

Level 1 – 2: Repeatable

- Work is conducted through defined, documented, and repeatable processes.
- Most critical processes are flowcharted.
- Some risks and controls are identified.
- Process and project stakeholders are identified.
- Customer requirements are identified and plans are developed to incorporate them into the product and processes.
- Policies and procedures have begun to be developed.
- Realistic cost, schedule, cope and quality measurements are developed and most of the time attained.
- Process management and control are used in a few documented processes.
- Corrective action is sometimes implemented and slowly moving toward preventive action process.
- Some work is teamed.

Level 2 – 3: Defined

- More work is conducted through defined, documented, repeatable, and capable processes.
- Process and project stakeholders are identified and satisfied through feedback loops.
- Critical process and product risks are identified and controlled.
- Customer satisfaction is progressively measured.
- Plans, policies and procedure have been developed for most critical processes as well as new product development processes.
- Realistic and aggressive cost, schedule, scope, and quality measurements are consistently attained.
- Process management, assurance, and control are tightened around measured, performance targets.
- Preventive processes and trending are often implemented.
- More work is collaborative and focus on process improvement.

Level 3 – 4: Managed

- All critical work activities are managed for improvement.
- Process and project stakeholders are consistently satisfied through enterprise and supplier wide process improvement.
- Risk management and control are integrated throughout the organization.
- Customer satisfaction trends upwards.
- Plans, policies and procedures focus on new work simplification methodologies, such as lean, six sigma, and other best practices.
- Aggressive cost, schedule, scope and quality metrics are consistently defined and attained.
- Process management, assurance and control focus on risk identification, mitigation, and management.
- Preventive processes and trending are used enterprise wide.
- Virtual teaming and other collaborative methods are widely used.

Level 4 – 5: Optimized

- Processes are tightly process controlled and capable using six sigma techniques.
- New processes, projects, and products are developed to create innovative products.
- Risk management and control are integrated throughout the value chain and into the supply base.
- Customer satisfaction is everyone's responsibility.
- Innovative work strategies are successfully implemented.
- Goal is six sigma consistency.

CMM IS AN AUDIT GUIDELINE

Cams are based on the concept of maturity. This begs the question: 'What's the difference between a mature and immature organization?' Consulting organizations have developed robust questionnaires consisting of 1000's of questions broken down into various categories.

In an immature, Level 1 organization, work is done in a loose, sometimes disorganized manner. Work is focused on activities or products. Work protocols are based on relationships and an agreed understanding on how to do work. Work is done through improvised, ad hoc activities. There are few structured processes. If there are procedures or work instructions, these are not seriously followed or enforced. Work is also inspection driven. Problem solving is reactive, after the fact, fire-fighting mode. Nonconformances, late schedules, over budgets, out of scope work, and poor customer satisfaction are routinely accepted. There are no metrics. Customer satisfaction and quality are not measured and improved.

CONDUCTING STEP 1 – ASSESS ORGANIZATIONAL MATURITY

CMM Figure

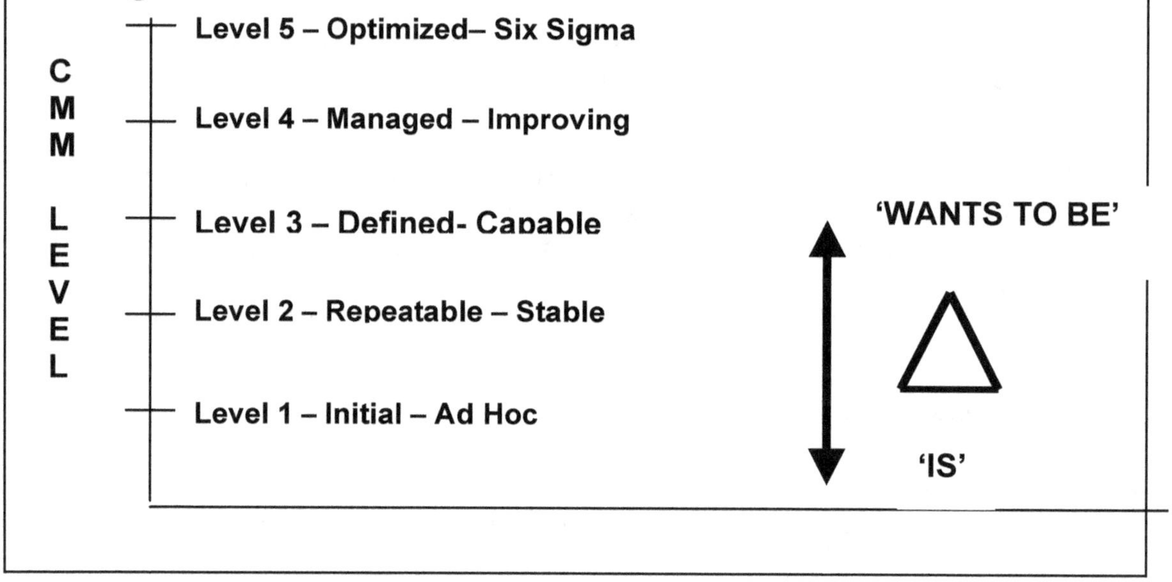

In a mature organization, work is process based and managed. Processes are documented, stable, capable and continually improving. Risks are managed and controlled. People are engaged and own their work processes. Roles, authorities, and responsibilities are well defined. Value added auditing is well accepted and routinely used. Value added auditors also serve as internal business consultants.

MOVING FROM LOW (PRODUCT INSPECTION) TO HIGH MATURITY (PREVENTION AND PROCESS MANAGEMENT)
Product inspection is a type of Level 1 CMM control and relatively low on the capability and maturity curve. Inspection is after the fact and does not add significant product or organizational value. For example, quality as 'conformance to specifications' cannot be assured or controlled by inspecting quality into the product. Inspectors check, sort, rework, rehandle, and reinsert defective products. This is wasteful, because it consumes time, effort, expense, material, personnel, and facilities.

On the other hand, defect prevention, risk management, and waste elimination are process objectives of a mature organization. This is a critical value added auditing concept. As a product is made or a service is delivered, value is added. Value is the labor, machinery, material, or facilities required to do work. For example, it is expensive to scrap a porous casting after it has gone through 10 or more manufacturing steps. Or, it is counter productive to manufacture a highly reliable product if the delivery or service system is inefficient. For example, customer satisfaction cannot be guaranteed by instructing a salesperson in courtesy after the sale. The salesperson or receptionist should have been instructed in policies and procedures before the assignment.

NORTEL NETWORKS 'PROCESS OWNER' RESPONSIBILITIES

- Owning the process so that their SUCCESS and other incentives are based on the effectiveness and efficiency of that business process.
- Ensuring standardization during development and implementation of common processes.
- Implementing effective and efficient processes for all products and markets.
- Addressing process changes needed to improve the business.
- Approving process modifications based on business case value.
- Implementing metrics for the process to support business strategies and align to Business Performance Indicators.
- Participating actively in the process to identify tools to enable the process and to integrate it across other processes.
- Communicating on-going and planned process/tools initiatives (upward and downward the organization).
- Utilizing common methodologies to document, define, and manage the process, which will facilitate horizontal integration across processes.

An underlying premise of process management is that the quality of the process outputs is enhanced through controlling and improving the processes that produce the products. This applies through the spectrum of different industry Cams, in software development, manufacturing, service delivery, and product development.

WHAT DOES THIS HAVE TO DO WITH VALUE ADDED AUDITING?

Most large consulting companies have developed generic or specific Cams. They use their proprietary CMM to differentiate themselves from the competition and to add customer value. They have developed a series of questions they use to evaluate a customer's processes. In some cases, I've seen up to a 1000 questions to determine where a client's maturity and capability. Then the consultant will develop a plan for the client to move up the maturity curve. At critical steps in the journey, the company is audited or assessed on their maturity or capability.

Common sense should dictate how Cam's are used. This book is called the **Value Added Auditing**. If the audit customer or sponsor expects a value added audit of a Level 1 organization based on Level 5 criteria, then there will be many findings based on invalid Level 5 audit assumptions and criteria. Common sense should dictate the use of value added audits of immature organizations.

Each organization has an optimal CMM. The long-term goal of an organization is to discover and move to its appropriate level of process capability and maturity. Not all organizations should pursue Level 5 processes. It may be appropriate for an organization to establish Level 2 or Level 3 CMM processes because Level 4 or Level 5 processes don't result in increasing returns.

It's important for the value added auditor to understand that as an organization moves up the CMM curve, sources of risk and variation are defined and controlled. As discussed, these processes are in control (stabilized) and capable of meeting requirements. As an organization moves up the capability curve, there are higher levels of risk man-

NORTEL NETWORKS 'PROCESS PRIME' RESPONSIBILITIES

- Enforcing process use.
- Improving the process (Benchmark Best in Class).
- Defining the scope of the process (coverage).
- Establishing project priorities and align them with other primes.
- Defining and implementing customer value chain metrics.
- Identifying appropriate IT requirements.
- Identifying critical success factors.
- Establishing clearly defined measurable goals.
- Establishing appropriate process sub-teams.
- Ensuring third party process alignment.
- Acting as a point-of-contact for other process primes.
- Offering hand-in-hand assistance to evolve to the new process with affected groups.
- Ensuring sustainability.
- Reviewing and driving appropriate policy changes.
- Recognizing and escalating problems for resolution.

agement, assurance and control. Variation is minimized. Risks are identified and mitigated.

CMM is a powerful way to look at how organizations do their work. Since, most companies organize their work around core processes, the CMM concept is a transferable and adaptable methodology to many organizational processes. As an organization matures, processes are defined, documented, stabilized, capable of meeting requirements, and improved throughout the value chain. Consistency around process targets as well as process improvement results in lower deficiencies, lower risks, higher operational effectiveness/efficiency, and satisfied customers.

NORTEL PROCESS LIFECYCLE ASSESSMENT

Technology or perish.
John Pierce

Nortel Networks is a global company. It operates in many countries. The challenge to Nortel and to other global organizations is how to satisfy customers in different countries through standardized and common processes.

Nortel Networks developed a generic process lifecycle assessment methodology based on maturity and capability that value added auditors can use to evaluate organizational processes. To clarify, Nortel calls its compliance assessments 'audits' and calls its process reviews 'assessments.' This assessment is conducted along with regular quality management system compliance audits and provides an additional service to the organization.

CONDUCTING STEP 1 – ASSESS ORGANIZATIONAL MATURITY

GOAL/OBJECTIVES OF NORTEL ASSESSMENT

The objective of the Nortel assessment is to identify opportunities for improvement (Obi) to Nortel Network process primes and process owners. Process primes are the primary process sponsors that often include senior management. Process owners are responsible for improving, controlling and putting improvement plans into effect. See above 'Nortel Networks Process Prime Responsibilities' and 'Nortel Networks Process Owner Responsibilities' sidebars.

The Nortel process assessment evaluates quality issues and specific business imperatives based on common process checklists, similar to those discussed in the *Develop Audit Survey* chapter. This assessment may look for evidence of:

- Management review.
- Business continuity planning.
- Customer focus.
- Root cause analysis.
- Corrective action/preventive action (CAPA).
- Defect prevention.
- Reliability & dependability.
- Lessons learned.
- Documentation control.
- Metrics.
- Customer improvement.
- Regulatory management.

NORTEL PROCESS LEVEL HIERARCHY

Nortel Networks created a hierarchy of 5 process levels. See 'Nortel Networks Process Level Hierarchy' figure on the next page. Nortel first identified overall business function processes (Level 1). Nortel Level 1 processes include core business functions as defined by its corporate strategic direction and value chain. This level captures the corporate vision, principles, target customers, business architecture and performance targets.

Nortel Level 2 processes translate business functions into major process functions. This level defines the scope and scale of essential process components.

Nortel Level 3 processes are deployed across all of Nortel Networks and are similar across the organization. These processes are detailed and documented through process maps, quality manual, and procedures. Level 3 processes have the following common elements:

- Identified components of the process (all process flows, triggers, input/outputs, roles).
- Detailed integration points including gaps and bottlenecks.

NORTEL NETWORKS PROCESS LEVEL HIERARCHY[3]		
Process Level	**Definition**	**Attributes**
Level 1: Business Functions	Identifies mega business functions defined by strategic direction and value chain	• Captures vision, principles, target customer, business architecture and performance targets
Level 2: Major Process Functions	Translates business functions into major process functions	• Defines scope and scale of essential process components (plan, acquire, transform, deliver, monitor).
Level 3: Main Process	Defines and sequences the main events within a major process function	• Identifies components of the main process (all process flows, triggers, input / outputs, roles). • Identifies integration points to highlight gaps and bottlenecks. • Defines major metrics such as cycle time, cost, frequency, quality, and volume.
Level 4: Sub Process	Defines the sub processes within a main process.	• Defines sequence of activities, role definition, and tool(s) used. • Expands definition of the performance metrics to identify performance issues for process redesign and tools implementation. • Defines data elements within Integration Points
Level 5: Procedures	Defines the procedures and reference materials used to guide, clarify, or instruct how tasks are performed.	• Defines methods, practices, and work instructions required for consistent process execution. • Defines detailed requirements for tools implementation (screens, data elements, worksheets, inputs, sequencing).

- Identified major metrics, such as cycle time, cost, frequency, quality, and volume are identified.

Nortel Level 4 processes are more detailed and have the following characteristics:

- Defined sequence of activities, role definitions and tools used.
- Expanded definition of the performance metrics to identify diagnostics for process redesign and tools implementation.
- Defined data elements with integration points.
- Detailed local or regional requirements.

Nortel Level 5 is the lowest and most detailed level. This level defines the procedures and reference materials used to guide, clarify, or instruct people how specific tasks are performed. Level 5 documentation include 'work instructions,' and 'work diagrams/schematics.'

Several examples may clarify the differences between a Level 3 and Level 4. A Level 3 process would flowchart a telecom installation, while a Level 4 subprocess would flowchart the specific steps for installing the test equipment. Or, a Level 3 process would flowchart a customer order service while the Level 4 subprocess details the steps from receiving the order to logging and acknowledging the finished order.

PROCESS LIFECYCLE METHODOLOGY

The power of the Nortel assessment methodology lies in its Process Lifecycle survey approach, which consists of the following five stages:

- Process Design.
- Process Deployment.
- Process Implementation.
- Process Management.
- Process Improvement.[4]

These five elements form the basis of a process lifecycle and are detailed in generic questionnaires or surveys. Quality and business imperatives are process goals and objectives, which are incorporated into each process stage.

Each stage of the process lifecycle methodology has process criteria/objectives, which are detailed in a survey or questionnaire. The general process criteria/objective are first described and then a survey details the process questions that a value added auditor would ask, analyze or determine in each lifecycle stage.

The assessor starts at Stage #1 - Process Design and moves to the next step assessing 'opportunities for improvement.' If there are negative responses to critical questions, then the auditor may stop the assessment at this point. The survey objectives and questions for the five stages are listed in the following pages:

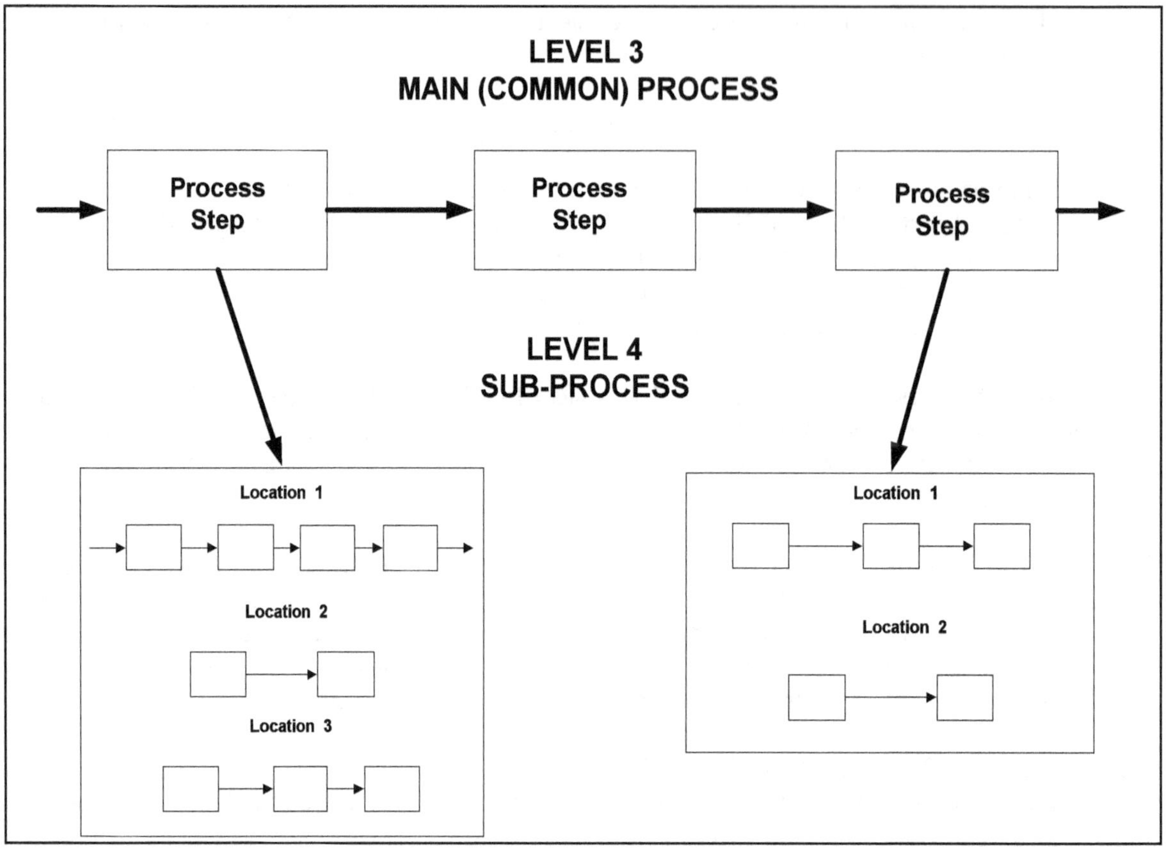

#1) Process Design Criteria/Objectives

Process design criteria are standardized across the main processes (Level 3) and are based on the following criteria/objectives:

- When designing processes, the same set of process criteria and objectives is used.
- Design criteria include quality and business imperatives.
- Customer and supplier input is included when designing a process (these are called integration points).
- Capability of obtaining metrics is designed into the process. In-process as well as output metrics are available for review. These include customer as well as supplier related metrics.
- Process ownership and boundaries are an outcome of process design.
- Assessment determines if imperatives, customer and supplier input and metrics are designed into the process.
- Information related to the design of the process is obtained through a review of Level 3 process documentation and through interviews with the design team or design team prime.

I. PROCESS DESIGN (DESIGN TEAM OR DESIGN TEAM PRIMES) SURVEY

- What is the process designed to deliver?
- Is the process documented?
- Is the process designed around the process boundaries? Are integration points (inputs and outputs) identified?
- Are roles and responsibilities defined?
- Did the process design incorporate management systems such as TL 9000, ISO 9001, ISO14000?
- Does the process enable management to meet current and future Nortel regulatory requirements?
- Does the process incorporate new/changing technology?
- Does the process ensure trouble-free and timely introduction of products/services (e.g., software tools, equipment)?
- Are business imperatives considered during the design of the process?
- Is there a 'business continuity plan' in place? Is 'business continuity planning' designed into the process?
- Does the process include methods for addressing changes in customer and market requirements?
- Are key customers, suppliers and partners involved in the design of the process?
- Is the process designed with the supplier and customer requirements in mind? Are these requirements identified and communicated to the organization?
- Are key input measures (e.g., on-time delivery) determined and documented? Are they specified or set locally?
- Are suppliers and partners involved in determining the measures?
- Does the process design include methods for monitoring key input measures?
- Does the process design assure that products and services received are from supplier's approved sites (if applicable)?
- Does the process include a method for communicating performance results to suppliers and partners?
- Are key output measures (e.g., on-time delivery) determined and documented? Are they specified or set locally?
- Are customers involved in determining the measures?
- Does process design include methods for monitoring key output measures?
- Is there a mechanism to ensure the process meets key performance requirements?
- Within the process, are quality, cycle time, cost, and lessons learned addressed?
- Does the process include reliability and dependability models, if applicable?
- Has the process defined the key operational performance metrics? (ie. quality, cost, on time delivery)? Are these specified or set locally?
- Does the process enable the organization to set and monitor targets for customer requirements?
- Is there a mechanism in place for feedback from suppliers and customers?
- Does the process design include a root cause analysis program?
- Does the process design include a defect prevention model (ie., problem solving prevention system PSPS)? `
- Does the process include periodic reviews for process effectiveness, efficiency, and improvement?
- Was the process designed with adequate document control methods?

CONDUCTING STEP 1 – ASSESS ORGANIZATIONAL MATURITY

#2) Process Deployment Criteria/Objectives

Process deployment criteria are standardized across the main processes (Level 3). Depending on the maturity and capability of the process, the process deployment stage may or may not be assessed. Local process primes and process owners are primarily interviewed in this stage. When deploying processes, the same set of criteria/objectives should be used, specifically:

- Each global process is to be deployed across representative organizations.
- Assessment confirms if the designed process is communicated throughout the organization.
- Sampling of geographical locations is conducted to determine deployment status.
- Information relating to the deployment of the process is obtained through the review of communication type documents (ie.: training materials, presentations, meetings etc) and interviews with local primes and subject matter experts (Sees).
- If deployment is determined to have not occurred, the assessment concludes at that time.

II. PROCESS DEPLOYMENT (LOCAL PROCESS PRIME AND PROCESS USERS) SURVEY
• Is the process consistently deployed? • Is there a process for process deployment? • Was process deployment communicated to your organization? • Are roles and responsibilities communicated to all involved? • Is formal training utilized to support deployment of the process? • Is process training conducted in each location? • Are local process primes and users trained?

#3) Process Implementation Criteria/Objectives

Process implementation criteria again are standardized across the main processes (Level 3). When implementing processes, the same set of criteria should be used. If it is determined that the global process is deployed throughout the organization, the next step in the assessment looks for evidence of implementation, specifically:

- Assessment confirms whether the process is implemented as the process design dictates.
- Assessment highlights where implementation is evident and where it is not.
- As in the deployment step, sampling is conducted to determine the depth of implementation across the business.
- Information relating to the implementation of the process is obtained through a review of the process documentation (Levels 3 and 4 primarily) and through interviews with local primes and process subject matter experts.
- If implementation is determined to have not occurred, the assessment concludes at that time.

III. PROCESS IMPLEMENTATION (PROCESS USERS) SURVEY

- Are local process users aware of the business process?
- Is the process implemented consistently at this location? If yes, how long has the process been in place?
- Was the local procedure modified as a result of the new process implementation? If not, was the local procedure reviewed to determine gaps?
- Do local procedures support the business processes and agree with the process maps documented?
- Do local process users know who is responsible for the local business process?

#4) Process Management Criteria/Objectives

Management criteria are also standardized across the main processes (Level 3). When managing processes, the same set of criteria should be used. Once it is determined that the process is deployed and implemented, the assessment looks for evidence that key criteria/objectives for managing a process is in place and is consistently utilized, specifically:

- Assessment looks for metrics and how the metrics are used.
- Assessment determines if communication between customers, suppliers and partners is in place.
- Assessment looks for evidence regarding the Management Review aspect of process management.
- Information relating to the management of the process is obtained through a review of records (ie. metrics, reports, meeting minutes, corrective and preventive actions, etc.) and through interviews with local primes and subject matter experts (SME's).

IV. PROCESS MANAGEMENT (LOCAL PROCESS PRIME & PROCESS USERS) SURVEY

- Are products and services received from supplier's approved sites (if applicable)?
- Are key input measures tracked, reviewed, and used?
- Are results provided to suppliers/partners to help improve their overall performance and to improve their ability to contribute to long-term performance?
- Is there evidence that the process has incorporated new/changing technology?
- Are new products/services (ie. software tools, equipment) introduced timely and trouble-free?
- Are changes in customer and market requirements addressed, (if applicable)?
- Are quality, cycle time, and cost issues being addressed?
- Are reliability and dependability models being used, (if applicable)?
- Is a root cause analysis program in place and used?
- Is a defect prevention model being used (ie., problem solving prevention system PSPS)?
- Are documents adequately controlled?
- Does the local process owner obtain real time input from process users?
- Are key output measures tracked, reviewed, and used?
- Are there key performance measurements to determine if the process is in control and operating efficiently?

- Are management reviews conducted? If so, by who and how often?
- Do improvement activities/actions result from the management reviews?
- Are business imperatives considered during the management review? Are processes adjusted as a result?
- Are supplier/customer requirements reviewed and managed?
- Are current and future Nortel regulatory requirements reviewed and acted upon?
- Is historical performance data available?
- Is there evidence of positive trends for the key operational performance metrics (ie., quality, cost, on time delivery)?
- Are lessons learned from preview processes/products/services identified and used for future improvements?

#5) Process Improvement Criteria/Objectives
Process improvement criteria are also standardized across the main processes (Level 3). When improving processes, the same set of criteria/objectives should be used, specifically:

- Assessment looks for evidence that shows mechanisms exist to improve the process.
- Information relating to process improvement is also obtained through a review of records (i.e.: metrics, reports, meeting minutes, corrective and preventive actions, etc.) and through interviews with local primes and subject matter experts.

V. PROCESS IMPROVEMENT (LOCAL PROCESS PRIME) SURVEY

- Is there a systematic method to evaluate the process and make improvements?
- Are process improvements implemented as a result of the evaluation process?
- Is feedback provided to the process owner when change to the process is required? Are the changes acknowledged? Are they acted upon in an appropriate time frame?
- Are improvements shared on a global basis?

HOW THE ASSESSMENT IS CONDUCTED
A typical Nortel assessment involves collecting and evaluating information through:

- Review of process maps (flowcharts).
- Interview of the process owner (s) using the previously discussed process assessment criteria and survey questionnaires.
- Interviews of key individuals (process primes, owners, users, suppliers, and stakeholders) within the organization and as appropriate outside the organization using the process assessment criteria.
- Observation of how the processes function.
- Examination of pertinent documents and data.

CONDUCTING STEP 1 – ASSESS ORGANIZATIONAL MATURITY

The organization, function, process, or project being assessed is encouraged to provide guides to help assist the assessment with such things as:

- Locating the people to be interviewed.
- Identifying the correct person when specific information is needed.

The auditee is as responsible for the success of the assessment as is the value added auditor team. This activity is a collaborative effort. The design team member or the design team prime describes the process at a high level, such as inputs, outputs, suppliers, customers, metrics, action plans, and actions taken. The process operator demonstrates one or more steps in how the process is used and provides evidence that the process is followed. Process suppliers (where applicable) provide evidence of interaction with the process.

During an interview, the assessor asks questions and gathers evidence to:

- Identify strengths and areas of best practice within the organization.
- Identify areas of weakness or process roadblocks to operating more effectively and to evaluate their seriousness and pervasiveness.
- Verify that the process has been effectively implemented, maintained and managed.
- Verify global deployment of the process across the business.

The assessors report on system and process gaps and strengths. The final report emphasizes what's done right and identifies opportunities for improvement (OFIs).

SUMMARY

The Capability Maturity Model (CMM) is a generic model for determining the process maturity of almost any organization. All Cams focus on the capability and maturity of an organization to develop and deliver high quality products and services on time and on budget that meet stakeholder requirements. The value added auditor shouldn't apply 'world-class' auditing techniques to an organization low on the CMM curve.

The CMM is a process concept that should be part of the toolbox of every value added auditor and process consultant. The concept allows a value added auditor to examine the processes and practices of an organization and to see how they compare against a generic model.

A significant feature of value added auditing is that it has a business-process focus. It becomes critical that value added auditors understand the concepts of process control, capability, improvement, and maturity. A major challenge to value added auditing is that many organizations have not mapped their core processes and their major subprocesses, which may include their ISO 9001 Quality Management System process.

CHAPTER 13
STEP 2: ASSESS PROCESS CAPABILITIES

A hunch is creativity trying to tell you something.
Frank Capra

As mentioned, value added auditors may also serve as internal consultants. This is a departure from the traditional arms length relationship between auditor and auditee. External accounting and financial firms have traditionally consulted with their clients and are now statutorily restricted from providing many consulting services to the companies they audit. This gap can be filled by value added auditors who assume the additional role of business process management consultants.

AUDITORS AS BUSINESS PROCESS CONSULTANTS

Give a man a fish and you feed him for one day. Teach a man to fish and you feed him for a lifetime.
Chinese proverb

Most types of auditing are maturing to become less punitive and less policing to become more collaborative. Why? Self-management and self-control are key principles these days. The IIA standards allow an auditor to conduct objective audits in situations where they may not be independent. What does this mean? The auditor conducts objective assessments but may also work with the auditee or process owners to implement the recommendations.

This is a major change in value added auditing from the traditional independent and objective, arms-length relationship between the auditor and auditee. The auditor in some organizations may evolve into a process consultant or a 'business generalist.'

COLLABORATIVE AUDITING

As has been stressed throughout this book, the value added auditor may work with senior management, customer, stakeholders, and even the auditee in a collaborative role. In some cases, the value added auditor may conduct an arms length assessment. The IIA practice guide recommends:

> "auditors should assist both management and the audit committee by examining, evaluating, reporting, and recommending improvements on the adequacy and effectiveness of management's risk processes. Management and the board are re-

> sponsible for their organization's risk management and control processes. However, internal auditors acting in a consulting role can assist the organization in identifying, evaluating, and implementing risk management methodologies and controls to address those risks."[1]

Collaborative auditing may seem like an oxymoron, an idea that is internally inconsistent or even meaningless. But, this is a significant result, outcome, and benefit of value added auditing. Traditional auditing was an adversarial relationship. Some called it policing or 'gotcha' auditing. The value addition of this process is negligible over the long term. It results at best in some fixes but doesn't result in long-term improvement.

Collaborative auditing changes the traditional adversarial auditee-auditor relationship. In this atmosphere, all relationships change. The customer or the person requesting the audit evolves into an audit sponsor or even an audit partner. The auditee, the person or activity being audited may become the sponsor, client, or the person requesting the audit. Why? The audit group brings consultative services that the internal process doesn't have. The auditor becomes a problem solver and partner with the auditee to solve problems, find better ways of controlling the process, and find ways of adding more organizational value. The results of the audit become an opportunity for improvement instead of a finding or nonconformance.

This change is significant for internal, operational, customer-supplier, ISO 9001, internal, and other auditors. These people provide management, assurance, consulting, training, and other consulting services to their clients in an objective, independent and systematic manner that adds organizational value. This is a stretch for many organizations as well as many traditional auditors.

USING VALUE ADDED AUDITORS AS BUSINESS CONSULTANTS

Using value added auditors as internal consultants is gaining many adherents. Internal customers may use value added auditors to study, evaluate, and recommend improvements. There are many reasons for using value added auditors as internal consultants. Value added auditors know the overall organization's operations, culture, policies, procedures, people, and style. External consultants would have to go through a learning curve. Value added auditors often have extensive experience, training, and exposure to different disciplines and approaches.

Value added consultants have established the trust and goodwill that are required for studying problems and recommending solutions. People know the value added auditors have a stake in implementing recommendations and solutions.

Value added auditors may offer a menu of business services, including:

- Business performance improvement.
- Risk analysts.
- Nonconformance intervention, correction, and prevention advice.

- Internal control assessments.
- Consulting and training.
- Customer-supplier audits.
- Risk mitigation, control, and prevention.

CHANGE AUDIT SCOPE

A value added auditor can provide assessment, assurance, management, and analytical services. To avoid confusion, the value added auditor's role should be spelled out in an expanded audit brief or scope of work. As discussed, value added auditors can conduct audits as well provide process assessment and consulting services. Many of the Imia's standards and guidelines offer suggestions for conducting consulting engagements. For example, the 'IIA Implementation Standard 2120.C2 - (Consulting Engagements)' states:

> "Internal auditors should incorporate knowledge of controls gained from consulting engagements into the process of identifying and evaluating significant risk exposures of the organization."

If a value added auditor provides consulting services, the auditor needs to develop what W. Edwards Deming, the quality guru, called 'profound' system, process, and product understanding. For example, major recurring risk events or critical isolated problems require immediate attention and resolution. These problems can compromise operational or product integrity. For example, a major problem may involve product safety or reliability such as automobile brake reliability. Otherwise, an accident due to faulty brakes can result in costly litigation. Value added auditors need deep product and broad process knowledge to solve these risk problems.

TYPES OF PROBLEMS

Less critical, minor recurring or isolated material problems may also require immediate attention. These problems do not affect operating integrity. More often, they are nuisances that lower operational efficiency or effectiveness. Minor recurring or isolated problems are usually fixed as they occur. However, they can lower operational efficiency and become a part of a suboptimal operating process.

Many of us take the path of least resistance when analyzing problems or situations. What is comfortable, known, successful, and reinforcing tends to be repeated. This is the same with many process owners. As well, most organizational policies, procedures, culture, principles, objectives, and even appraisal systems tend to promote status quo systems. Sometimes, radical change is required to break an organization of self-defeating patterns.

How can an organization break out of its patterns, which may challenge status quo systems or the conventional wisdom of the organization? Insiders who are subject and some would say trapped in the existing systems may not have the perspective or the impetus to transform the system. This is a major reason why more companies are engaging value added auditors to conduct risk management and process improvement assessments because they can provide 'out of the box' thinking and risk analysis.

'OUT OF THE BOX' THINKING AND ANALYSIS

'Out of box' thinking means seeing things differently. Tools for out of box thinking include: scenario building, 'what if' analysis, simulation, contingency planning, or brainstorming. The power of these tools rests with their ability to look at problems differently, with different people using different tools. Value added auditors serving as internal consultants can provide 'out of the box' process thinking and analysis. This knowledge and exposure to the entire organization is a significant value-add for auditors.

'Out of box' thinking and analysis is especially critical for functional managers and process owners because they own their systems and processes. When processes have been stabilized, made capable, and proceduralized, there is a tendency to be content with the status quo. A fundamental shift may be required to pursue breakthrough improvement or process reengineering. These examples may illustrate this point:

- Instead of minimizing process variation, what may be the value of substantially improving this process?
- Instead of documenting all processes, what are critical quality related processes that require documentation?
- Instead of auditing all systems and processes equally, what are the critical systems processes which have the most risk of failure?
- Instead of correcting deficiencies, what can be done to prevent deficiencies from occurring or recurring?

TYPES OF CONSULTING AND PROBLEM SOLVING

Anticipating, controlling, and correcting problems are essential to all risk management and process improvement. Value added consulting may involve:

- **Proactive value added auditing.** Compliance auditing can be a passive activity. Forms and procedures are checked for compliance. Products are inspected to ensure compliance with specifications. However, a more important objective of value added auditing is to actively search for opportunities for improvement (OFIs). This is called proactive auditing.
- **Intervention/interdiction.** If a problem or risk event is discovered, then immediate intervention may be required. Intervention or interdiction is the process of discovering or anticipating problems then containing and eliminating them so they don't recur.
- **Containment.** If a critical problem or risk event occurs, such as a critical product failure that can result in a loss of life, then a team is immediately assembled to contain the damage, fix the immediate problem and ensure that it does not spread. Containment of a nonconforming shipment of products would involve segregating the shipment until permanent corrective action can be accomplished.
- **Corrective action.** Corrective action is after-the-fact actions or activities to fix a problem.

SEVEN SINS OF BUSINESS PROCESS MANAGEMENT[2]

Business Process Management is the hot buzz-word of this decade. Process defines how organizations are structured, how work gets done, and how products are developed. Every value added auditor should be aware of these 7 sins of process management and improvement:

- Process improvement is not tied to the strategic issues the business faces.
- Process improvement initiative does not involve the right people.
- Process improvement teams are not provided with clear directions and are not held accountable for their objectives.
- Senior managers think if they are not carving away at the existing organization, they are not making significant improvements.
- Process designers do not consider how the changes will affect people.
- Organization focuses more on process redesign than on improvement implementation.
- Teams fail to develop a measurement system and other elements of a continual improvement infrastructure.

- **Corrective action audit.** This evaluates the effectiveness of the corrective action to eliminate the symptom and root-cause of the problem.
- **Preventive action.** Preventive action encompasses the set of activities to ensure that areas of risk are monitored and there are sufficient internal process controls to prevent deficiencies from occurring.

ROOT CAUSE DEFICIENCY AND/OR RISK ANALYSIS

If deficiencies or problems continue, this indicates that the symptom was fixed but the root cause of the problem was not discovered and eliminated. When a risk, deficiency, or nonconformance is discovered, the problem and data should be entered into a database for future examination and trend analysis. Was this a recurrence? How often and in what area did it recur? Could the recurrence be due to multiple factors?

A number of additional questions may arise when a problem occurs or recurs:

- **Type and number of deficiencies.** Deficiencies can be a customer complaint, cost overrun, ISO 9001 noncompliance, or product field failure. This data may indicate if the problem is isolated or if it is an early warning of a widespread problem.
- **Location of risk or deficiency.** The location of the deficiency also reveals if the problem is isolated to a specific physical location, instrument, or use. This data when collated or analyzed can provide more clues for isolating the problem and finding the root-cause.
- **Corrective action analysis.** Corrective action analysis focuses on investigating the deficiency and determining actions to eliminate the cause of the deficiency.
- **Preventive action analysis.** Preventive analysis is a proactive analysis of data and information such as of records, results, service reports and customer complaints to detect, analyze, and eliminate potential causes of deficiencies. Once

these are identified, then steps are identified to implement internal controls to prevent their occurrence or recurrence.

- **Cost analysis.** Quality, process, product, customer, and operational cost analysis indicate where there are existing and/or potential problems waiting to occur.
- **Risk analysis.** Areas of high risk are identified and internal controls are developed to minimize or eliminate the risk.
- **Trend analysis.** Trend analysis indicates if the deficiency or problem recurred. If it did then it may indicate the problem was not totally understood and there were multiple causes that were not eliminated. Trend analysis is important because it can be used to establish quality baselines, indicate process improvement, make comparisons, use for tactical quality planning, and indicate historical patterns.
- **Management review.** Management review is the periodic review of the suitability and effectiveness of a company's quality systems, including ISO 9001 compliance and process improvement.

PROCESS CONTROL, MANAGEMENT, AND IMPROVEMENT EVALUATION

The difficult we do immediately. The impossible takes a little longer.
Anonymous

The evaluation of process control and management follows a linear process. The value added auditor may assess, lead, coach, and assist an organization to evolve from ad hoc processes (CMM Level 1) to formal, controlled, operational processes (CMM Level 3 and higher) that are capable of meeting internal and external customer requirements. We previously described this process in terms of the Capability and Maturity Model (CMM).

PROCESS CONTROL, MANAGEMENT, AND IMPROVEMENT

Process control, management and improvement are fundamental capabilities of value added auditors. They may need to assess, assure, or even assist in process control, management, and improvement.

Usually process owners are responsible for proceduralizing, controlling, and stabilizing their operations. If there is an unusual occurrence or risk event then the process is stopped and the source of the problem is discovered and eliminated. Internal control depends on owners using their best judgment to make the right decisions to solve the problem and minimize risks. Value added auditors need to understand process control, management, and improvement to assess the effectiveness of their decisions.

Process control, management and improvement may involve a number of service activities including decreasing process variation around operational targets, improving staff responsiveness, increasing member satisfaction, minimizing risks, or lowering process nonconformances.

CONDUCTING STEP 2 – ASSESS PROCESS CAPABILITIES

PROCESS CONTROL, MANAGEMENT, AND IMPROVEMENT METHODOLOGY

Value added auditors may need to serve as consultants to assess, assure, and even implement risk management and process management practices. Remember variation off a target or a business objective is a possible risk waiting to occur. It may be worthwhile to review the *Process Management 101* chapter again. While there is no one commonly approved methodology for evaluating processes, the following lists many of the steps found in most methodologies:

- **Determine the maturity and capability of the organization.** This may well be the most critical element of all risk management and process management. If the value added auditor tries to impose or adapt management processes that are beyond the maturity and capability level of the organization, then there is a high probability of failure. The organization will reject the risk management or process management initiative because it is beyond its capability to understand and adopt the ideas in the audit report.
- **Identify and prioritize improvement opportunities.** Organizational opportunities for improvement (OFIs) are identified and prioritized through Pareto analysis, cost of quality, trend analysis, risk analysis, and force field analysis. Projects with the highest return, quickest return, or the highest risk are chosen and audited first
- **Understand process culture.** Each process has stakeholders, process owners, and a culture. The organizational culture at the process, function, or team level says how things are done. This culture has to be understood before an assessment or improvement initiative can be implemented. Process owners are critical to any assessment and corrective action, because they will be responsible for ensuring long-term process success.
- **Assess customer needs.** Each overall process and each process step has an internal customer whose needs must be satisfied. Process customer satisfaction determines if the process is doing what it is supposed to be doing. Customer or member satisfaction may be determined by satisfying a number of process requirements such as speed, quality, cost, efficiency, or effectiveness. Each process step also adds product value. Process and product value attributes were discussed in the 'Understand Audit and Business Objectives.' Processes can be controlled and proceduralized. Specific process parameters can be controlled, capability can be ensured, and the process improved. Specific techniques to use include statistical methods, inspection, sampling, and capability analysis.
- **Flow chart process.** A flowchart shows the process as a series of steps. Each step has a customer and supplier. Process owners are responsible for flowcharting their processes. The flowchart can be used to understand the process and to identify redundancies, waste, or other non-value added activities. Specific techniques to use include: block diagrams, input/output analysis, benchmarking, and reengineering.
- **Identify process steps.** A process may consist of one or a hundred steps in a series or series-parallel configuration. More commonly a process will consist of about 10 service steps.

CONDUCTING STEP 2 – ASSESS PROCESS CAPABILITIES

- **Identify inputs and outputs of each process step.** Each process step has inputs from a supplier and each step has outputs sent to a customer.
- **Develop and establish process measures.** The inputs, process, and outputs of each process can be measured. Measures may be process or product specific. Process measures include capability indices, costs, or time related measures. Product measures include critical dimensions, service characteristics, or member requirements.
- **Identify the resources spent in each process step.** Each step in the process consumes resources, including time, monies, or materials. Are these resources being consumed in the most efficient and effective manner possible?
- **Determine costs and other value subtractions.** This analysis determines the cost of all resources being expended.
- **Redesign process to add the most value. If costs or risks are beyond the risk appetite or tolerance of the organization, then t**he process is redesigned to add the most value to external and then to internal customers. Process redesign may mean adding more process steps, eliminating steps, or redistributing the flow of resources or work.
- **Specify performance standards and requirements.** Once the process has been redesigned then new performance standards are developed for the entire process, for each process step and for discrete variables within each process step including for people, resources, suppliers, etc.
- **Examine variation and stabilize each process step.** If a process has to be redesigned, there is a higher chance of process instability and special cause variation. Each process step has to be examined, understood, proceduralized, controlled, stabilized, and be capable. At a minimum, the process owners have to understand the sources of risks and variation and be able to minimize variation around process targets.
- **Understand process variation.** The measurement and control of variation is fundamental to all continual process control, management, and improvement. Once data is collected and reviewed, then special causes of variation involving people, suppliers, equipment, or products can be eliminated. Once the process is in control, then common causes of variation can be identified and eliminated.
- **Implement improvement.** Process improvement involves several objectives. One major objective of improvement is to establish controlled and capable processes. Another objective is to control process variation around process targets. Another objective is to minimize risk events or special cause variation.
- **Measure improvement.** Measurement, monitoring, and evaluation of results determine whether process controls work. Measurement identifies the effectiveness, efficiency, and economy of the improvement. From a business decision, it determines if the investment was worth it. Results may involve lowered costs, reduced cycle times, improved productivity, improved customer satisfaction, lowered risk, or other specific factors. All measurements should be meaningful to the process team, plant/business unit stakeholders, and support the corporate mission.

- **Evaluate customer satisfaction.** A number of process stakeholders may have to be satisfied. The overall process has stakeholders as well as each process step has an internal customer. The downstream step of each process step has an internal customer who receives the service or product from the prior step. If there has been a history of problems, deficiencies, rework, missing documentation, or other value losing activities, the internal process step customer can bring these up
- **Pursue continual process improvement.** Start over again. Continual or breakthrough improvement is ensured through establishing and meeting process benchmarks and improvement targets. One of W. Edward Deming's 14 points is 'constancy of purpose' which means that improvement is not instant pudding or a silver bullet but requires continual attention and application. Any type of improvement required long-term commitment and effort from senior management and day-to-day commitment from each employee. Continual improvement implies the process is never ending, which may involve reducing costs, improving customer service, or decreasing cycle times.

CONTINUAL OR BREAKTHROUGH IMPROVEMENT

Fundamental progress has to do with the reinterpretation of basic ideas.
Alfred North Whitehead

An organization can pursue 'continual or incremental improvement' or 'breakthrough or dramatic improvement'. 'Continual improvement' is often characterized by a series of singles, while 'breakthrough improvement' is characterized by homeruns. Most organizations pursue a combination of both approaches depending on their maturity and capability (CMM) goals.

BREAKTHROUGH IMPROVEMENT

Some organizations are changing and evolving through tweaking existing systems/processes but are not transforming themselves. Continual improvement is often associated with 10% or 20% improvements while transformational change may mean 10 fold improvements. Incremental change focuses on tweaking processes, reducing waste or reducing costs. Transformational change challenges the basic assumptions of the organization and establishes new processes.

Some organizations are more conducive to radical or breakthrough improvement change such as Six Sigma than others. Mature organizations are less likely to change rapidly because of entrenched practices, top-heavy hierarchies, and inflexible cultures. Breakthrough improvement tends to flourish in organizations where the marketplace requires speed and flexibility. Organizational systems and processes must be flexible to adapt to dynamic marketplace and customers. Value added consulting or auditing must be congruent, supportive, and reinforce the culture, maturity and capabilities of the organization.

An example may illustrate the importance of 'breakthrough improvement.' An organization may have accepted chronic problems and must now focus on improving core processes. A development team may discover a process only has a 80% yield, which is clearly unacceptable. As a result, the development team may have to redesign the product in order to eliminate the defect, which could not be achieved by simply adjusting or designing the production process.

BREAKTHROUGH INNOVATION

Innovation is another example of 'breakthrough improvement.' The Malcolm Baldrige National Quality Award defines innovation as:

> "Innovation refers to making meaningful change to improve products, services, and/or processes and create new value for stakeholders. Innovation involves the adoption of an idea, process, technology, or product that is considered new or new to its proposed application. Successful organizational innovation is a multi-step process that involves development and knowledge sharing, a decision for implementation, evaluation, and learning. Although innovation is often associated with technological innovation, it is applicable to all key organizational processes that would benefit from breakthrough improvement and/or change."[3]

SUMMARY

Major changes are occurring in auditing. These changes are driving internal, quality, customer-supplier, and other types of auditing into value added auditing. One change is there is no longer a purely independent and objective, arms-length relationship between the auditor and auditee. The auditor in some organizations may evolve into a process or risk consultant.

Traditionally, auditors were independent and objective. Independence implies that there is an arms length relationship between the auditor and auditee. The challenge is that if the auditor provides the auditee consulting assistance, the auditor's independence may be impaired while the auditor's objectivity to the auditee still provides value.

CHAPTER 14
STEP 3: ASSESS SYSTEM/PROCESS RISKS

Once process maturity, stability, and capabilities are understood, then the value added auditor can evaluate organizational system, and process risks. All business revolves around opportunities and risk. Any time there is a business opportunity, there is an associated risk. In **Against the Gods: The Remarkable Story of Risk**, the author says the mastery of risk is the foundation of modern life and is what divides modern from ancient times. By consciously or unconsciously calculating probabilities, auditors make intelligent decisions about business processes.[1] Risk management in the last year has become the critical filter by which management makes its decisions. Senior and middle management used to make decisions based on cost, schedule, or quality. Now, risk is the first filter by which all decisions are made.

Smart management is risk management. Smart managers want to avoid crisis, firefighting, or reactive management. It's much smarter decision making to have stable and capable processes. If there is a system or process spike, it is easier to have systems that are designed to correct and even to anticipate the problem so that it doesn't spin out of control. This is the fundamental idea behind 'risk management' and 'management by exception.'

EVALUATION OF RISKS

You've removed most of the roadblocks to success when you've learnt the difference between motion and direction.
Bill Copeland

Different level and types of risk can occur in a business. In this section, we discuss 4 types of risk that should be managed:

- Enterprise/business risks.
- Process risks.
- Product risks.
- People risks.

When approaching an engagement, the value added auditor may want to identify overall business risks. The value added auditor would work with the board of directors, senior management, process owners, or other stakeholders to determine overall risks. In this section, we offer questions that the value added auditor may want to ask to understand the above risks.

ENTERPRISE/BUSINESS RISKS

To assess business risks, the value added auditor and/or process owner may ask:

- Who are new global competitors?
- Are new competing products being introduced?
- Are there terrorism or security concerns?
- Are current customers satisfied?
- Are new or existing customer requirements identified?
- Are there legal constraints to product introduction or development, such as new laws or new regulations?
- Are there new safety, environmental, or other considerations that haven't been identified?
- Are new technologies being introduced?
- Is the present business model still viable and profitable?
- Is there a possibility of currency revaluations?

PROCESS RISKS

When evaluating process risk, the value added auditor may ask:

- What are the business objectives for this process?
- Do specific process steps have business objectives, requirements, or metrics?
- Is the process or process step aligned with the overall business objectives?
- Are critical process variables identified?
- Are critical process variables controlled?
- Is the process in control?
- Is the process capable of meeting requirements?
- Is the process improving?
- What is the capability of the process?
- Are process risks identified?
- What are the process risks or what can go wrong?
- What are the constraints or restraints to achieve business objectives in the process?
- What is the likelihood that the specific risk will occur?
- What are the impacts on business or process objectives if things go wrong?
- What and where are the process management and controls implemented to correct and prevent this?
- How much can and do process owners control or influence risk events?

ENTERPRISE/BUSINESS RISK QUESTIONS

- Does the board of directors understand critical organizational risks?
- Is the business model identified and understood?
- Are core customers identified and ability to satisfy them analyzed?
- Are core stakeholders identified and ability to satisfy them analyzed?
- Is risk a strategic issue?
- Are core processes identified, process, and risk-control mapped?
- Are risks of the business model identified and sourced?
- Are the magnitude and likelihood of critical risks identified?
- Are the financial and cost impacts of critical risk identified?
- Are there risks of:
  a. Business interruption?
  b. Catastrophic events?
  c. Health, safety, and environmental events?
  d. Reputation and image issues?
- Is the upside and down side of each risk event identified?
- Is there a process for determining risk exposures and risk relevance?
- Does the organization have capabilities to manage risks?
- Are risks of the business model, critical initiatives and/or projects discussed and determined?
- Are critical disclosures, operational, supplier and other risks identified?
- Are accountabilities and authorities determined based on risk?
- Is there a process for escalating and expanding risk accountability?
- Do performance measures include risk?
- Is performance tied to risk acceptance and control effectiveness?
- Is compensation tied to acceptance and control of risk?
- Is there a process for capturing, measuring, monitoring, controlling and reporting on new risks that may arise?
- Does senior management and/or the board of directors have plans regarding managing risk:
  a. Corrective action?
  b. Preventive action?
  c. Proactive action?
  d. Preemptive action?

PRODUCT RISKS

At each process step, value is added. The entire process results in a product. The product can be a tangible product or intangible service. At this point, product risks can be evaluated by addressing the following questions:

- Are all product stakeholders identified?
- Are stakeholder requirements identified?
- Are stakeholder requirements understood and how they will be satisfied?
- Are all the critical and major product characteristics identified?
- Are product characteristics process controlled?
- Are product characteristics prioritized in terms of their risk of failure?
- Are process capability measures assigned to critical product characteristics?
- Has the product been properly tested?
- Are possible types of failure and risk identified?

PEOPLE RISKS

Of the various types of risk, people are often a significant source of variation and of risk. When evaluating people, the following are factors to consider:

- Is senior management committed to risk management and control?
- Are people trained in risk management and control?
- Are there sufficient people to get the job done?
- Do they understand business objectives and process metrics?
- Is there sufficient time for people to do their jobs?
- Do they understand what needs to be done?
- Are they engaged with the process?
- Is training sufficient for the tasks at hand?
- Are there sufficient tools to get the work done?
- Is supervision adequate, responsive, and responsible for getting the work done?
- Are people remunerated competitively?
- Is there high turnover?
- Are people controls understood and maintained?
- Are people dedicated and honest?

ASSESS RISK MANAGEMENT MATURITY

A pinch of probability is worth a pound of perhaps.
James Thurber

The role of value added auditing has recently changed. Before the September 11, 2001 New York Trade Towers catastrophe, most organizations did not have a risk management function. Now, this is changing for even the smallest organization. So, what does a value added auditor do if the organization has no risk management process as it may be required by ISO 9001: 2015? Value added auditors may assist management by examining, evaluating, reporting and recommending risk management practices. Then once these are in place, then the value added auditor can work with senior management and process owners to determine their adequacy and effectiveness.

RISK MANAGEMENT RESPONSIBILITIES

The Institute of Internal Auditors standards place the ultimate responsibility for risk management on senior management:

> "Risk management is a key responsibility of management. To achieve its business objectives, management should ensure that sound risk management processes are in place and functioning. Boards and audit committees have an oversight role to determine that appropriate risk management processes are in place and that these processes are adequate and effective."[2]

CONDUCTING STEP 3 – ASSESS SYSTEM AND PROCESS RISKS

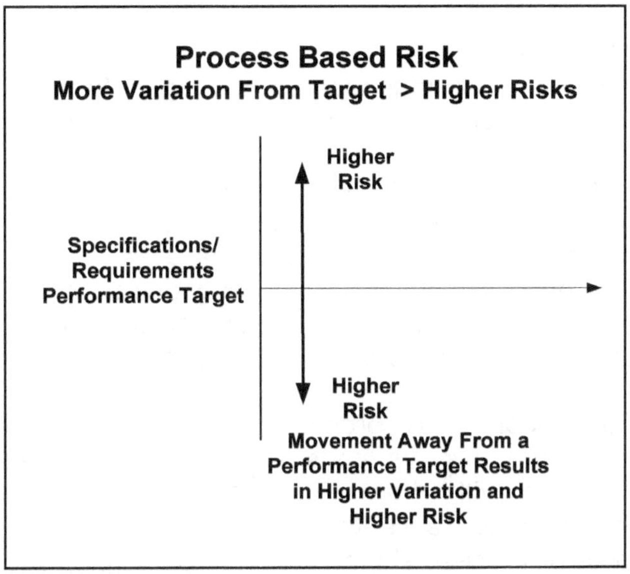

Who is ultimately responsible for controlling and managing operational risks? Process owners are responsible for the quality and risk of their day-to-day work. The board of directors and senior management are responsible for developing risk policies. Middle management is responsible for ensuring risk management procedures are applied and implemented consistently. 'After the fact' auditors can't be responsible for risk management on a daily basis.

In the last few years, we've seen that self-management has become more prevalent. Self-management infers risk management. What does this mean? Organizations and stakeholders believe that people are responsible for their work and have the authority to change their processes. This is what underlies the concept of process ownership. If this concept is carried a little further, then the process owners are also responsible for identifying, controlling, and managing process and product risks.

At the enterprise or organizational level, senior management should have processes and procedures to manage overall business risks, which may incorporate elements of the following:

- Identify mission critical objectives.
- Align processes to mission critical objectives.
- Determine risks or what can go wrong in each business unit, function, or process.
- Determine the extent of the exposure or risk.
- Determine if possible, the probability of the exposure or risk.
- Examine the current level of overall controls.
- Review applicable residual risks.
- Close the gaps to cover the shortfalls.
- Start the cycle over again looking for system, process, or product gaps.

RISK MANAGEMENT STRATEGIES

The organization should have risk management strategies that the value added auditor can assess. For example, the following discusses several 'make or buy' risk strategies:

- **Avoid risk.** This requires the root cause of the risk is eliminated. This may mean finding a new supplier or developing new processes.
- **Mitigate risk.** This requires the probability of the event occurring is reduced. Risks can be mitigated through transferring, deflecting or reducing them. Risks may be mitigated through contingency plans, and insurance. Examples of risk mitigation include using redundant suppliers, cross training existing personnel, or using backup hardware.
- **Accept risk.** This means the process owner has identified the risk, its consequences, and can live with them. The process owner or customer accepts the fact that the supplier or process shipments have an acceptable number of nonconforming products. If the quality level of received products needs to be improved then the customer or supplier can invest in additional machinery to improve capability levels.
- **Share risk.** This means the customer and supplier may assume and share the risk. The customer may need improved quality from a captive or sole-source supplier and will invest in capital equipment at the supplier's location.
- **Diversify risk.** This means the risk is unacceptable and must be spread out. The process owners may have a sole source supplier and the risk of an event such as a strike implies that shipments may be disrupted. The customer diversifies the risk by finding acceptable, alternate suppliers.
- **Control risk.** This means the occurrence or recurrence of the risk can be monitored and even prevented. The process owner has sufficient trend data or other information to predict when a shipment may contain nonconforming products, when a machine's output may be out of specification, or when machines should be preventively maintained.
- **Increase risk.** This means the customer will increase risks because the potential returns outweigh the risks. For example, a company may choose a new supplier to reduce cycle times or increase product development opportunities. The future potential returns outweigh present perceived risks.

OVERALL RISKS

- Lack of business effectiveness.
- No systems or processes.
- No business model.
- Poorly deployed business model.
- Errors and omissions.
- Violations.
- Disasters.
- Fraud.
- No ethics.
- Poor corporate reputation.
- Systemic/chronic problems.

CONDUCTING STEP 3 – ASSESS SYSTEM AND PROCESS RISKS

MANAGING TACTICAL RISKS

Hopefully, process owners have quantified the risks in their processes. If not, auditee management may have to introduce these concepts and work with the process owners to establish process risk controls.

Management or the process owners should understand the following in evaluating tactical risks:

- Identify risks for the particular business unit and appropriate process level.
- Identify mission critical objectives.
- Understand what and how the process works.
- Identify critical input process variables.
- Prioritize input risk variables.
- Identify possible and probable risk events for the input process variables.
- Understand the probability of exposure or events.
- Understand the impact or consequences of an event occurring.
- Develop quantitative or qualitative process and risk analysis.
- Develop process risk profile.
- Understand the output process variables.
- Identify stakeholder risk tolerances.
- Be able to put cost estimates of an event occurring.
- Evaluate process outputs for conformance to standards.

RISK MANAGEMENT TACTICS

A simple but effective way to look at risk is in terms of movement or variation away from a specification, requirement, or performance target. This is shown in the 'More Variation From Target > Higher Risks' figure on page 273. Since process management is fundamentally the study of variation, process management is a critical means to manage risks.

Each process consists of a number of steps. Each process step has a number of input variables, such operator, methods, machines, materials, and environment that were discussed in the *Process Management* 101 chapter. Some input variables are more critical (potential risk) than others. Process owners should be able to identify, specify, control, and improve these critical process variables.

As we discussed, Nortel Networks and many large companies have multiple process levels. Nortel Networks had 5 levels of processes. Each level may have a number of high-risk process steps. Process owners should identify specific high risk steps and develop commensurate controls to manage these risks. This is shown in the 'Location of High Risk Process Steps' figure on the next page.

Once the high-risk, process steps have been identified, then critical process variables can be identified. These are the process variables that can create unwanted variation

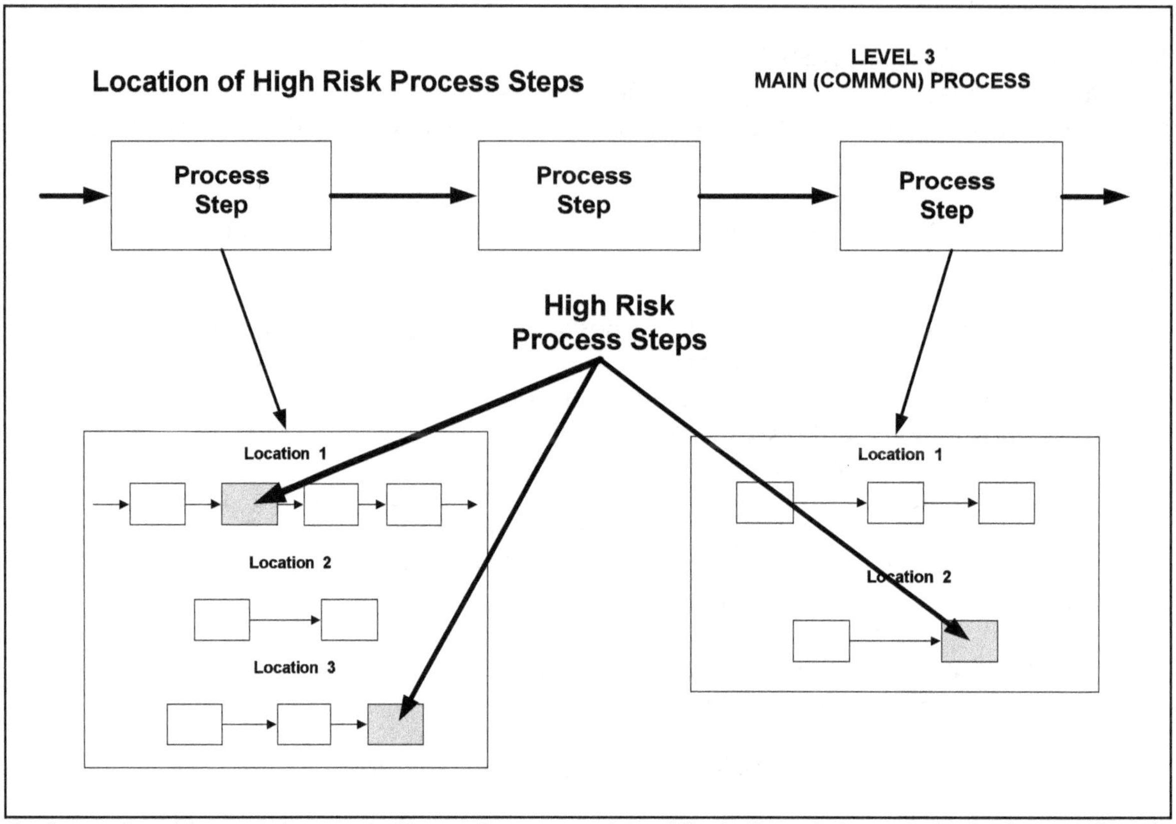

that can lead to a risk event. Once these are identified, then internal controls are developed to minimize and control the variable. This is fundamental to all risk management.

Risks also may not be quantifiable. The customer or auditor must determine whether they are high, moderate, or low. It is largely a subjective judgment. There are models of risk assessment that quantify risk in terms of weighted criteria, but these models are beyond the scope of this book. In some cases, external regulatory authorities, such as the Nuclear Regulatory Commission, may determine risk levels and controls in the construction and operation of nuclear power plants.

There can also be significant differences in risk management methodologies used by an organization. Depending on the organizational size, products, processes, and complexity, risk management processes can be:

- Formal.
- Informal.
- Quantitative.
- Qualitative.

The type of risk management approach reflects the organizational processes and products. Large organizations can adopt a more sophisticated and quantitative risk manage-

PROCESS FMEA TABLE

System | Process Name
Process Product/Service | Prepared By

Functional Risks	Failure Mode	Possible/ Probable Causes	Effects		Failure Detection Techniques	Failure Criticality	Symptom/ Root Cause Analysis	Counter Measures	Com-ments
			Process	Sub Process					

ment approach than a smaller company. Smaller, less complex organizations may follow an informal and qualitative approach in developing risk profiles.

PARETO PROBABLE RISKS

High risks imply high internal controls. So, an important function of value added auditing is to determine the level of risk and then evaluate the effectiveness of compensating controls.

The reality is that only one or several input variables is a course of a critical or major process risk. The others are negligible. To use Joseph Juran's term, process owners and value added auditors should be able to identify the 'critical few' risk variables from the 'insignificant many.' This is basically a rewording of the Pareto Principle that says that 20% of the process variables can produce 80% of the process variation and produce 80% of the risks. The trick, of course, is to identify the critical 20%. This isn't easy in technical processes. Design of experiments, FMEA, and other statistical tools may have to be used.

FAILURE MODE EFFECTS ANALYSIS (FMEA)

FMEA is a commonly used tool/tactic for identifying, controlling, and preventing risks. There are two types of FMEAs: product and process. FMEAs can be used to:

- Identify process and product risks.
- Anticipate risk and nonconformances during the design stage of the product lifecycle.
- Identify probable and possible causes of failure.
- Develop countermeasures to prevent the occurrence and recurrence of the failure or risk.

The power of FMEA is to anticipate and predict failures. For example, FMEA analysis occurs when engineers or process stakeholders identify all the potential causes of critical or high-risk failures and develop plans to prevent the failures.

AUDIT RISK

GAGAS 6.05 Audit risk is the possibility that the auditors' findings, conclusions, recommendations, or assurance may be improper or incomplete, as a result of factors such as evidence that is not sufficient and/or appropriate, an inadequate audit process, or intentional omissions or misleading information due to misrepresentation or fraud. The assessment of audit risk involves both qualitative and quantitative considerations. Factors impacting audit risk include the time frames, complexity, or sensitivity of the work; size of the program in terms of dollar amounts and number of citizens served; adequacy of the audited entity's systems and processes to detect inconsistencies, significant errors, or fraud; and auditors' access to records. Audit risk includes the risk that auditors will not detect a mistake, inconsistency, significant error, or fraud in the evidence supporting the audit. Audit risk can be reduced by taking actions such as increasing the scope of work; adding specialists, additional reviewers, and other resources to perform the audit; changing the methodology to obtain additional evidence, higher quality evidence, or alternative forms of corroborating evidence; or aligning the findings and conclusions to reflect the evidence obtained.

FMEA analysis can be conducted at various levels of a process. For example, global risks and possible failures can be identified for core processes. Then, this analysis can cascade or be deployed to business unit, plant, functional, and work area processes.

Also, this analysis can be applied to a product. Potential risks and failures can be identified at the product, subassembly, and component levels. Again is it the responsibility of the value added auditor to conduct this analysis? No. The process owners have the primary responsibility for this. The value added auditor may assist or preferably review the effectiveness of this process.

IDENTIFYING HIGH RISK AREAS

Keep in mind the fact that Ralph Nader could be the first customer for your new product.
Anonymous

As discussed, all processes and all process steps are not equal in importance. The Pareto Principle basically says the 20% of the areas will cause 80% of the problems. Probably more accurately, 5% of the process variables are the most critical to control. Read the above sidebar on 'Audit Risk.' So, the challenge for the value added auditor is to focus his or her attention on the 'critical few' processes not the 'insignificant many.' How does the value added auditor know how to do this?

FOCUS ON HIGH RISK AREAS

The key to value added auditing is to understand the consequences of failure caused by the specific risk variables and to evaluate the effectiveness of the controls used to manage these critical risks. Value added auditors may have to work with process owners to develop controls to manage risk exposure to a product failure, product recall, lawsuits, and customer dissatisfaction.

CONDUCTING STEP 3 – ASSESS SYSTEM AND PROCESS RISKS

EVALUATING DETECTION RISKS
1. Is the value added auditor properly trained?
2. Are value added auditors tested and certified?
3. Does the value added auditor have specific industry, process, and product knowledge and background?
4. When was the last audit?
5. What were the results of the last audit's assessment?
6. What reliance do parties put on the value added audit?
7. What are the consequences of deficiency findings in an audit?
8. What was the scope of the last audit?
9. How extensive are value added audits?
10. How intensive are value added audits?
11. What was the cost of the audit and who paid for it?
12. Is there a quality manual?
13. Is it followed?
14. What are the operational constraints to a value added audit?
15. What specialists are used in the audit?
16. Who conducted the last audit?
17. What types of tests were conducted?
18. When was a post audit conducted?
19. What is the quality of the permanent files?
20. Is the auditee cooperative?
21. Is the customer cooperative?
22. What is the statistical confidence of the internal tests?
23. Did they miss anything?

The customer, the auditee, and the value added auditor all involve a level of risk. The value added auditor should understand these. The auditor's risk is that he or she may make a mistake in terms of not understanding risks, not finding a deficiency, mistaking a situation, overstating a situation, misstating a situation, miscalculating, or any other number of errors. The customer's risk is that he or she may obtain incorrect information that results in poor business decisions with potentially high risk consequences. Finally, the auditee's risk that he or she may be accused of producing poor quality goods or services or have ineffective controls. The threat of litigation and high damage rewards raises the stakes for each party.

The value added auditor should be familiar with these auditing risks:

- Size risk.
- Control risk.
- Detection risk.
- Inherent risk.

SIZE RISK

Size involves the magnitude of the risk or exposure. Size risk implies that an occurrence may have catastrophic or critical implications. On the other hand, these events usually

EVALUATING SIZE (MAGNITUDE) RISKS

1. What is the size of the auditee?
2. What is the size of the auditee's budget?
3. How many employees does the auditee employ?
4. What types of products does the auditee produce?
5. How many products does the auditee produce?
6. What is the cost of each product?
7. Does the auditee function in a special regulatory environment?
8. What is the auditee's industry?
9. Who are the auditee's customers?
10. How much capital is committed to the project?
11. What is the number of customer transactions?
12. What is the average value of each transaction?
13. What are the political risks?
14. What are the environmental risks?
15. Are there risky conditions or penalty clauses in any contracts?
16. Is there a risk of customer dissatisfaction?
17. How are the elements of the cost of quality distributed?
18. Are there speculative or high-risk projects?
19. Is there a chance of bankruptcy?
20. What are major external influences, such as political, technological, social, cultural, etc.?
21. What are the related party transactions?
22. Are there special purpose entities?
23. How are these controlled and disclosed?
24. What is the level of organizational candor and disclosure?
25. What is the vulnerability to fraud and illegal acts?

have a low probability of occurrence. For example given a large number of items, statistically there will always be some random items outside specification limits.

Risk may also involve a limited number of items, but each has a high value or high associated risk if failure occurs. Nuclear power plants and space shuttles are examples of such failures. Size risk can involve the number, value, importance, or other factors. A list of questions detailing inherent risk is shown in the above 'Evaluating Size Risks' sidebar.

CONTROL RISK

Much of this discussion in this book is devoted to discussing how to minimize control risk and to optimize the effectiveness of internal process controls. Control risk is the risk that the internal control system will not detect or prevent a nonconformance or risk event in a timely manner.

CONDUCTING STEP 3 – ASSESS SYSTEM AND PROCESS RISKS

EVALUATING CONTROL RISKS

1. Are process controls used company wide?
2. Are statistical controls used company wide?
3. Are major processes in statistical control?
4. Is process capability determined for most processes?
5. Is the type of internal control suitable to the process risk?
6. How long have processes been operating and in control?
7. What are special control factors?
8. Are corrective/preventive actions effective?
9. What is the evidence of continual improvement?
10. Are the inputs, process, and outputs analyzed in terms of statistical controls?
11. Are projects effectively controlled?
12. What are recent system and process changes?
13. What is the extent of interaction or influence by the outside world?
14. Are customer concerns communicated quickly?
15. Is corrective action prompt and effective?
16. What is the quality of the auditee's management?
17. What is the quality of the auditee's employees?
18. What is the quality of the auditee's suppliers?
19. Is employee training and development effective?
20. Does the legal department evaluate contracts?
21. What is employee turnover?
22. What is the degree of company wide automation?
23. Are employees technically literate?
24. Are operating procedures and quality manuals developed?
25. Do actions reflect the information in the manuals?
26. Do customers or other parties express concerns?
27. How are the concerns resolved?
28. Is quality measured throughout operations?
29. Are world-class benchmarks established?
30. Do measurement systems produce accurate and precise data?
31. Are internal value added audits extensively used?
32. What are known losses and exposures?
33. What are the results of the last audit?

As discussed, process owners and management own their risks and their controls. The value added auditor primarily assures that process owners identify and control their risks. The value added auditor may assess that process owners:

- Analyze the process or project.
- Identify the triggers of a deviation, nonconformance, or variance.
- Trend or track the status of process risk variables.
- Monitor out of control or incapable process situations.
- Determine frequency of monitoring the internal control system.
- Test the system for risk management effectiveness.
- Review their risk plans after each risk occurrence.
- Reassess the probability and impact for each risk event.

- Document each risk event.
- Conduct a causal analysis of the risk, threat, or vulnerability.
- Create lessons learned file.
- Communicate results to the project team or process owners.
- Develop corrective and preventive action plans to mitigate risks.
- Integrate risk management into each meeting.

DETECTION RISK

Detection factors are issues that deal with the auditor's due care and professional proficiency. If both are properly exercised, this risk should be low. A list of questions is shown in the 'Evaluating Detection Risks' figure on page 279.

Professional proficiency and due care are defining attributes of value added auditors. As discussed, due professional care implies that the value added auditor seeks a level of assurance that organizational and process controls are in place, operating properly, detecting risks and preventing nonconformances. In performing this risk assessment, the auditor is part of the organization's overall internal control mechanism to ensure that risks are managed, deficiencies are minimized and continual process improvement is pursued.

Detection risk is the risk that the value added auditor didn't exercise sufficient due care or in other words the auditor may have made a mistake. The value added auditor may conclude there is no error or nonconformance when there is one. The value added auditor can minimize this risk by having the requisite professional proficiency and exercising due professional care.

INHERENT RISK

Inherent risk is the natural risk in a process or system that can result in a nonconformance regardless of internal control procedures or the ability of the value added auditor. The following are examples of inherent risk:

- Complex engineering design calculations have a higher inherent risk of errors than adding up a grocery bill.
- Complex products have a higher probability of failing than similar simpler products.

The value added auditor usually can't reduce inherent audit risk. The organization itself must change its basic risk management system to reduce inherent risk and this is usually beyond the auditor's scope and capability, unless this is part of the auditor's consulting engagement activities. The value added auditor can only recommend system or process improvements that change the basic system. If the auditor believes there is a risk of a miscalculation, the value added auditor can recommend that engineering conducts design and calculation reviews. If engineering has to purchase computer-aided design tools to simulate different conditions and check design calculations, then management must authorize the design review and the purchase of a new machine.

OPEN ENDED QUESTIONS

The questions in the detection, size and control risk figures can be open-ended. They reveal the possibility and potential of different types of failure. The risk questions deal with 'what if' types of issues, such as:

- What are possible types of process or product failure?
- What are the possible consequences, magnitude, and probability of each type of failure?
- What are the estimated costs of each failure?
- What internal controls seem to prevent failures?
- What corrective actions may be implemented?
- What FMEAs have been implemented?
- What history of success is there with FMEAs?

When the value added auditor comes across one of the situations addressed in the above questions, this should raise a red flag and caution the value added auditor to a possibly damaging or risky situation, which may require gathering more evidence before issuing a finding or opinion. The value added auditor may notice that some questions fall into different risk categories.

VALUE ADDED AUDITOR RESPONSIBILITIES

There's no such thing as "zero risk."
William Driver

Value added auditors can serve various risk management roles from evaluator/auditor to even consultant.

VALUE ADDED AUDITOR RESPONSIBILITIES

Depending on how senior management and the board of directors want to manage risks, value added auditors can serve different roles in the risk assessment and management process from:

- Have no role in risk management.
- Consider risk management as part of every value added audit plan and auditing process.
- Participate actively and continually support the organizational risk management process.
- Manage and coordinate the organizational risk management and assessment process.

Hopefully, process owners manage risks and have developed the appropriate control – control systems. The value added auditor can then validate and verify the right things have been done to analyze and manage the risks.

CONDUCTING STEP 3 – ASSESS SYSTEM AND PROCESS RISKS

DISCUSS ROLES AND RESPONSIBILITIES

In determining the appropriate risk management role, authorities, responsibilities, and expectations, the value added auditor may discuss with the board, senior management, and process owners the following questions:

- What risk management role does the board, senior management, or process owners want the value added auditor to assume?
- How risk sensitive is the board, senior management, and process owner?
- What risks can the board, senior management, and process owners live with?
- What are specific risks of the enterprise system, process, or product?
- What can be done to mitigate each risk variable?
- What risk approach does the owner want to follow?
- What risk approach should be followed, ie. risk transfer, risk deflection, or risk reduction?
- What happens if risks are accepted?
- What type of risk contingency plans are developed?
- Can the risk management role be defined in the value added auditing charter?

THE MOST CRITICAL RISK MANAGEMENT QUESTION

One of the first questions the value added auditor should ask is: 'Does the auditee have a process of managing and controlling risks.' An organization can have different risk management processes or methodologies. The value added auditor should assess whether the risk management methodology is appropriate to the organization. The value added auditor may want to:

- Review industry trends.
- Understand strategic mission, strategies, goals, and objectives.
- Identify stakeholder requirements.
- Determine risks and exposures and that may influence the organization.
- Review internal control procedures.
- Review corporate policies, procedures, and work instructions.
- Review previous risk evaluations by senior management, internal auditors, quality auditors, value added auditors, third party auditors, or second party auditors.
- Interview senior and operational management and related stakeholders regarding risks, risk mitigation strategies and control activities.
- Assess effectiveness of risk mitigation, initiation, monitoring, control, and communication activities.
- Assess adequate and timeliness of reporting of risk management audits.
- Review the completeness of management's risk analysis.
- Review management's action to remedy issues raised by risk assessments.
- Observe efficiency and effectiveness of management's self-assessment processes.
- Test internal controls and monitoring procedures.
- Test accuracy of information.

CONDUCTING STEP 3 – ASSESS SYSTEM AND PROCESS RISKS

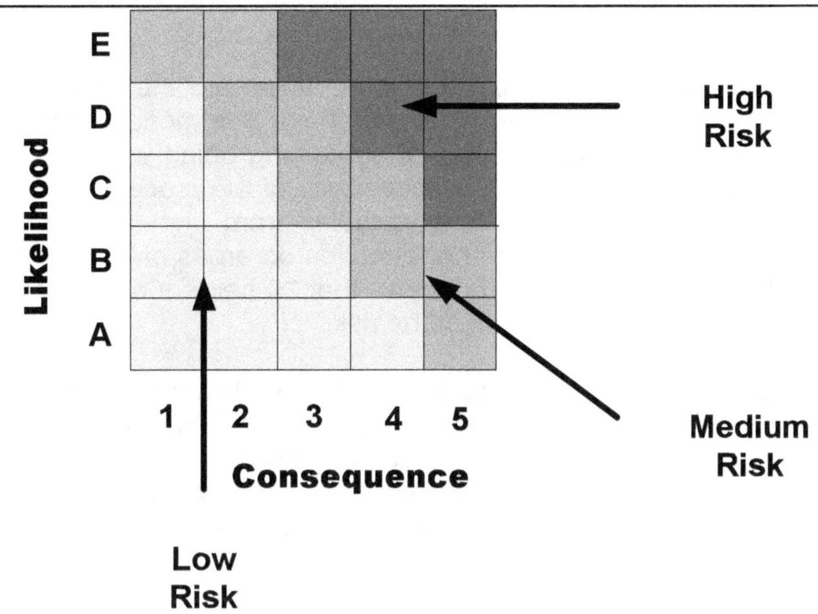

Risk Profile Matrix

- Review weak risk management controls.
- Review accepted level of risk and controls.
- Review risk management policies and strategies.

MINIMUM REQUIREMENTS FOR RISK ANALYSIS

At a minimum, the value added auditor should have the following to identify the possible risk on a project or process:

- Flow diagram of the process or project.
- Work breakdown structure of a project.
- Description of the management and controls for the overall process as well as of each process step.
- Risk profile matrix for process or product (see above 'risk profile matrix').
- Expert knowledge of the system, process, and products.
- Process procedures, work instructions, trending data, and files.

What does the value added auditor do if he or she doesn't have sufficient knowledge or information of the process? The auditor then must get qualified experts or the process owners preferably to identify the process risks and controls.

CONDUCTING STEP 3 – ASSESS SYSTEM AND PROCESS RISKS

SUMMARY

The critical question of value added auditors is: 'What are the critical risks to manage and control?' Or expressed another way, 'what are the most critical variables to the process and which pose the highest risk of something going wrong?' Should the value added auditor do this? No. This is the responsibility of the process owners. The process owners should prioritize risk areas and variables from highest to lowest. The value added auditor would then focus his or her attention on areas of highest risk. Again using the 80-20 rule, the value added auditor knows that 20 percent of the variables create 80 percent of the variation and 80 percent of the risk.

The value added auditor may be asked to initially establish a risk management process, and then become an internal consultant. The auditor should not lose his/her objectivity. The auditor would ensure that he/she does not assume 'ownership of the risks' role. Auditors can assist, coach, facilitate, design, and even design risk management processes as long as the auditor does not 'assume' or 'own' the risk.

One of the biggest business risks is the unknown and the unknowable. There is no such thing as perfect knowledge and information. Murphies, the unexpected, can happen at the worst possible moment. These are the most difficult to control. The business can only do something. It can develop a contingency plan if something like this may occur. For example, it can purchase insurance.

Risk analysis is not easy. A risk can have many dimensions and causes. It can interact and interrelate with others risk. Different risks can come from the same source.

CHAPTER 15
STEP 4: EVALUATE CONTROL EFFECTIVENESS

Controls can be internal or external to an organization. Controls should focus on reducing risk to levels that are acceptable to the organization. This is always a cost-benefit decision. It may mean that the cost or implementing additional risk controls offers diminishing returns.

INTRODUCTION TO INTERNAL CONTROLS

We think in generalities, but we live in detail.
Alfred North Whitehead

In engineering terms, internal control is a continual feedback, self-adjusting mechanism. Some machines are able to machine, inspect, adjust, and segregate nonconforming material automatically. If measurements are on the specification target, measurement variation is random and the machine or process keeps on running. If unnatural variation or unwanted measurements are outside the specification limits, the machine automatically adjusts itself or stops. Review the figures in the 'Process Management 101' chapter for an explanation of this process.

WHAT IS RISK - CONTROL?

The purpose of internal control is to ensure that management systems are in place and working properly to prevent deficiencies and manage risks. Internal control may also be used to evaluate the level of performance in a machine, administrative, or service process. At a macro level, the value added auditor evaluates the internal control structure to determine that organizational and other risks are controlled. At a more detailed level, the value added auditor determines that deficiencies don't occur and the process is continually improving.

In the same manner, an administrative process can go out of control. A receptionist is abrupt and impolite and external customers comment. An administrative assistant makes numerous typos and internal customers pick up on it. In the inspection mode, external and internal customers would detect and may accept poor service. If the administrative assistant kept making the same mistake, customer complaints would indicate there is a control flaw. The root cause problem might be poor training, poor supervision, family problems, poor methods, too much work, or poor attitude. Each of these variables is correctable.

CONTROL DEFINITIONS

There are a number of definitions of control and we present these three:

- ISO 9001 – 2000.
- COSO (Treadway Commission).
- Institute of Internal Auditors.

And, control has been a key element of most ISO 9001 documents. For example, ISO 9001 – 2000 defined quality control as:

> '... part of quality management focusing on fulfilling quality requirements."[1]

The American Institute of Certified Public Accountants, American Accounting Association, Financial Executives International, Institute of Internal Auditors, and Institute of Management Accountants (Committee of Sponsoring Organizations of the Treadway Commission) developed the following definition of 'internal control':

> "Internal control is a process, effected by an entity's board of directors, management, and other personnel, designed to provide reasonable assurance regarding the achievement of objectives in the following categories:

- Effectiveness and efficiency of operations.
- Reliability of financial reporting.
- Compliance with applicable laws and regulations".[2]

The Institute of Internal Auditors defines several types of internal control (illustrated in the figure on the next page):

> "...any action taken by management to enhance the likelihood that established objectives and goals will be achieved. Controls may be **preventive** (to deter undesirable events from occurring), **detective** (to detect and correct undesirable events which have occurred), or **directive** (to cause or encourage a desirable event to occur). The concept of a system of control is the integrated collection of control components and activities that are used by an organization to achieve its objectives and goals."[3]

There are some important outcomes of the above definition, including:

- The five components of internal control according to the Committee of Sponsoring Organizations (COSO) include control environment, risk assessment, control activities, information and communication, and monitoring.
- Internal control is a process. It is a means to an end, not an end in itself.

CONDUCTING STEP 4 – EVALUATE CONTROL EFFECTIVENESS

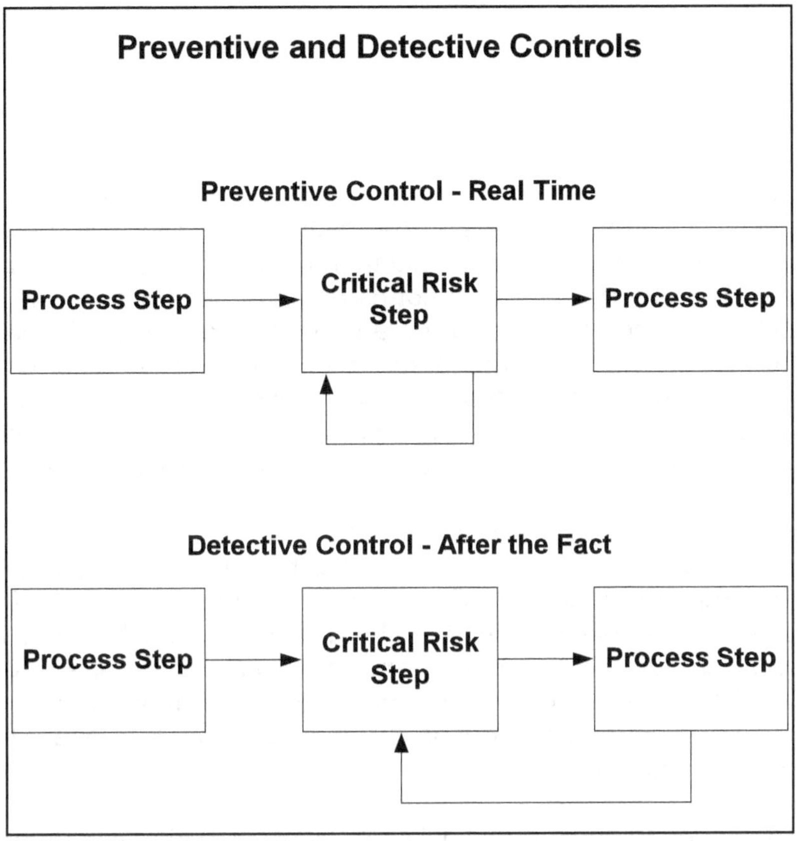

- People impact internal control. It's not merely policy manuals and forms, but people at every level of an organization are controlling their work processes.
- Internal control provides only 'reasonable assurance', not absolute assurance to an entity's management and board.
- Internal control focuses on the achievement of business objectives in one or more separate but overlapping categories.[4]

MEETING BUSINESS OBJECTIVES

Expressed another way, another purpose of internal control is to lower the risk of not meeting a business requirement or objective. This may involve detecting and preventing unusual occurrences. The unusual occurrence can be an exception, risk, excessive variation, variance, abnormality, deficiency, irregularity, nonconformance, waste, or system weakness. The value added auditor may encounter several types of irregularities, which are usually understood in context. However, the following have technical or special meanings:

- Nonconformance.
- Defects/defectives.
- Flaws.
- Deficiencies.

- Deviations.
- Mistakes.
- Misstatements.
- Irregularities.
- Errors.

Nonconformances or nonconformities are the favored terms for describing a deviation from a specification or desired process objective. Nonconformance can be a defect or defective product. Nonconformance does not have the onus of blame, like defects or defectives, which imply that someone willfully made a mistake. Nonconformance is a more benign term, implying that a measurement did not comply with a specification. If a case ends up in court, a lawyer can argue that the specification was incorrect. Deviations refer to differences between what is and what should be. Misstatements are errors or irregularities. The difference is important. Errors are unintentional; irregularities are intentional.

IDENTIFY THE CRITICAL RISKS TO CONTROL

There are two fundamental elements of risk: 1. the probability of occurrence and 2. the impact of the occurrence or exposure. For example, there may be a 25% probability or chance of occurrence of a relatively minor risk. This occurrence has little impact on the process, project, or product. On the other hand, a 1% probability of a major occurrence may be critical to the project. For example, this occurrence may involve life threatening health, safety, or environment degradation. In this case, a 1% probability may be too high and should be mitigated further.

What does the value added auditor do if risks have not been identified or quantified? The value added auditor should then spend time understanding the auditee and then identify risk areas and controls. There are a few common sense rules regarding risk and what areas the value added auditor may have to audit:

- High probability of occurrence or exposure.
- High impact of the occurrence or exposure.
- Life, safety, or environmental threats.
- Customer or market conditions change.
- Major changes in the input variables in a process, including changes in personnel, management, environment, materials, suppliers, etc.
- Process instability or lack of capability of meeting requirements.
- Significant understanding, variables, or factors that change the process.

TYPES OF CONTROLS

An organization usually has a hierarchal system of controls consisting of enterprise governance controls, process management controls, process assurance controls, and activity level controls. At the activity level, there may be controls at each critical process step.

Internal controls is a broad concept including hard and soft controls such as:

CONDUCTING STEP 4 – EVALUATE CONTROL EFFECTIVENESS

Control of Process Variables

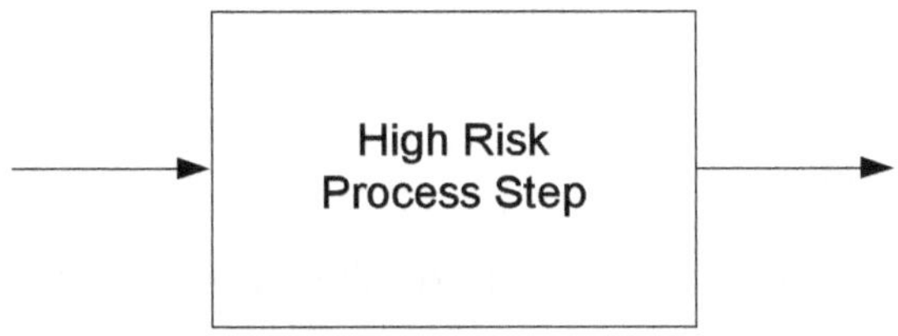

Process Variables	Internal Controls
Operator	Training
Methods	Common procedures
Machine	Statistically controlled
Materials	Single source suppliers
Environment	Safe, clean, and monitored

- People (operator).
- Methods.
- Machines.
- Materials.
- Environment.
- Culture.
- Leadership and management.
- Soft (relationships).

As discussed in the previous chapter, some processes and certain process steps have higher risk than others. These process steps have process variables, such as operator, methods, machines, materials and environment. Each of these is a source of variation and risk, which can be controlled. See the above 'Control of Process Variables.'

And, there are individual controls. There is much more emphasis on managing, controlling, and owning one's processes. As discussed, each person is responsible for developing, and managing internal controls within his/her control area.

EVALUATING THE INTERNAL CONTROL SYSTEM

At best, extensive internal process controls exist, operate properly, and improve the process continually. These may be false assumptions. Controls may not exist, may be weak, or may control unimportant variables. To evaluate an internal control system, the value added auditor can:

CONDUCTING STEP 4 – EVALUATE CONTROL EFFECTIVENESS

- Interview customers.
- Interview personnel.
- Interview area supervisors.
- Interview operating personnel.
- Interview technical and operating personnel.
- Analyze customer surveys.
- Analyze flowcharts.
- Evaluate the quality manual.
- Evaluate the workmanship manual.
- Evaluate control charts for critical and major quality characteristics.
- Evaluate training procedures.
- Observe upstream operations.
- Observe downstream operations.
- Verify supply and raw material controls.

The 'objective – risk – controls – assessment' methodology is an easy and systematic method for evaluating the internal control system. The 'objective' of the audit is first identified. 'Risk' events or process variables are then identified. 'Controls' are then identified and then 'evaluated.' This is a fundamental to all value added auditing.

ENTERPRISE/ORGANIZATIONAL CONTROLS

An organizational environment that emphasizes soft controls (tone at the top), such as integrity, accountability and responsibility is a major control element that can ensure consistency, improvement, and value added activities. The tone starts at the top, cascades down and permeates all elements of the organizational and value stream.

The organizational environment is key to process management and control. The environment sets the tone and attitudes for people to do their work. The environment deals with integrity, values, attitudes, consistency, honesty, feedback, and attention to detail.

A simple example may illustrate this. I'm often consulting and coaching senior management. They will often say: "I rely on my people to do their jobs and control their processes. I trust them." That's a great call. People are the biggest source of variation thereby the biggest source of risk. But... A senior manager may have 5 or more levels of reporting relationships. How does he or she know that what immediate reports believe, say or do cascades down the organization? Each manager may have 7 or more direct reports. We've all played the game where we're seated around a table. One person whispers something to a downstream person and after five or more people, the story as initially told is totally distorted.

CONDUCTING STEP 4 – EVALUATE CONTROL EFFECTIVENESS

STRATEGIC AND BUSINESS ENTITY CONTROL QUESTIONS
1. What are the key and core processes? 2. Have key processes been flowcharted at these levels: a. Enterprise? b. Business entity? c. Plant? d. Area? e. Work activity? 3. Have process primes and process owners been identified? 4. Have controls been identified at these levels: a. Enterprise? b. Business entity? c. Plant? d. Area? e. Work activity? 5. Are internal controls based on an integrated structure consisting of: a. Control environment? b. Risk assessment? c. Control activities? d. Information and communication? e. Monitoring? 6. Do controls ensure: a. Process stability? b. Process capability? c. Process improvement? 7. Are key controls mapped to the appropriate risks? a. Are the process risks and controls properly designed for effectiveness? b. Are the controls successful in managing risks? c. Do the process consistently meet business objectives?

PEOPLE CONTROLS

Of the factors involved in a process - the environment, operator, material, method, or machine - the operator is often the most important element of internal control because the operator has the most control over a process. A person can:

- Cause the most unwanted variation.
- Identify and eliminate unwanted variation.
- Prevent defects or risks from occurring.
- Improve the process over time.
- Ensure process stability and capability.
- Manage and control risks.

CONDUCTING STEP 4 – EVALUATE CONTROL EFFECTIVENESS

PROCESS CONTROL QUESTIONS

1. Is the process aligned with strategic business objectives?
2. Is the process a core organizational process?
3. Does the process support the core process?
4. Are critical processes defined and documented?
5. Are critical process mapped (flowcharted)?
6. Are the following defined for the process:
   a. Metrics/objectives?
   b. Accountabilities?
   c. Deliverables/outputs?
   d. Inputs?
   e. Inputs?
   f. Systems?
   g. Policies/procedures/work instructions?
7. Are critical processes designed properly?
8. Are critical processes:
   a. In control (stable)?
   b. Capable of meeting requirements and objectives?
   c. Improving?
9. Are there procedures and other documentation to support the processes?
10. Are process risk points identified in:
    a. Systems?
    b. Processes?
    c. Products?
    d. Activities?
11. Are internal controls of above risks based on:
    a. Control environment?
    b. Risk assessment?
    c. Control activities?
    d. Information and communication?
    e. Monitoring?
12. Are risk-mitigating controls identified?
13. Is there information from pervious audits indicating control effectiveness?
14. Are corrective and prevention actions fed back into the process to improve it?
15. Have problems or deficiencies recurred?

The person or persons responsible for the activity can also weaken strong internal controls or strengthen weak controls. If the operator is competent, honest, and has been properly trained, then other elements of internal control are less critical for high quality goods and services to be delivered. Good intentions, however, are not sufficient. The operator should be trained, guided, and reinforced on controlling and improving the process. If the other factors of the process, methods, machine, materials, and environment show excessive variation, the operator can:

- Use another method to machine parts.
- Take action to exchange faulty material.

CONDUCTING STEP 4 – EVALUATE CONTROL EFFECTIVENESS

INTERNAL CONTROL EVALUATION
Audit objective: Machine process has internal controls to limit nonconformances. **Risk:** Machine produces nonconforming products. **Internal Control System:** **Operators** go through intensive 6-month training program. One of every five applicants is hired. Each applicant goes through three sets of interviews by supervision and team members. Pay is 25 percent above industry average. Two years of technical college and a 3.25 Grade Point Average (GPA) is required. **Machine** is a flexible manufacturing system (FMS) from Sweden. Process controls are integrated into the machine. Specifications are downloaded directly from computer-aided design (CAD) equipment. Critical and major quality characteristics are process-controlled as verified by SPC charts, first article analysis, and capability studies. Process is control and capable to Cpk = 1.33. **Material** is purchased from single-source or single/acceptable-alternate sources. Suppliers have gone through an extensive certification program based on progressively higher Cpk values. Suppliers have been audited. Suppliers have been selected carefully and have improved over the years. **Methods** of operation are detailed and documented. Each operator is trained in own operation as well as cross-trained in upstream and downstream operations. Each operator receives 40 hours of process- and product-specific training. **Environment** is particulate, temperature, humidity, and airflow controlled. Automatic measurement is conducted every 15 minutes, and automatic controls adjust heating, ventilation, and air conditioning systems. Clean rooms meet minimum industry standards.

- Adjust the heating and ventilation system.
- Tweak the process to the target dimension.
- Carry out other activities to prevent nonconformances.

People are the biggest source of variation. People can bypass a control system. So, if the operator does not care or is incompetent, strong controls in other factors won't guarantee defect-free products or manage risks.

CHANGES IN PROCESSES

If there is a change in a process variable, it is always a good idea to revisit the process to ensure the process is stable and capable of meeting requirements. The new process variable can destabilize the process by creating more or unexpected variation, which can produce unexpected circumstances and higher risk. Unexpected or new variables can include:

- New source of materials.
- New workers.

CONDUCTING STEP 4 – EVALUATE CONTROL EFFECTIVENESS

INTERNAL CONTROL OBJECTIVES

GAGAS 6.16 Auditors should obtain an understanding of internal control that is significant within the context of the audit objectives. For internal control that is significant within the context of the audit objectives, auditors should assess whether internal control has been properly designed and implemented and should perform procedures designed to obtain sufficient, appropriate evidence to support their assessment about the effectiveness of those controls. Information systems controls are often an integral part of an entity's internal control. The effectiveness of significant internal controls is frequently dependent on the effectiveness of information systems controls. Thus, when obtaining an understanding of internal control significant to the audit objectives, auditors should also determine whether it is necessary to evaluate information systems controls.

6.17 The effectiveness of internal control that is significant within the context of the audit objectives can affect audit risk. Consequently, auditors may determine that it is necessary to modify the nature, timing, or extent of the audit procedures based on the auditors' assessment of internal control and the results of internal control testing. For example, poorly controlled aspects of a program have a higher risk of failure, so auditors may choose to focus more efforts in these areas. Conversely, effective controls at the audited entity may enable the auditors to limit the extent and type of audit testing needed.

- Market place changes.
- New processes and practices.

CRITICAL INTERNAL CONTROL QUESTIONS

Problem solving and constructive innovation are what business is all about.
Randall Meyer

If the value added audit objective is to evaluate internal control effectiveness, how does the value added auditor accomplish this? Internal controls can be organizational system, process, supply chain, people, documentation, and other controls. Not all internal controls have to be deployed for the auditor to be confident of the effectiveness of the auditee's process controls. Confidence is based on the level of assurance and risk the customer and auditee want to assume.

The internal control system ultimately reduces risks and increases the level of assurance. The value added auditor can assess the internal control structure by addressing the following questions:

1. Do the appropriate controls exist?
2. Are there informal or soft controls?
3. Do controls monitor the major process variables?
4. Are controls the most effective controls?
5. Are controls working properly?
6. Do controls work to improve the process?[5]

CONDUCTING STEP 4 – EVALUATE CONTROL EFFECTIVENESS

INTERNAL CONTROL EVALUATION

GAGAS 6.18 Auditors may obtain an understanding of internal control through inquiries, observations, inspection of documents and records, review of other auditors' reports, or direct tests. The nature and extent of procedures auditors perform to obtain an understanding of internal control may vary among audits based on audit objectives, audit risk, known or potential internal control deficiencies, and the auditors' knowledge about internal control gained in prior audits.

6.19 The following discussion of the principal types of internal control objectives is intended to help auditors better understand internal controls and determine whether or to what extent they are significant to the audit objectives.
a. Effectiveness and efficiency of program operations: Controls over program operations include policies and procedures that the audited entity has implemented to provide reasonable assurance that a program meets its objectives, while considering cost-effectiveness and efficiency. Understanding these controls can help auditors understand the program operations that convert inputs to outputs and outcomes.
b. Relevance and reliability of information: Controls over the relevance and reliability of information include policies and procedures that officials of the audited entity have implemented to provide themselves reasonable assurance that operational and financial information they use for decision making and reporting externally is relevant and reliable and fairly disclosed in reports. Understanding these controls can help auditors (1) assess the risk that the information gathered by the entity may not be relevant or reliable and (2) design appropriate tests of the information considering the audit objectives.
c. Compliance with applicable laws, regulations, contracts, and grant agreements: Controls over compliance include policies and procedures that the audited entity has implemented to provide reasonable assurance that program implementation is in accordance with provisions of laws, regulations, contracts, and grant agreements. Understanding the relevant controls concerning compliance with those laws, regulations, contracts or grant agreements that the auditors have determined are significant within the context of the audit objectives can help them assess the risk of noncompliance with provisions of laws, regulations, contracts, or grant agreements, or abuse.

6.20 A subset of these categories of internal control objectives is the safeguarding of assets and resources. Controls over the safeguarding of assets and resources include policies and procedures that the audited entity has implemented to reasonably prevent or promptly detect unauthorized acquisition, use, or disposition of assets and resources.

1. DO THE APPROPRIATE CONTROLS EXIST?

The appropriate controls should be used to control risks and processes. If internal controls exist, then the value added auditor can feel some assurance and confidence that the process is monitored and there is some level of prevention. If there are no manifest internal controls, the value added auditor should inquire with the operator, supervisor, or process engineer if there are compensating and overlapping controls. Another option is to inquire with internal stakeholders to see how satisfied they are with a process's output or with external customers to determine how happy they are with the product or service.

While there may be no preventive internal controls, there may be other types of controls, such as incoming, in process, final inspection or testing controls. While inspection is not preferable to a prevention control, it may be warranted under certain conditions. For example, a product may be one-of-a-kind or come under regulatory oversight that specifies that products will be inspected as well as controlled during inspection. In this situation, the value added auditor would evaluate the effectiveness of the inspection and testing. If products are sampled, the value added auditor may test the system by pulling a representative sample and testing measuring products in the sample.

The major problem with inspection is that it is after the fact. Inspection occurs after a process has produced parts. There is no way to go back to the process to adjust the machine, eliminate the cause of the defects, or improve the process.

2. ARE THERE INFORMAL OR SOFT CONTROLS?

There are two types of internal control: formal and informal. The preceding discussion focused on formal internal controls. These are relatively easy to recognize. They are specified, standardized, prescribed and proceduralized. They are often found in standards or procedures in operational manuals.

Informal control systems are more difficult to identify and evaluate. Often, these are based on social networks. Many times, informal control systems are based on preexisting personal relationships. For example, managers who joined the organization at the same time form a supportive relationship of advice and counsel, forming an informal control system. If one manager has a problem, personal or otherwise, or requires resources, money, equipment, or advice, the informal support network supplies these.

These informal or soft control systems are often outside the organizational chart. They are in what is sometimes called the 'organizational white spaces.' One way to determine the extent of informal controls is to check how things work compared to organizational policies, procedures, and the organizational chart. If things work according to policies and procedures, then informal controls are subtle and well hidden. If things work much differently than what's detailed in procedures, there may be active and informal controls. The value added auditor checks the process by interviewing process owners, interviewing stakeholders, analyzing trend charts, or performing system tests. If these don't exist, the auditors should check whether informal controls meet required performance levels. And if standards and specifications are maintained, then the auditor can recommend that procedures be adopted to reflect how things are done.

3. DO CONTROLS MONITOR THE MAJOR PROCESS VARIABLES?

There are many product and process characteristics that can be monitored and controlled. Each is a possible source of variation, which can cause product or service deterioration if the variation is abnormal or outside the specification target. For example, a product may have 200 dimensions. A service may have 5 service attributes or characteristics. The question is which dimension or process variable to control and how to do it. The quick answer is to control the variables and product characteristics that relate to ex-

INTERNAL CONTROL DEFICIENCIES

GAGAS 6.21 In performance audits, a deficiency in internal control exists when the design or operation of a control does not allow management or employees, in the normal course of performing their assigned functions, to prevent, or detect and correct
(1) impairments of effectiveness or efficiency of operations, (2) misstatements in financial or performance information, or (3) noncompliance with provisions of laws, regulations, contracts, or grant agreements on a timely basis. A deficiency in design exists when (a) a control necessary to meet the control objective is missing or (b) an existing control is not properly designed so that, even if the control operates as designed, the control objective is not met. A deficiency in operation exists when a properly designed control does not operate as designed, or when the person performing the control does not possess the necessary authority or qualifications to perform the control effectively.

6.22 Internal auditing is an important part of overall governance, accountability, and internal control. A key role of many internal audit organizations is to provide assurance that internal controls are in place to adequately mitigate risks and achieve program goals and objectives. The auditor may determine that it is appropriate to use the work of the internal auditors in the auditor's assessment of the effectiveness of design or operation of internal controls that are significant within the context of the audit objectives.

ternal customer satisfaction, business objectives, and present the most risk. These should be identified on engineering drawings, specifications, operations manual, or a business mission critical statement. External customer satisfaction is the end user. Internal customer satisfaction is the next or downstream user of the products or services.

There are formal mechanisms for communicating and deploying 'voice of the customer' requirements into internal processes. These requirements are the critical or major process variables that deal with customer satisfaction. In addition, there are other product characteristics that the customer cares about and assumes are satisfactory, such as safety, reliability, maintainability, and durability. If any of these critical product characteristics fail, the manufacturer can be held liable.

4. ARE CONTROLS THE MOST EFFECTIVE CONTROLS?

There can be different and overlapping internal controls. Operator, machine, method, material, and environmental variables can be controlled in several ways. For example, incoming raw materials can be controlled by obtaining them from certified suppliers that have extensive management systems and process controls. Or, the incoming materials can be inspected.

5. ARE THE CONTROLS WORKING PROPERLY?

It's not sufficient just to have internal controls - they must be working properly to prevent defects in customer-defined characteristics or to reduce risks. History of performance can indicate this. The process has been in statistical control and has been capable over a period of time. This has been verified through inspection and there are no reported field failures. The existence of process control charts (SPC) with corrective action is another indicator.

CONDUCTING STEP 4 – EVALUATE CONTROL EFFECTIVENESS

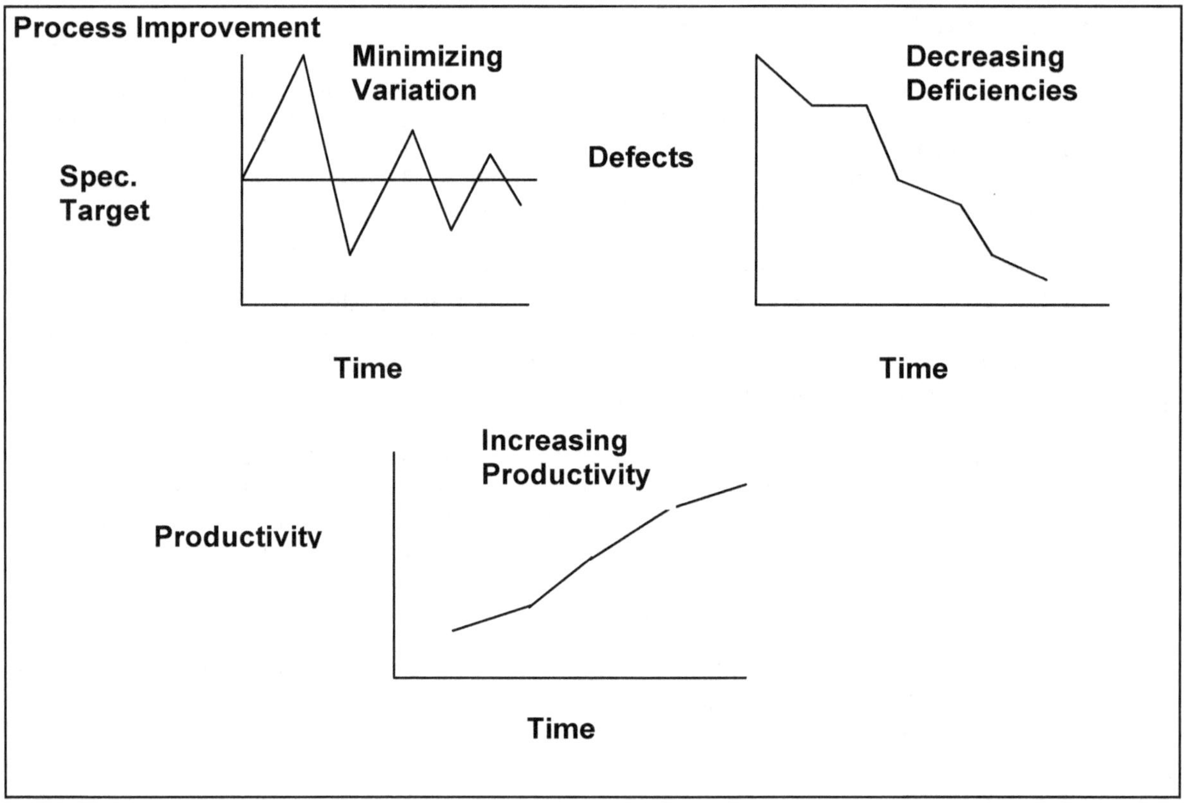

Other indicators are:

- No reported problems or risk events.
- Low internal failure cost of quality.
- Low external failure cost of quality.
- Low warranty costs.
- No customer complaints.
- Higher capability indices.
- No deviations or waivers.

6. DO CONTROLS WORK TO IMPROVE THE PROCESS?

Having internal controls to detect and prevent defectives is important, but are they used to improve the process? Improvement can be demonstrated in three ways as demonstrated in the above 'Process Improvement' figure:

- Variation is minimized around the specification target (top left graph).
- Level of defects is decreased (top right graph).
- Level of performance is increased (bottom graph).

CRITICAL CONTROL ISSUES

Many types of risk are unknowable or unquantifiable.
Anonymous

The risk concept can be applied to all stages of value added auditing from planning to conducting to reporting the value added audit. Its worth is especially high in the planning stage of the audit when determining customer requirements and evaluating the effectiveness of the auditee's process controls. The value added auditor can use the concept to determine if a value added audit is feasible, where to focus attention, and how to develop the audit plan. The value added auditor can also use it to understand the relationships among the various types of risk and the amount of evidence to gather in each area of the audit as a result of risk levels.

RISK CAN'T BE COMPLETELY ELIMINATED OR CONTROLLED

There are some fundamental laws about risk. First, some risks can't be completely managed, eliminated, and detected such as inherent risks discussed in the last chapter. The value added auditor, unless requested or instructed to construct a risk management system, must rely on process owner's risk management and control system. The value added auditor then evaluates the 'effectiveness' of the process owner's risk management process.

The IIA states the following about controlling risks:

> "... risk management, control, and governance processes is present if management directs processes in such a manner as to provide reasonable assurance that the organization's objectives and goals will be achieved. In addition to accomplishing the objectives and planned activities, management directs by authorizing activities and transactions, monitoring resulting performance, and verifying that the organization's processes are operating as designed."[6]

PROVIDE 'REASONABLE ASSURANCE'

And then, the value added auditor can only evaluate the 'adequacy' and provide only 'reasonable assurance' that processes are risk managed, controls are in place, and controls are working properly.

The threshold of 'reasonable assurance' is indicated when:

> "...the most cost-effective measures are taken in the design and implementation stages to reduce risks and restrict expected deviations to a tolerable level."[7]

Adequacy is indicated when:

> "... risk management, control, and governance processes are present if management has planned and designed them in a manner, which provides reason-

able assurance that the organization's objectives and goals will be achieved efficiently and economically."[8]

RISK EVALUATION CAVEATS

Process management and project risk analysis are still in its infancy. A process owner or auditor can get hard numbers and think there is certainty in the third decimal point of a number. The value added auditor should be careful! Process, project and product risk analysis makes it easier to understand the dynamics of what, who, how, when, and where things are going on. It also reveals what may and can go wrong. In general, the value added auditor should be aware of the following:

- Risks are sometimes unknown and unknowable.
- Risks interact in unexpected ways.
- Single variable can interact with others that can result in multiple effects.
- Process owners may be too close, (staring at the knothole in the tree) and may not have the context of the forest.
- Mathematical models provide a false sense of security and accuracy.

DETERMINING CONTROLS OF A 200-STEP PROCESS

What does the value added auditor do if a process consists of 200 process steps? Depending on the required level of assurance, the value added auditor has several options for evaluating a production process:

.

- **Assess policies, procedures, and work instructions, including the manuals, standard operating procedures**. If these exist and are being followed, then there is some assurance that the organization believes in and documents value added processes.
- **Assess major assembly and fabrication areas**. If work areas have posted control charts, work areas are clean, people are concerned about work, risks are managed, and material is from certified suppliers, then the auditor can be assured that control procedures are being implemented.
- **Assess individual processes and machines**. At this detailed level, the value added auditor assesses machines in terms of possible sources of variation, including process inputs, process controls, and conformance of the outputs.
- **Test the final product by conducting performance tests**. The assumption is that if the final product performs satisfactorily under extreme or exaggerated conditions then the auditor can assume that all the elements that went into the product are satisfactory. Performance tests include salt water, accelerated stress, and other reliability tests.

If the value added audit customer wants very high or near 100% assurance or confidence, then each process step would be assessed in terms of inputs, the process, and outputs. In addition, supplied material and interaction of the many parts at each level of the product from component, subassembly, assembly, and final product may be as-

Reporting of Critical Risk/Control Information

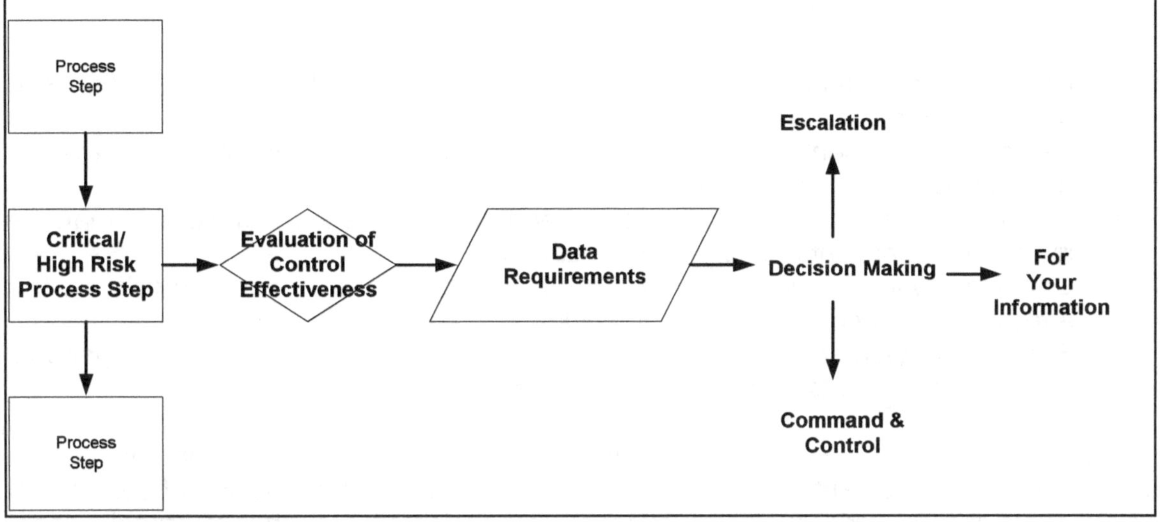

sessed. While the above analyzes a production process, the same analysis can be conducted of a service.

Should the value added auditor do this? Probably not, because it would be beyond the point of diminishing return. The threshold of a value added audit is 'reasonable assurance.' Costs would outweigh benefits. Again, the level of assurance the customer wants from the audit determines this. If a critical product failure occurs in the field, then a team of auditors might go to this effort to determine the root cause of the problem and eliminate it.

REPORTING RISK/CONTROLS

An organization should develop a risk management process, proceduralize it, and consistently follow it. A typical risk/control reporting process is shown in the above 'Reporting of Critical Risk/Control Information' figure. The process is pretty straightforward. The value added auditor or preferably the process owner has identified critical or high-risk process steps. The auditor evaluates internal control effectiveness based on data derived from the process. The data is then processed in one of three ways: specifically 1. escalated for further advise or comment; 2. sent across the organization 'for your information' (FYI) or 3. sent down the organization as part of a 'command/control' decision. The value added auditor would then evaluate the entire risk/control effectiveness of the control system.

CONDUCTING STEP 4 – EVALUATE CONTROL EFFECTIVENESS

SUMMARY

Value added auditors evaluate the organization's risk management process by addressing five key objectives in developing an overall opinion of the adequacy of risk management. Key objectives to be evaluated:

- Risks arising from business strategies and mission critical activities are identified and prioritized.
- Senior management and the board of directors should determine the level of risks acceptable to the organization.
- Risk mitigation activities are designed and implemented to reduce and manage risks that are acceptable to the organization.
- Ongoing monitoring activities are conducted to periodically reassess risk and the effectiveness of controls to manage risks.
- Senior management and the organization's board of directors receive periodic reports of the results of the risk management processes.[9]

The overall goal of a value added audit can be business improvement, operational effectiveness, risk management, or internal control assessment. The purpose of the control assessments is to provide business managers an educated snapshot of organizational risks and determine the type and effectiveness of internal controls to minimize the risks. Internal control assessments can be performed at the corporate/organizational level, business unit, process level, or process step level.

A critical, some would say the most critical valued function of a value added auditor is to evaluate the effectiveness of risk controls. To do this, the value added auditor may:

- Highlight areas of concern or hazards that may be a failure waiting to occur.
- Focus attention on important matters, the 20% of the factors that cause 80% of the problems.
- Help to eliminate ineffective process controls and assist in improving processes.
- Provide a means to support management decisions.
- Aid in the detection and prevention of process and product deficiencies.
- Determine how much evidence should be gathered to support a finding or an opinion – high risk requires more evidence.

The value added auditor should focus his or her time in understanding and analyzing critical process variables and their risks/controls, specifically those that have a high probability of occurring, have major product impacts, or can affect quality, time, scope, cost adversely. The goal of risk identification and control is to ensure that critical processes are in control, stable, capable, and improving.

CHAPTER 16
STEP 5: ASSESS EVIDENCE

In this step, the value added auditor gathers sufficient evidence and conducts tests at the system, process, or product levels to be able to make a 'reasonable determination' in support of an audit judgment.

Testing a process or system may be necessary in order to gather and evaluate evidence. Testing may involve tracking products or tracing items through a process or system to determine if process controls exist, are working properly, are in control, and are capable. Testing assumes that there is an internal control system, records and documentation exist; people are concerned about business performance and risk management; and people have been trained extensively. Testing must be carefully designed and conducted. It is time consuming, but it may be necessary if the value added auditor has to obtain a high level of assurance as in the case of catastrophic failure, liability, costly failure, fraud, or other high risks.

GATHERING EVIDENCE

Facts apart from their relationships are like labels on empty bottles.
Sve Halla

Gathering evidence is the process of obtaining data and information on critical and major processes to support an audit decision or opinion. The judgment may be to prove or disprove a statement, condition, situation, test, premise, or assertion. Gathering evidence is also part of fieldwork to obtain sufficient understanding of organizational culture, processes, products, and internal control effectiveness.

TYPES OF EVIDENCE

In developing value added audit opinions, the value added auditor must rely on evidence obtained by conducting tests or through some other means. Evidence is data and information that provide proof for an audit determination. Various types of evidence are covered below and in the sidebar on the next page:

- **Primary evidence.** This is original data or information. It indicates the highest level of fact. An original, signed certification, contract, purchase order, or test results are examples of primary evidence.
- **Secondary evidence.** This is weaker than primary evidence and has a lower level of reliability and assurance. This type of evidence would include oral con-

EVIDENCE QUESTIONS
I. Are internal controls in place and being used?
2. Do internal controls manage critical risks?
3. Are processes in control and capable?
4. What types of evidence exist?
5. When should the evidence be obtained?
6. Who should obtain the evidence?
7. How much evidence should be gathered?
8. What type of evidence should be gathered?
9. What test procedures should be used?
10. Should and can the process or the product audit be conducted?
11. How should it be tested?
12. Are inspection procedures written?
13. What is the population?
14. Is sampling necessary?
15. Is sampling 100 percent or statistical?
16. What sample size should be selected?
17. Which items should be selected from the population?
18. Are people properly trained in sampling and testing?

firmation of the above test data. This evidence is acceptable as long as secondary evidence reflects or represents the primary evidence.

- **Direct evidence.** This evidence proves a fact or information without inference. If people who have actual knowledge by directly witnessing events or asserting facts, then evidence is direct.
- **Circumstantial evidence.** This evidence establishes one fact by proving another collateral fact but it does not directly establish the fact. It is inferred from other information or data.
- **Corroborative evidence.** This evidence is gathered from another reputable source that supports or verifies the original information. An example would be an independent testing lab confirming the auditee's test results.
- **Conclusive evidence.** This evidence cannot be questioned or contradicted. It does not need corroboration.

PROCESS OF FORMING A FINDING OR AN OPINION

Much of the value added auditor's work is forming an opinion about the auditee's systems, processes, and products by conducting tests and following test procedures. As a general rule, sufficient evidence should be gathered to form the reasonable basis for an audit opinion. The value added auditor may test a product to destruction, beyond its limits, or within defined parameters. The many types of product tests are beyond the scope of this book.

Tests are conducted by strictly defined rules. Standards or specifications usually specify explicit requirements. They deal with issues of quality, cost, productivity, effectiveness, time, technology, service, and delivery. The standards may state objectives, goals, and

OBTAINING SUFFICIENT, APPROPRIATE EVIDENCE

GAGAS 6.56 Auditors must obtain sufficient, appropriate evidence to provide a reasonable basis for their findings and conclusions.

6.57 The concept of sufficient, appropriate evidence is integral to an audit. Appropriateness is the measure of the quality of evidence that encompasses its relevance, validity, and reliability in providing support for findings and conclusions related to the audit objectives. In assessing the overall appropriateness of evidence, auditors should assess whether the evidence is relevant, valid, and reliable. Sufficiency is a measure of the quantity of evidence used to support the findings and conclusions related to the audit objectives. In assessing the sufficiency of evidence, auditors should determine whether enough evidence has been obtained to persuade a knowledgeable person that the findings are reasonable.

6.58 In assessing evidence, auditors should evaluate whether the evidence taken as a whole is sufficient and appropriate for addressing the audit objectives and supporting findings and conclusions. Audit objectives may vary widely, as may the level of work necessary to assess the sufficiency and appropriateness of evidence to address the objectives. For example, in establishing the appropriateness of evidence, auditors may test its reliability by obtaining supporting evidence, using statistical testing, or obtaining corroborating evidence. The concepts of audit risk and significance assist auditors with evaluating the audit evidence.

6.59 Professional judgment assists auditors in determining the sufficiency and appropriateness of evidence taken as a whole. Interpreting, summarizing, or analyzing evidence is typically used in the process of determining the sufficiency and appropriateness of evidence and in reporting the results of the audit work. When appropriate, auditors may use statistical methods to analyze and interpret evidence to assess its sufficiency.

accountabilities; timelines and benchmarks; how requirements will be met and by whom; and types of tests and measurements.

COLLECT SUFFICIENT EVIDENCE BASED ON ASSURANCE/RISK PROFILE

The value added auditor should obtain only the amount of evidence necessary to support the level of assurance required by the value added audit customer. How does the value added auditor determine this? This should be part of the audit interview and was discussed in the planning phase of the audit. Further discussion can be found in the above sidebar.

For example, the auditor obtains evidence that internal controls relative to a critical product characteristic will prevent or detect nonconformances. The amount of evidence or information to gather is based on the customer's assurance requirements and the auditee's risk profile.

For example, the amount of evidence to gather can be based on:

- Strong internal controls-less evidence.
- Moderate internal controls - more evidence.

CONDUCTING STEP 5 – ASSESS EVIDENCE

APPROPRIATENESS OF EVIDENCE

GAGAS 6.60 Appropriateness is the measure of the quality of evidence that encompasses the relevance, validity, and reliability of evidence used for addressing the audit objectives and supporting findings and conclusions.

a. Relevance refers to the extent to which evidence has a logical relationship with, and importance to, the issue being addressed.
b. Validity refers to the extent to which evidence is a meaningful or reasonable basis for measuring what is being evaluated. In other words, validity refers to the extent to which evidence represents what it is purported to represent.
c. Reliability refers to the consistency of results when information is measured or tested and includes the concepts of being verifiable or supported.

- Weak internal controls-even more evidence.
- No internal controls-note this in audit report and require corrective action.

This determines the level of assurance the auditor can have in assessing the auditee. Questions to consider when evaluating evidence can be found in the above sidebar.

DETERMINING ASSURANCE LEVEL
Higher levels of assurance require higher levels of internal control. The levels and types of internal control also answer the following questions:

- How much evidence based on assurance/risk has to be gathered?
- What type of tests should be conducted?
- Can an audit even be conducted?

The above questions assume a static level of assurance. The value added audit customer may want to have a lower level of assurance, which means that the auditor may gather less evidence in order to support a reasonable decision.

The value added auditor has a number of methods for gathering data, information and evidence. For example, the auditor may:

- Trace the process.
- Check process results against performance targets.
- Evaluate risks and controls.
- Check policy compliance.
- Observe quality or performance tests.
- Interview personnel.
- Analyze documents.
- Analyze work instructions.
- Test prototypes.
- Conduct reliability tests.
- Analyze contracts and purchase orders.

APPROPRIATENESS OF EVIDENCE – TYPES OF EVIDENCE

GAGAS 6.61 There are different types and sources of evidence that auditors may use, depending on the audit objectives. Evidence may be obtained by observation, inquiry, or inspection. Each type of evidence has its own strengths and weaknesses. The following contrasts are useful in judging the appropriateness of evidence. However, these contrasts are not adequate in themselves to determine appropriateness. The nature and types of evidence to support auditors' findings and conclusions are matters of the auditors' professional judgment based on the audit objectives and audit risk.

a. Evidence obtained when internal control is effective is generally more reliable than evidence obtained when internal control is weak or nonexistent.
b. Evidence obtained through the auditors' direct physical examination, observation, computation, and inspection is generally more reliable than evidence obtained indirectly.
c. Examination of original documents is generally more reliable than examination of copies.
d. Testimonial evidence obtained under conditions in which persons may speak freely is generally more reliable than evidence obtained under circumstances in which the persons may be intimidated.
e. Testimonial evidence obtained from an individual who is not biased and has direct knowledge about the area is generally more reliable than testimonial evidence obtained from an individual who is biased or has indirect or partial knowledge about the area.
f. Evidence obtained from a knowledgeable, credible, and unbiased third party is generally more reliable than evidence obtained from management of the audited entity or others who have a direct interest in the audited entity.

- Verify engineering calculations.
- Verify written representations.
- Investigate field failures.

If audit results do not confirm or support the audit objective, then the value added auditor may evaluate other elements of the control structure; for example, interview more personnel or check more training records.

GATHERING EVIDENCE METHODS

It is a capital mistake to theorize before one has data. Insensibly one begins to twist facts to suit theories, instead of theories to suit facts.
Sir Arthur Conan Doyle, writer

Evidence is gathered from a process or system to reach the threshold of 'reasonable assurance.' Gathering evidence to reach a compliance decision may entail checking documentation against a standard. Or, gathering evidence may mean determining the degree of correspondence or adherence between what a process is doing and comparing it against an operational or business objective. Or gathering evidence may mean checking that critical risks are identified and controlled in the proper manner.

In general, there are six techniques that the value added auditor can use to gather evidence:

CONDUCTING STEP 5 – ASSESS EVIDENCE

APPROPRIATENESS OF EVIDENCE – TYPES OF EVIDENCE

GAGAS 6.62 Testimonial evidence may be useful in interpreting or corroborating documentary or physical information. Auditors should evaluate the objectivity, credibility, and reliability of the testimonial evidence. Documentary evidence may be used to help verify, support, or challenge testimonial evidence.

6.63 Surveys generally provide self-reported information about existing conditions or programs. Evaluation of the survey design and administration assists auditors in evaluating the objectivity, credibility, and reliability of the self-reported information.

6.64 When sampling is used, the method of selection that is appropriate will depend on the audit objectives. When a representative sample is needed, the use of statistical sampling approaches generally results in stronger evidence than that obtained from nonstatistical techniques. When a representative sample is not needed, a targeted selection may be effective if the auditors have isolated risk factors or other criteria to target the selection.

6.65 When auditors use information provided by officials of the audited entity as part of their evidence, they should determine what the officials of the audited entity or other auditors did to obtain assurance over the reliability of the information. The auditor may find it necessary to perform testing of management's procedures to obtain assurance or perform direct testing of the information. The nature and extent of the auditors' procedures will depend on the significance of the information to the audit objectives and the nature of the information being used.

6.66 Auditors should assess the sufficiency and appropriateness of computer-processed information regardless of whether this information is provided to auditors or auditors independently extract it. The nature, timing, and extent of audit procedures to assess sufficiency and appropriateness is affected by the effectiveness of the audited entity's internal controls over the information, including information systems controls, and the significance of the information and the level of detail presented in the auditors' findings and conclusions in light of the audit objectives. The assessment of the sufficiency and appropriateness of computer-processed information includes considerations regarding the completeness and accuracy of the data for the intended purposes.

1. Observe conditions.
2. Interview people.
3. Analyze information.
4. Verify information.
5. Investigate systems, processes, and products.
6. Evaluate situations, data, or information.[1]

I. OBSERVE CONDITIONS

Observing involves looking for risks, unusual conditions, variances, unstable processes, incapable processes and anything that may not lead to the fulfillment of business objectives, process requirements or other documents. A walkthrough of a plant is an example of using this technique and is commonly used for gathering information and making mental notes. In the walkthrough, the value added auditor inspects operational cleanliness, looks for unusual conditions, checks posted process control charts, and other factors.

In some ways, observing is one of the more difficult value added auditing techniques because it assumes that the auditor has an educated eye and knows what to look for. Deming the quality expert called this 'profound knowledge.' A novice auditor may go to an auditee's operations and be overwhelmed by the mass of data and activity. An experienced auditor would focus on specific unstable conditions, risks, threats, vulnerabilities, exceptions, or unusual occurrences.

Something critical may have changed since the previous audit. The experienced value added auditor observes the changes to determine how they affect the process and the quality of the output. The experienced auditor checks compliance with safety and environmental factors. The auditor looks at process control charts to determine whether processes are in control, capable, and improving.

The assessment tools of value added auditing are similar to those used in internal auditing. Sometimes the deficiencies observed lead to a further investigation. If the auditor notices deficiencies, nonconformances, risks, or unusual situations, the auditor investigates these in the audit and may recommend how they can be corrected.

2. INTERVIEW PEOPLE

Interviewing is one of the more frequently used value added auditing techniques. It is sometimes called inquiry or questioning. Responses may be written or oral. The purpose of interviewing is to obtain impressions and other specific information. Interviewing may start early when the auditee is asked to fill out the preliminary audit.

Interviewing requires tact and human relations skills otherwise, the value added auditor may offend the people whose cooperation is required. If the auditor is perceived as interrogating the auditee then defenses rise, resistance increases, and information is harder to obtain.

It is important to understand the auditee's environment. If the environment is participatory, then interviews may be conducted with the team leader or the entire team. These interviews are more of a dialogue in risk management and continual improvement. The value added auditor and the auditee's team discuss problems and jointly try to understand what is occurring. The value added auditor may assist the team in developing solutions. The auditor may serve as a means for the team to obtain additional resources. For example, if the process owners have an intractable problem and need additional resources to eliminate it, the value added auditor may listen to the team, distill what is said, inform management of the problem, and request additional resources.

CONDUCTING STEP 5 – ASSESS EVIDENCE

SUFFICIENCY OF EVIDENCE

GAGAS 6.67 Sufficiency is a measure of the quantity of evidence used for addressing the audit objectives and supporting findings and conclusions. Sufficiency also depends on the appropriateness of the evidence. In determining the sufficiency of evidence, auditors should determine whether enough appropriate evidence exists to address the audit objectives and support the findings and conclusions.

6.68 The following presumptions are useful in judging the sufficiency of evidence. The sufficiency of evidence required to support the auditors' findings and conclusions is a matter of the auditors' professional judgment.
a. The greater the audit risk, the greater the quantity and quality of evidence required.
b. Stronger evidence may allow less evidence to be used.
c. Having a large volume of audit evidence does not compensate for a lack of relevance, validity, or reliability.

The manner in which questions are posed is important and can affect the audit. For example many questions are phrased in terms of non-threatening yes or no questions. 'Do you always store the sensitive measuring equipment in this unlocked dirty cabinet?' is a threatening type of question. This elicits a different response than 'How do you store measuring equipment?' Many people don't realize how they come across. Facial expressions and body posture offer nonverbal cues that may communicate unintended communicate information. For example, if the auditor has his arms crossed, rocks his body, and tenses facial muscles, then the auditee will be on the defensive.

Questions of internal customers reveal how well an upstream activity is being performed. This is especially useful in service organizations where the work is information. Questions the value added auditor could ask are:

- Do you receive information on time?
- Is it presented in a useful and understandable fashion?
- How is the information used?
- Is the information inaccurate or not useful?
- If yes, how is the information corrected?
- What do you do if the problem recurs?

3. ANALYZE INFORMATION

Analysis is a systematic assessment of data so it becomes understandable. Analysis may involve:

- Comparing and contrasting data.
- Sampling, inspecting, and testing products.
- Calculating and simulating operating conditions.
- Understanding human motives.
- Understanding cause and effect relationships.
- Interpreting performance or compliance documents.
- Performing complex calculations.

- Testing, inspecting, or sampling products.
- Understanding trends and patterns.
- Segmenting audits.

In terms of segmenting an audit, analysis involves taking a complex operation and breaking it up into understandable and auditable chunks. This is called segmenting or chunking the audit. Large audits are chunked into bite-sized, workable parts. The value added auditor would segment an audit to understand resource flows, to manage a large team of audit specialists, or to determine the relationship among parts.

4. VERIFY INFORMATION

Verifying involves attesting to the accuracy and confirming the validity of test procedures, test results, and training effectiveness. This means inspecting a procedure, product, or standard to ensure conformance or compliance. Verifying may also mean determining the correspondence, adherence, or alignment between requirements/expectations/objectives and factual/objective information. The value added auditor would establish the accuracy or correspondence by testing systems, processes, or products. The value added auditor may corroborate one person's statement by a discussion with another or by searching records for corroboration.

Interviewing is often coupled with verifying. One way to verify information is to interview one group of people and then interview another group that interfaces with the first group and check responses. The relationship between the groups can be superior-subordinate, customer-supplier, peer-peer, or user-salesperson. The purpose is to corroborate the information and focus on differences, variance, risks, or exceptions. This type of interviewing is more efficient than obtaining responses from only one party. If the auditor does not believe either group, then the auditor can pursue inquiries by other means.

Verification is used to establish accuracy, precision, existence, and other performance objectives that were discussed early in the chapter. In an administrative process, testing involves scrutinizing documents and records to confirm that procedures and policies are consistently followed. Verification is most effective when it is provided by people independent of the auditee. Verification may involve observation, interviewing, or physical identification.

5. INVESTIGATE SYSTEMS, PROCESSES, AND PRODUCTS

Investigation is an inquiry into an operation, usually for the purpose of uncovering something hidden, risky, unstable, unusual, subtle, or unexpected. For example, the value added auditor may encounter situations, circumstances, relationships or results that are anomalous. An investigation implies tracking down information to determine the truth. A number of instances may justify an investigation, such as:

- Risk event.
- Abnormal variation on a control chart.
- Major change in design, process, or system.

- Improprieties, such as conflicts or fraud.
- Weakness in controls.
- Anomalous conditions.
- Unexpected or unexplained results.
- Unexpected or unexplained relationships.
- Noncompliances or nonconformances.
- Illegal acts.
- Major gaps.
- Irregularities.
- Potential errors.
- Systemic or chronic conditions.

6. EVALUATE SITUATIONS, DATA OR INFORMATION

In a formal auditing sense, evaluation is the same as appraising or reaching a judgment. Once an operation has been surveyed and analyzed, the value added auditor would determine internal control effectiveness and efficiency. This could be considered the end result of a value added audit. In other words, the auditor evaluates evidence and information to reach an opinion that is based on professional proficiency and due care. The auditor would arrive at a conclusion that leads to a formal opinion of systems, processes and products. The auditor could evaluate information throughout the value added audit, such as:

- Determining which activities to audit.
- Determining risks, costs, and budgets.
- Determining tests.
- Determining accuracy, precision, and confidence levels.

REMEMBER THAT DATA IS NOT INFORMATION

The suggestions or recommendations in this book do not necessarily result in understandable information or audit evidence. The results of a test or interview are data, which may have to be analyzed so they result in information useful for decision-making and audit assurance. Information is characterized by facts that are understandable and can be used to make decisions. For example, raw test data are simply numbers. These numbers are graphed, compared, analyzed, or collated so that meaningful patterns and trends arise which may lead to information. For the value added auditor, the decision is whether this information has to be corroborated or analyzed to determine if it is sufficient and reliable evidence that can be used to satisfy the value added audit's objectives.

'Lessons learned' are a key closed-loop, feedback system for feeding data and information back into the organization so they can improve systems, processes, and products. As data and information are fed back into the system, knowledge and wisdom are created. See the 'Value Pyramid' figure on page 316. for the relative value of various types of information. These provide a critical soft control and a level of assurance to the value added auditor.

CONDUCTING STEP 5 – ASSESS EVIDENCE

TIPS FOR CONDUCTING AN INTERVIEW

There is no one set of tips for conducting an interview. However, keep the following in mind:

- Have and follow an agenda.
- Prepare for the interview.
- Know the objective of the interview.
- Be on time.
- Be cheerful and smile.
- Explain the purpose, background, and need for interview.
- Know process and product names.
- Know background and general information on the auditee.
- Establish rapport.
- Ask both open-ended and closed questions.
- Set a friendly tone.
- Allow person to finish answering questions.
- Slow down.
- Look attentive.
- Use an interpreter if necessary.
- Be an active listener.
- Speak carefully and clearly.
- Ensure that people understand you.
- Ensure that you understand people.
- Be firm when necessary.
- Set the interviewee at ease.
- Find the best location for the interview.
- Dress appropriately for the interview.
- Don’t visit.
- Don't interrogate.
- Don't be rude.
- Don’t challenge.
- Don't be sexist, homophobic, or racist.
- Don't discuss religion and politics.
- Don't intimidate.
- Don't fidget.
- Don't be accusatory.

Reliable data result in better information, which lead to better audit evidence. Are some data better or more reliable than others? Yes, data or evidence may be gathered in different ways or come of different sources:

- Data from an independent source outside the auditee's operations is usually more reliable than internal data.
- Data from different departments within the auditee's operations is usually more reliable than data from the audited area.

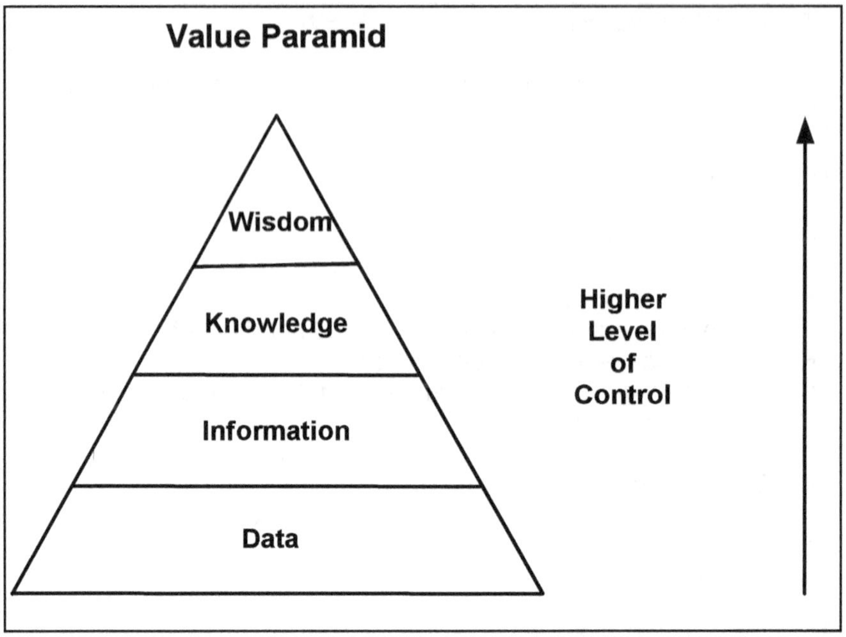

- Data from a process with effective controls is more reliable than data from a poorly controlled system or one with no controls.
- Data that was audited in the previous year or in the current year audit is more reliable than unaudited data.
- Corroborated data from a variety of sources is more reliable than data from one source.[2]

TESTING RISK - CONTROLS

Situations are rarely ever black or white; they are usually varying shades of gray.
Professor Ray Brown

How much evidence should the value added auditor gather? As discussed, the required levels of management's appetite for risk, customer assurance and auditee's risk/controls determine how much evidence to gather. So, one common method of gathering evidence is to evaluate transactions, processes, systems, activities, or records by sampling part of the population or by testing internal control effectiveness.

This review compares results of the evaluation against established business objectives, performance metrics, requirements, or standards. To review the entire population of items is often impossible, impractical, or costly, so the value added auditor selects a random, representative sample. The important point is that the auditor needs to obtain sufficient information to support the audit opinion.

TESTING PROCESS

The testing process accumulates evidence and information to support audit decisions. There can be up to eight steps in the testing process:

CONDUCTING STEP 5 – ASSESS EVIDENCE

- Identify standards, metrics, or objectives.
- Determine level of assurance and required controls.
- Identify testing resources.
- Define the population.
- Select the sample.
- Inspect, test, and/or assess sample items.
- Test control effectiveness.
- Decide on the strength or weakness of the internal controls.

I. IDENTIFY STANDARDS, METRICS, OR OBJECTIVES

Metrics or objectives detail the requirements a system, process, or product must satisfy or comply with. Requirements are found in::

- Process standards.
- Quality standards.
- Policy statements.
- Technical specifications.
- Industry standards.
- International standards.
- Trade practices.
- Service standards.
- Workmanship instructions.
- Work instructions.
- Purchase orders.
- Contracts.
- Company benchmarks.
- Company directives.
- Procedures.
- Budgets.
- Workmanship practices.
- Recommended practices.

What does the auditor do if standards, objectives, or other audit requirements don't exist? The value added auditor may:

- Work with the auditee to develop realistic standards, business objectives, or process metrics.
- Conduct a cursory evaluation of undocumented and uncontrolled systems and report results in the audit.
- Recommend to the auditee what steps to take to obtain the appropriate documentation and develop process controls.
- Proceed no further and inform the customer of the decision.

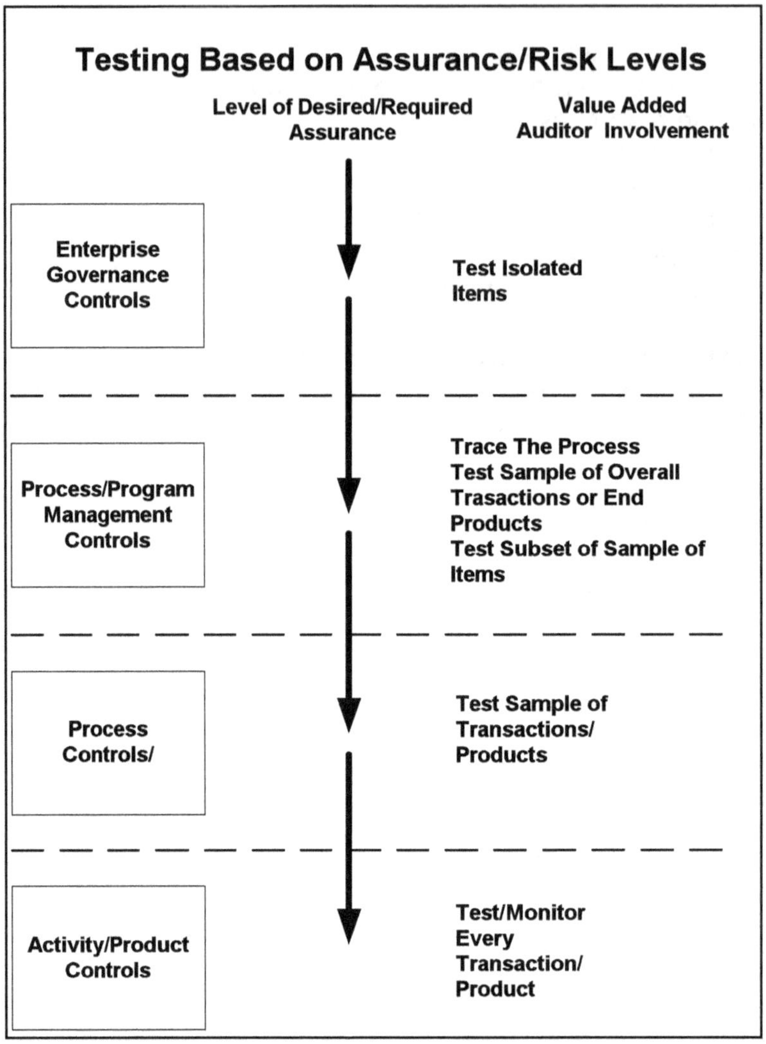

STEP 2: DETERMINE LEVEL OF ASSURANCE AND REQUIRED CONTROLS
In most organizations, there are levels of control, specifically (See above 'Testing Based on Assurance/Risk Level' sidebar):

- Enterprise 'Governance' controls.
- Process/Program management controls.
- Process controls.
- Activity controls.

Enterprise 'Governance' Controls
'Governance' controls are implemented at the 'value chain' or business unit levels normally on a scheduled basis. The company's board of directors, regulatory authority, third

party assessor, or some other authority may authorize these. Examples of 'governance controls' include:

- Traditional financial audits to GAAP (Generally Accepted Accounting Principles).
- Securities and Exchange Commission reviews.
- Regulatory compliance audits.
- Management reviews.
- External stakeholder satisfaction benchmarking.
- 'Effectiveness/efficiency' reviews by Internal Auditing.
- Portfolio project reviews.

Process/Program Management Controls
Process management or program management controls are oversight controls over a group of processes, projects, and functions. Senior management and process sponsors evaluate the existence, effectiveness, and economy of the processes within their purview. Examples of process management controls include:

- Budget vs. actual comparisons.
- Policies/procedures implementation.
- Trend/variance analysis.
- Internal customer satisfaction reviews.
- Complaint analysis.
- Peer reviews.
- Product development reviews.
- Project reviews.

Process Controls
Process controls provide a level of confidence that the right controls exist, are effective, are written, and are followed consistently. Process owners monitor and control their process to ensure stability, capability, and improvement. Examples process assurance include audits and analysis of:

- Process stability.
- Process capability.
- Process improvement.
- Trend analysis.
- Corrective/preventive action application.
- Exception reporting.

Activity/Product Controls
Activity controls are the day-to-day controls embedded in processes. Day to day process owners use these controls to ensure that products, both services and tangible products, coming out of these processes satisfy technical requirements, business objectives, and other metrics. Examples of process activity controls include:

- Product inspection and testing.
- Inspection.
- Segregation of duties.
- Verification of corrective/preventive actions.
- Written instructions.
- Authorizations and approvals.

3. IDENTIFY TESTING RESOURCES

Resources, such as personnel, facilities, and equipment have to be allocated for inspection, test, or analysis. Inspectors, process engineers, or R&D specialists may be required to perform tests and analyze results. For example, chemical analysis requires Ph.D. chemists or process engineers. These people are trained in specialized testing, standards, regulatory requirements, and acceptance standards. Facilities are required for receiving, handling, or storing products. Calibrated measurement equipment is required for conducting tests, typically, destructive, nondestructive, dimensional, optical, mechanical, and pneumatic equipment. As specified in the standard, measurement equipment is specially calibrated, stored, and handled.

4. DEFINE THE POPULATION

In general, a population consists of all items in a group that is being studied. A sample is a subgroup of the population. A population of items can be specified differently. For example, in terms of determining dimensional conformance of products from a manufacturing process, the population can be:

- All products from similar machines from all shifts since the last audit.
- All products from similar machines from one shift since the last audit.
- All products from one machine from one shift since the last audit.
- All products from one machine from one shift since last week.

Preferably, the population should be homogeneous from one machine, one operator, one shift, over a short period of time. As the population gets smaller and more specific, the value added auditor can target and identify the causes of variation and deficiencies. Of course this is the optimum and may be difficult to achieve.

5. SELECT THE SAMPLE

So when conducting an audit, the value added auditor pulls a representative sample of the population of data and tries to determine the effectiveness of the system of controls. This could be compared with looking at selected frames of a movie to see what has occurred throughout the picture. If enough movie frames are examined, the value added auditor can understand the movie's plot. However, if the frames are not representative of all the scenes in the picture, the value added auditor would then lose the plot line or a sense of what occurred. The auditor may want to see the whole picture, not just one frame, and in many circumstances, this is possible. For example, a process may continually monitor and measure a product characteristic. For each production run, there is a control chart showing variation of each characteristic. The value added auditor should

'OBJECTIVE – RISK – CONTROL – TEST' METHOD

Audit objective: Critical quality attributes are identified on engineering prints.
Risks: Supplier or manufacturing doesn't know what to focus its attention on.
Controls: Drawings identify quality attributes by accepted symbols.
Audit tests: Sample and inspect engineering drawings to ensure critical quality characteristics are identified.

Audit objective: Quality characteristics are prioritized.
Risks: Supplier or manufacturing gives all product dimensions equal priority.
Controls: Drawings prioritize quality characteristics in terms of critical, major, and minor.
Audit tests: Sample and inspect engineering drawings to ensure critical quality characteristics are prioritized.

Audit objective: Dimensions and tolerances are specified on engineering drawings.
Risks: Wrong or no dimensions are specified on drawings.
Controls: Drawings are approved and signed by authorized persons.
Audit test: Engineering drawings are sampled and inspected for proper dimensions and tolerances. Are unusual conditions or instructions specified in notes?

ask the auditee to call up the information on the computer to obtain the production history of the characteristic. Now, the value added auditor can view the whole reel of film.

How to select the sample

The sample size can be 100 percent of the population or a statistical sample. Two popular sampling tables are: Sampling By Attributes and Sampling By Variables.

The following are basic statistical sampling rules:

- **Population is homogeneous.** If there is substantial variation in the population, it is difficult to pull a sample that reflects the characteristics of the population.
- **Samples are randomly selected.** A sample should be selected randomly and represent the characteristics of the population of items. This is done by using random numbers from a table or computer.
- **Samples are representative of the population.** A sample should be representative of the parts designed or produced during the time in which the assessment is conducted.
- **Auditor follows prescribed procedure.** Sampling tables require that the person conducting the test follows directions. Otherwise, errors will arise and results are skewed.

6. INSPECT, TEST, AND/OR ASSESS SAMPLE ITEMS

The audit may have a number of objectives, such as evaluating conformance to a procedure or evaluating the effectiveness of a help desk based on a performance objective. The above 'Objective – Risk – Control – Tests Model' is a common method of evaluating and reporting risk-control effectiveness.

Items, products, or services can be inspected and assessed in terms of the following criteria that are spelled out in a standard or similar document:

- Conformance/compliance.
- Improvement.
- Efficiency.
- Effectiveness.
- Economy.
- Existence.
- Occurrence.
- Completeness.
- Accuracy.
- Precision.
- Valuation.
- Disclosure.
- Allocation.
- Reasonableness.
- Sufficiency.
- Timing.
- Validity.
- Ownership.
- Classification.

In general, if the value added auditor wants higher levels of assurance of control effectiveness, then he or she will evaluate more items in the sample as indicated by the 'Testing Based on Assurance/Risk Levels' sidebar. To obtain an overview of enterprise governance controls, the value added auditor would test selected items. To obtain a higher level of assurance or confidence, the auditor may evaluate process management controls by tracing the process, testing a sample of the end product items and/or test sample of the transactions or products. The auditor can also drill down to the process assurance or process activity level and test more items.

The value added auditor should use common sense when evaluating the items. In a large operation, there can be many minor risks, exceptions, variables, nonconformances, or deficiencies. The auditor then asks the question: 'Do these deficiencies show lax control or are they minor exceptions to an otherwise excellent internal control structure?' For example, if during an assessment of corrective action requests, two or three were not implemented within the time stipulated in the request and there have been no recurrences or other major problems, the two or three late corrective actions may not be important.

7. TEST CONTROL EFFECTIVENESS

The purpose of testing a process is to validate that internal controls are effectively managing risks (process stability), process is consistently meeting business objectives (capability), and determine the process is consistently improving.

CONDUCTING STEP 5 – ASSESS EVIDENCE

Organizational, system, or process control validation is applied to critical processes an process steps, those that are most significant to overall decision-making. Validation is a process that should continuously monitor the organizational system, and process controls.

The following should be considered when assessing the effectiveness of the process controls:

- Evaluate the materiality, significance, and type of the risks involved in the process.
- Determine the nature, magnitude, and likelihood of the risk.
- Evaluate the effectiveness of 1. control environment; 2. risk assessment; 3. control activities; 4. information and communication; 5. monitoring.
- Evaluate process 1. control (stability); 2. capability of meeting requirements; 3. improvement.
- Determine if the controls are enterprise governance controls, program/process management controls, process controls, or product level controls.
- Determine if the controls are detective, preventive, preemptive, corrective, or predictive.

8. DECIDE ON THE STRENGTH OR WEAKNESS OF INTERNAL CONTROLS

After the preliminary review of process samples, the value added auditor may have sufficient information to determine if the results of the process meet business objectives, standards, requirements, or metrics. At this point, the auditor can evaluate the effectiveness of internal controls and address the following questions, which were examined in a previous section:

- Do internal controls exist?
- Are there informal controls?
- Do they control the correct process variables?
- Are they the most effective controls?
- Are they working properly?
- Do they work to improve the process?

In value added auditing, this process step evaluates the strengths and weaknesses of the internal control system. Strengths refer to the existence and effectiveness of system, process, and product controls to ensure that defects are prevented, business objectives are met, risks are controlled, and customers are satisfied. If the system has extensive controls, less testing is conducted. Testing refers to corroborating the existence and effectiveness of the controls. If sufficient controls exist and have effectively operated over a period of time to eliminate deficiencies, the value added auditor may conclude that the system does not have to be tested.

INTERNAL CONTROL AUDIT OBJECTIVES

GAGAS A2.03 Examples of audit objectives related to internal control include an assessment of the extent to which internal control provides reasonable assurance about whether

a. Organizational missions, goals, and objectives are achieved effectively and efficiently.
b. Resources are used in compliance with laws, regulations, or other requirements.
c. Resources, including sensitive information accessed or stored outside the organization's physical perimeter, are safeguarded against unauthorized acquisition, use, or disposition.
d. Management information, such as performance measures, and public reports are complete, accurate, and consistent to support performance and decision making.
e. The integrity of information from computerized systems is achieved.
f. Contingency planning for information systems provides essential back-up to prevent unwarranted disruption of the activities and functions that the systems support.

Weaknesses refer to a lack of adequate risk - control or to no control. This situation automatically raises a red flag. No controls, not even some form of inspection, indicate there is a high probability of a systemic problem, risk event, process instability, or product nonconformance. Also, no controls may even result in making the audit impossible.

If controls are weak, then there is a higher probability that risks or nonconformances will result. For example, statistical process control is a strong internal control to prevent product nonconformances and to improve a process. Inspection is a weak internal control because it guarantees a set defect level and can't be used to improve a process. Thus if a machine process is not statistically controlled or not capable of meeting specifications, the risk of nonconforming products is high. Or, if engineering has no design review or calculation review, the probability of design error also increases.

Reporting Strength or Weakness of Internal Controls
The strength or weakness of internal controls can be tested in several ways:

- 'Objective – Risk - Controls – Test' method.
- 'Criteria – Condition – Cause – Effect' method.

One method for evaluating risks and their controls is shown in the 'Objective – Risk - Control – Tests' sidebar on page 321. This method evaluates the audit objective directly against its risks and its controls. Audit tests are then conducted to determine effectiveness of the controls. This method consists of the following:

- **Objective.** Audit objectives, standards, measures, or expectations are identified.
- **Risk.** The organizational, process, project, product, or process risks of not being able to meet the audit objective are identified.
- **Control.** The effectiveness of internal controls to mitigate or manage the risks relating to the audit objective is investigated.

'CRITERIA – CONDITION – CAUSE – EFFECT' METHOD

Criteria. Compliance to ISO 9001 requirement of 'internal auditing' procedural requirement.
Condition. Preliminary surveillance assessment indicated 'Internal Auditing' procedure is incomplete and does not address all of the ISO 9001 requirements.
Cause. New Quality Management System manager was hired to register the organization to ISO 9001. This person is unfamiliar with the new standard.
Effect. ISO 9001 registration may require internal training and may take longer than expected.

- **Test.** The value added auditor conducts tests to determine if the risks are being managed effectively and to obtain sufficient evidence to reach a reasonable value added audit determination.

Another method for evaluating the effectiveness of risk controls uses a 'gap analysis' and 'cause-effect' approach. See above 'Criteria – Condition – Cause – Effect' sidebar for an example of this method. This method consists of the following:

- **Criteria:** The audit objectives, standards, measures, or expectations are identified in making an evaluation and/or verification (what should exist).
- **Condition:** This consists of the evidence the value added auditor found while conducting the audit (what does exist).
- **Cause:** The reason for the difference (the gap) between the expected and actual conditions (why the difference exists) is investigated.
- **Effect:** The risk or exposure (the impact of the difference) to the organization and stakeholders is examined. In determining the degree of gap, risk or exposure, the value added auditor considers the effect audit observations and recommendations may have on the organization's operations, risks, and financial statements.[3]

EVALUATE EVIDENCE

Never take anything for granted.
Benjamin Disraeli

Once evidence and information have been gathered from tests, observations, interviews and other sources, then the value added auditor distills and summarizes the data. The auditor's objective is to have sufficient data to make an educated and 'reasonable assurance' assessment. The auditor wants to gather sufficient evidence to determine whether systems and controls are working effectively and efficiently.

ART AND SCIENCE OF VALUE ADDED AUDITING

In many ways, value added auditing is both art and science. The art of auditing involves evaluating operational systems, processes, and controls. An inexperienced process or value added auditor may come in and properly note each deficiency, regardless of its criticality. The novice auditor goes for the quantity of deficiencies, not understanding their underlying process or risk importance, causes, and consequences. This often re-

DEVELOPING ELEMENTS OF A FINDING – CRITERIA/CONDITION/CAUSE/EFFECT

GAGAS 6.37 Auditors should identify criteria. Criteria represent the laws, regulations, contracts, grant agreements, standards, specific requirements, measures, expected performance, defined business practices, and benchmarks against which performance is compared or evaluated. Criteria identify the required or desired state or expectation with respect to the program or operation. Criteria provide a context for evaluating evidence and understanding the findings, conclusions, and recommendations included in the report. Auditors should use criteria that are relevant to the audit objectives and permit consistent assessment of the subject matter.

6.75 Condition: Condition is a situation that exists. The condition is determined and documented during the audit.

6.76 Cause: The cause identifies the reason or explanation for the condition or the factor or factors responsible for the difference between the situation that exists (condition) and the required or desired state (criteria), which may also serve as a basis for recommendations for corrective actions. Common factors include poorly designed policies, procedures, or criteria; inconsistent, incomplete, or incorrect implementation; or factors beyond the control of program management. Auditors may assess whether the evidence provides a reasonable and convincing argument for why the stated cause is the key factor or factors contributing to the difference between the condition and the criteria.

6.77 Effect or potential effect: The effect is a clear, logical link to establish the impact or potential impact of the difference between the situation that exists (condition) and the required or desired state (criteria). The effect or potential effect identifies the outcomes or consequences of the condition. When the audit objectives include identifying the actual or potential consequences of a condition that varies (either positively or negatively) from the criteria identified in the audit, "effect" is a measure of those consequences. Effect or potential effect may be used to demonstrate the need for corrective action in response to identified problems or relevant risks.

sults in the auditee being resentful and angry. The participatory, problem-solving culture that organizations are promoting for continual process improvement can be destroyed by insensitive actions.

The science of auditing is the ability to conduct and analyze different processes, tests, and products. In compliance auditing, evaluation may simply be comparing what is done against what is required in performance documents. If the value added auditor can interpret standards or specification documents strictly, then compliance auditing is relatively straightforward. This, of course, assumes that the documents are comprehensive and well written.

HOW MUCH INFORMATION TO GATHER

Information or evidence gathered in an audit should be objective, independent, sufficient, competent, accurate, factual, suitable, adequate, and convincing. Again, the IIA provides some direction on this. The value added auditor should gather evidence that is:

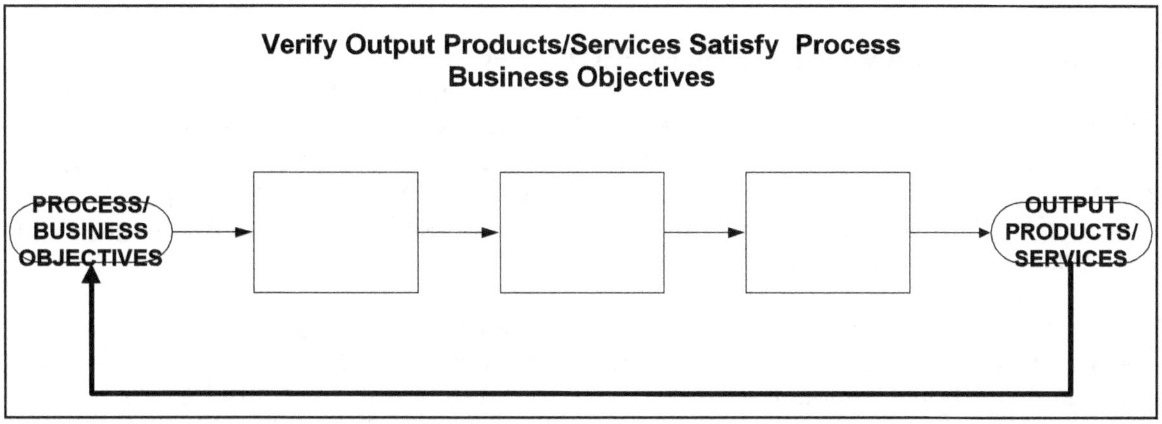

" ... sufficient, competent, relevant, and useful to provide a sound basis for engagement observations and recommendations. Sufficient information is factual, adequate, and convincing so that a prudent, informed person would reach the same conclusions as the auditor. Competent information is reliable and the best attainable through the use of appropriate engagement techniques. Relevant information supports engagement observations and recommendations and is consistent with the objectives for the engagement. Useful information helps the organization meet its goals.[4]

VERIFY OUTPUTS SATISFY PROCESS/BUSINESS OBJECTIVES

The value added auditor should also verify that process outputs satisfy process business objectives. For example, the business or process objective may be 95% customer satisfaction with a help-desk or 99.7% conforming products from a production line. Both of these statements are process objectives. It's critical that added auditors check outputs against business objectives or requirements. Value added auditors may focus exclusively on the process and forget that a process has a business objective such as the following discussed earlier in this book:

- Performance.
- Reliability.
- Conformance.
- Usability.
- Maintainability.
- Customer service.
- Aesthetics.
- Environmental.
- Software quality.

For example, if the process output or product was performance or reliability related, the value added auditor should verify that process outputs satisfy these process and business objectives.

CONDUCTING STEP 5 – ASSESS EVIDENCE

EVALUATE RISKS AND CONTROL EFFECTIVENESS

Identifying and evaluating risks are among the most important tasks of a value added auditor. The value added auditor assesses the importance and the risk of internal controls not detecting and preventing nonconformances and determines the amount of evidence to obtain to make a reliable decision. The value added auditor may debrief the auditee of control weaknesses during the audit, at scheduled progress meetings, in the post audit meeting, or in the audit report.

As defined earlier, control risk is the probability that the internal control system is able to prevent a risk, nonconformance or deficiency from occurring on a timely basis. The value added auditor should conduct sufficient tests to determine if the control system meets the business and audit objectives. In general, the value added auditor identifies a specific audit objective and then identifies the internal controls that relate to the value added audit objectives.

The amount of evidence to gather depends on the following risk factors:

- **Level of assurance.** High levels of risk require more controls and higher levels of assurance.
- **Cost.** Low risk implies low costs to obtain the proper amount of evidence from the right sources.
- **Reliability of evidence.** Reliability of evidence depends on the risk that senior management, customer and auditor may want to assume. High assurance implies highly reliable evidence.
- **Availability of evidence.** Evidence may be impossible to gather quickly because onsite analysis is required and the supplier may be located halfway around the world.
- **Time constraints.** If the customer wanted an audit report yesterday, then a rush audit requires more resources to overcome time constraints.
- **Scope.** Extensive or intensive audits in terms of time and location require more resources and evidence to obtain uniform reliable evidence.
- **Availability of resources.** Resources to conduct an audit may not be available. Equipment, specialists, or auditors may be unavailable.
- **Special circumstances, situations, or conditions.** Unusual or special situations can destabilize a controlled process. In terms of Statistical Process Control (SPC), this would show on the chart as an abnormal condition.
- **Recurring problems.** Recurring problems mean that the original fix did not eliminate the root cause of the problem. This is an example of symptomatic, fire-fighting mode of management.
- **Weak or no internal controls.** Weak or no internal controls will be highlighted in areas that show recurring problems or major types of failure. Often, the failure is caused by systemic or organizational factors.

OVERALL ASSESSMENT OF EVIDENCE

GAGAS 6.69 Auditors should determine the overall sufficiency and appropriateness of evidence to provide a reasonable basis for the findings and conclusions, within the context of the audit objectives. Professional judgments about the sufficiency and appropriateness of evidence are closely interrelated, as auditors interpret the results of audit testing and evaluate whether the nature and extent of the evidence obtained is sufficient and appropriate. Auditors should perform and document an overall assessment of the collective evidence used to support findings and conclusions, including the results of any specific assessments conducted to conclude on the validity and reliability of specific evidence.

6.70 Sufficiency and appropriateness of evidence are relative concepts, which may be thought of in terms of a continuum rather than as absolutes. Sufficiency and appropriateness are evaluated in the context of the related findings and conclusions. For example, even though the auditors may have some limitations or uncertainties about the sufficiency or appropriateness of some of the evidence, they may nonetheless determine that in total there is sufficient, appropriate evidence to support the findings and conclusions.

6.71 When assessing the sufficiency and appropriateness of evidence, auditors should evaluate the expected significance of evidence to the audit objectives, findings, and conclusions, available corroborating evidence, and the level of audit risk. The steps to assess evidence may depend on the nature of the evidence, how the evidence is used in the audit or report, and the audit objectives.

a. Evidence is sufficient and appropriate when it provides a reasonable basis for supporting the findings or conclusions within the context of the audit objectives.

b. Evidence is not sufficient or not appropriate when (1) using the evidence carries an unacceptably high risk that it could lead the auditor to reach an incorrect or improper conclusion, (2) the evidence has significant limitations, given the audit objectives and intended use of the evidence, or (3) the evidence does not provide an adequate basis for addressing the audit objectives or supporting the findings and conclusions. Auditors should not use such evidence as support for findings and conclusions.

- **Unstable and out of control conditions.** These conditions reveal themselves in an SPC chart or similar graph as a spike, trend line, points outside control limits, or similar conditions.
- **High risk areas.** High-risk areas may require a high level of assurance, which would mean the organization has an extensive internal control structure. One control system will not usually provide high levels of assurance. There usually are redundant human, procedural, and automated control systems that reinforce and back up each other.
- **High liability areas or products.** If failure results in critical nonconforming products, health or welfare are at risk and so is legal exposure.
- **Slow or lack of improvement.** Having value added systems and processes in control and capable is not sufficient. The rate of continual improvement must also accelerate to keep up with marketplace changes.

CONDUCTING STEP 5 – ASSESS EVIDENCE

OVERALL ASSESSMENT OF EVIDENCE - LIMITATIONS AND RISKS

GAGAS 6.72 Evidence has limitations or uncertainties when the validity or reliability of the evidence has not been assessed or cannot be assessed, given the audit objectives and the intended use of the evidence. Limitations also include errors identified by the auditors in their testing. When the auditors identify limitations or uncertainties in evidence that is significant to the audit findings and conclusions, they should apply additional procedures, as appropriate. Such procedures include:
a. Seeking independent, corroborating evidence from other sources;
b. Redefining the audit objectives or limiting the audit scope to eliminate the need to use the evidence
c. Presenting the findings and conclusions so that the supporting evidence is sufficient and appropriate and describing in the report the limitations or uncertainties with the validity or reliability of the evidence, if such disclosure is necessary to avoid misleading the report users about the findings or conclusions and
d. Determining whether to report the limitations or uncertainties as a finding, including any related, significant internal control deficiencies.

EXERCISE DUE CARE

As a rule, before issuing a finding or an opinion on the audit objectives, the value added auditor may want to reduce audit risk to a level that is acceptable to the customer. Thus if the auditor knows that an audit area has a high risk potential, there is a possibility of failure or high consequences as a result of a failure.

In high risk areas, the auditor must exercise higher professional due care. Due care requires that the value added auditor performs tests until sufficient evidence has been obtained for each system, process, or machine so that in the auditor's determination, there is reasonable assurance that processes are controlled and improved.

To conduct a value added audit with due professional care, the value added auditor can assess and reduce control and detection risks. The value added auditor can't eliminate inherent risk, but the auditor can reduce control risk by thoroughly understanding the business control environment including the inputs, outputs, and the process itself. The value added auditor can reduce detection risk by conducting sufficient tests on the system. Sufficiency is a subjective judgment based on the auditor's professional proficiency.

WHEN IS AN AUDIT COMPLETE?

This is a professional judgment call based on the preponderance of evidence based on the ability to meet the reasonable assurance threshold. The value added auditor should stop when the collective weight of the individual areas being assessed provides that reasonable assurance. This is often a professional judgment call.

A response of a 'nonconformance,' 'ineffective', 'not consistently applied, or 'not capable' with response to an isolated incident by itself would not warrant a ineffective conclusion regarding an audit objective. There may be soft or compensating controls in the process or in the area being audited. The value added auditor may recommend this is an 'opportunity for improvement.' If there are significant areas of deficiencies, ineffective controls, or high-risk areas, then these should be reported.

SUMMARY

The value added auditor, unless he or she has spent time working in the process, usually does not truly understand how the process or people really work. So, the value added auditor must rely on process artifacts, such as process reports, corrective action reports, process change documents, preventive action reports, and product tests. The value added auditor can then verify and validate existing process artifacts.

Value added auditors should consider several factors when evaluating the overall effectiveness of organizational control processes:

- Were significant discrepancies, risks, systemic weaknesses, or chronic conditions discovered from the audit work performed and other assessment information gathered?
- If so, were corrective actions, mitigations, preventive actions, or improvements made after the discoveries?
- Do the audit discoveries and their consequences lead to the conclusion that a pervasive, chronic, or systemic condition exists resulting in an unacceptable level of overall business, process, activity, or product risk?[5]

What happens if there is a significant control discrepancy, risk, deficiency, or weakness? The value added auditor may issue a corrective action request or a preventive action request. The significance, pattern, level of consequences, and type of exposure should be considered in evaluating the effectiveness of the entire systems of controls. If there are chronic, systemic, or unacceptable risks, then the auditing outcome would be different. Or, the discrepancy is a one-off, which is isolated and poses an acceptable residual risk. Or, the discrepancy is significant and poses an unacceptable residual risk.

CHAPTER 17
STEP 6: ISSUE OPINION/FINDINGS

After system or process tests have been conducted and sufficient evidence has been accumulated to provide reasonable assurance, then the value added auditor issues an opinion. If testing confirms that audit objectives are satisfied, then this is noted in the audit report. If there are significant errors, risks, flaws, no controls, ineffective controls, or deficiencies, then a finding is issued. In a finding, the value added auditor states that there is a large variance between 'what is' and 'what should be.'

FINDING/OPINION PROTOCOLS

Internal, quality, and operational audits usually did not require an opinion. Due to September 11 catastrophe and Sarbanes-Oxley 2002 law, there are higher requirements for the auditor or the chief audit executive to issue an opinion. Opinions are more often expressed at the 1. Enterprise level; 2. Programmatic/project/process level; and 3. Product/transaction level.

Opinions are more often requested or required as a result of:

- **Statutory compliance:** Sarbanes-Oxley Act of 2002 or the Federal Information System Management Act have increased the level of due diligence and due care in companies.
- **Internal control effectiveness:** Senior executives want assurance or 'skin in the game' as they are required to sign off on operational risks and on the financials.
- **Homeland security risks and assurance:** September 11 attacks have increased the need for security assurance and security controls. States are also passing laws requiring an audit of high asset facilities, such as oil refineries, etc.

WHAT IS AN OPINION?

An audit opinion is usually expressed because of a legal or high-risk need. The process of providing an opinion of the reliability of financial statement is called attestation. We are finding that more clients perceive an 'opinion' as a sufficient and necessary condition of a value added audit.

This presents a challenge. Most licensed professionals do not want to provide an opinion. Why? An opinion can result in legal exposure to the professional. Clients do not

DEVELOPING ELEMENTS OF A FINDING

GAGAS 7.73 Auditors should plan and perform procedures to develop the elements of a finding necessary to address the audit objectives. In addition, if auditors are able to sufficiently develop the elements of a finding, they should develop recommendations for corrective action if they are significant within the context of the audit objectives. The elements needed for a finding are related to the objectives of the audit. Thus, a finding or set of findings is complete to the extent that the audit objectives are addressed and the report clearly relates those objectives to the elements of a finding. For example, an audit objective may be to determine the current status or condition of program operations or progress in implementing legislative requirements, and not the related cause or effect. In this situation, developing the condition would address the audit objective and development of the other elements of a finding would not be necessary.

want to invest in the risk premium that an opinion requires. It is difficult to obtain 'errors an omission' insurance for some types of opinion, such as homeland security assessments.

OPINION TIPS

A professional can express an opinion, but it should be properly scoped and expressed. An opinion properly expressed should at a minimum consider the following:

- **Tight scope area of review:** Scope over which the opinion extends should include description of the organization, control components, audit objectives, time period, processes, etc.
- **Regulatory requirements:** Regulations may prescribe the type of work as well as the audit objectives with which that must be complied.
- **Based on defined criteria:** Criteria may be the audit objectives, requirements, etc.
- **Standard evaluation process was used:** Evaluation process should follow a prescribed approach, such as the 'Yellow Book' or 'Red Book.'
- **Sufficient evidence was examined:** Sufficient evidence should be gathered to support and form the basis for a finding or an opinion. Evidence is the basis for an informed evaluation, which a person in a similar position would reach.
- **Person providing opinion is trained:** Licensed or certificated professionals provide the opinion.

TYPES OF FINDING/OPINION – POSITIVE ASSURANCE

There are two different types of opinions: 1. Positive assurance and 2. Negative assurance. Each of these conveys different meaning and types of assurance. Positive assurance implies that sufficient and competent supporting evidence was gathered to support an opinion. The alternative is to formally disclaim an opinion because the auditor has not gathered or is unable to gather sufficient audit evidence to express an opinion. The auditor should clearly state that fact.[1]

Positive assurance is one of the strongest types of opinions. Positive assurance requires the highest level of evidence and testing by the value added auditor. A positive assurance opinion requirement is the highest level of evidence since sufficient data must be gathered and evaluated to determine where risks are located, how they are mitigated, and how effective controls are.

In providing a providing positive assurance, the auditor may provide the following:

- **Binary determination:** Internal controls are satisfactory or unsatisfactory, conforming or nonconforming, i.e. products are conforming or are not conforming to specifications; company's Quality Management System complies or does not comply with ISO 9001 requirements.
- **Variable (graded) determination:** Risks are graded; controls are evaluated for effectiveness on a scale from one (1) to five (5) or something similar, i.e. such as risk is evaluated for consequences (1-5) and likelihood (1-5).

How much evidence should be gathered for a positive assurance? This is based on the threshold of 'due care' and 'reasonable assurance.' A positive assurance for cyber security, nuclear facility, or a homeland security facility would require a high level of assurance, testing, and evidence, than for a commodity analytical evaluation.

TYPES OF OPINION – NEGATIVE ASSURANCE

"Negative assurance is a statement that nothing came to the auditor's attention that would indicate inadequate internal controls," according to the IIA.[2] Negative assurance opinion may indicate that auditor has not seen problems based on the auditor's work.

At the end of the day, the value added auditor is responsible that a methodology and procedures were consistently followed. The result is that a higher level of confidence or comfort in the accuracy of the opinion. This implies that the auditor has gathered sufficient amount of audit evidenced to express a finding or an opinion.

In general, a binary determination is a compliance finding and usually requires less evidence than a variable or graded determination. A compliance determination may require a review of documentation, while a graded determination may require extensive reviews of processes and projects. Sufficient evidence is required to support the final level of detail in a variable determination.

An opinion may be qualified with specific findings that contradict or scope the overall opinion. For example, an opinion may state that risks are mitigated and controls are appropriate, except for design controls for the Equicar product from Alpha supplier.

How much information should be gathered to support the opinion? This is based on the scope of the opinion, organizational controls, auditor capabilities, organizational risks, etc.

HIGHER STANDARD OF DUE CARE

During the last several years, certified public accounting (CPA) firms have been harvesting specialized technology credentials. It is not uncommon for CPAs and other professionals at these firms to possess Certified Information Systems Security Professional (CISSP) or Certified Information Systems Auditor (CISA) credentials for performing assessment, assurance an attestation services in cybersecurity and other technology-related areas. Addressing today's technology risks, however, requires expertise and extensive training beyond just certification. Professionals who provide technology or other specialized attestation need to possess appropriate, exert level qualification – especially where the risks can be truly catastrophic.

RISING STANDARD OF DUE CARE

In the United state, the standard of 'due care' has risen dramatically since events such as the September 11 attacks, the collapse of Enron, and the Iraq war. Greater consequences from risk events have led to increased control requirements and subsequently a higher standard of what truly constitutes due care. Cybersecurity, bio-terrorism, and other threats have raised 'proficiency' and assurance' thresholds as well.

Regulatory authorities – such as the US Department of Homeland Security and the Securities and Exchange Commission – are also raising the standard. The Sarbanes-Oxley Act of 2002 and Federal Information Security Management Act, for example, demand higher levels of organizational and professional knowledge. In light of the elevated regulatory climate, as well as increased nonregulatory threats, professionals who conduct work outside their areas of expertise can expose the organization to significant harm and even subject themselves to litigation risk.

To meet today's due care standards, organizations need to ensure that those who perform attestation or assurance possess the right level of expertise. To address heightened cybersecurity and other technology related threats, for example, those performing IT and security assessment should ideally possess a software engineering degree or comparable background. Similarly, a biologist with a doctorate level foundation of expertise should conduct a bio-terrorism assessment. Generally, technical or other highly specialized assessments should be handled by experts with appropriate academic training and professional licenses in the area under review.

CIO Magazine forewarned: "In 2010, information security will be much better than it is today. But between then and now, everything will get inconceivably worse."[3] Despite this, and other signs of increasing security challenges, one training firm currently offers to provide a cyber security certificate in just seven days. Can individuals who obtain this type of certification truly provide the requisite 'due diligence' and 'due care in areas where they are not degreed or licensed?

Special training and token credentials hardly seem adequate for the threats facing today's organizations. It is not sufficient to merely comply with established laws and requirements or give cursory treatment to risk areas. The security and overall health of the organization demand a much higher standard.[4]

SEEK REASONABLE ASSURANCE

The telephone book is full of facts, but it doesn't contain a single idea.
Mortimer Adler

There is seldom 100 percent assurance or 0 risk. So, the value added auditor must base his or her judgment on the 'reasonable assurance.'

ASSURANCE LEVELS

When conducting an audit, many questions arise about risk and control effectiveness, including:

- Is there 100% percent implementation in all areas, in all systems, and at all times?
- Is there partial implementation in some areas, in some systems, and at some times?
- Is there only partial implementation in one or two areas and at only specific times?
- Are good faith efforts being conducted?
- Is this only the start of implementation?

While 100 percent satisfaction of all process or product objectives in all areas, in all systems, and at all times is an optimum level of control performance, it may be unrealistic. As a rule, the value added auditor should seek reasonable assurance that the auditee has adopted sufficient internal process controls in specific areas to satisfy process or product objectives. Objectives can be minimizing defects, satisfying internal customers, providing prompt and courteous service, or designing reliable products.

Reasonable assurance is influenced by the value added auditor's assessment of the overall internal control structure and the environment in which it operates. For example, the auditee may have few or no automated manufacturing process controls. Process control charts are manually developed and maintained. In terms of the control system, would this be effective? This should be determined by evaluating the factors that most affect process variation and risk. These factors would be controlled more than lesser factors. For example, the value added auditor may determine that the human factor is the most important element of internal control and as long as human controls are strong, others can be weaker.

CONDUCTING STEP 6 – ISSUE OPINION/FINDINGS

LEVELS OF IMPORTANCE

Level O - No Risk
Level 1 - Level-Minor
- Internal controls do not focus on continual improvement.
- Sample Corrective Action Requests (CARs) do not indicate root cause correction.
- Some drawings difficult to access.
- Low warranty costs.

Level 2 - Major Risk
- Internal controls not used.
- Corrective action focuses on symptoms.
- Drawings do not specify process or product characteristics.
- No warranty costs.

Level 3 - Critical Risk
- No or limited internal controls.
- No or limited corrective action.
- Drawings out of date.
- High product failures.
- Level of product failures.

These questions deal with the efficiency and effectiveness issues. What does the auditor do if there is only 70 percent compliance in only part of the systems? The auditee does not have to be able to satisfy all the questions in a questionnaire or survey. This is where the value added auditor's abilities come in, specifically in terms of professional proficiency and the exercise of due care. An inexperienced auditor may categorically state that there is a major problem, cancel the audit and issue a negative finding. An experienced auditor would probably:

- Note the deficiencies in the audit report.
- Investigate compensating or overlapping controls.
- Check their effectiveness.
- Overview the entire process or system.
- Then deliver a finding or an opinion.

FINDING ISSUES

The list of items that can result in an audit finding may include:

- High areas of risk.
- Significant impacts of risk.
- Lack of or no conformance to requirements.
- No or ineffective controls.
- Low efficiencies and effectiveness.
- Calculation errors.
- Corrective actions not taken.
- No or poor documentation.
- No or poor measurement systems.

- Improper instructions.
- Lack of corrective action or preventive action.
- Lack of disclosure.
- No training.
- Imprecise and inaccurate measurement instruments.

PRIORITIZING RISKS

As mentioned, the value added auditor will seldom find 100 percent of the audit objectives being met l00 percent of the time. Deficiencies will arise in each case and the auditor must use his or her professional judgment to determine the level of importance of each deficiency. A simple method is to relate the level of deficiency to the degree of risk if internal controls did not detect and prevent a mishap or nonconformance. One method is to rate deficiencies in terms of a number that is correlated to risk and the criticality of the product. The rating system can be from 0 to 10,0 to 5, or 0 to 3. In our example, for simplicity we use 0 to 3, specifically:

- Level 0 - no risk.
- Level 1 - minor risk.
- Level 2- moderate risk.
- Level 3 - high risk.

The 'Levels of Importance' figure on the previous page illustrates the differences among the levels. Each level describes a deficiency in terms of:

- Existence and effectiveness of internal controls.
- Efficiency and effectiveness of corrective action.

It's fairly obvious that level 3 risks can result in major risk consequences or failure.

SUMMARY

Value added auditors should consider the following factors in issuing a finding or an opinion based on the type and level of information or evidence to gather:

- Complexity of the process.
- Availability, accuracy, and reliability of available information.
- Significant risks in the process.
- Type, depth, and extent of controls.
- Level of desired assurance.
- Precision and accuracy required.
- Extent or size of the population of items.
- Type of testing.

CHAPTER 18
STEP 7: CONDUCT EXIT MEETING

In this step, the value added auditor provides the auditee an initial impression of the results of the audit. This exit meeting can be formal or informal.

EXIT MEETING PROTOCOLS

A company in which anyone is afraid to speak up, to differ, to be daring and original, is closing the coffin door on itself.
Leo Burnett

The exit interview is the final meeting between the value added auditor and the auditee. It is sometimes called the post audit conference. The value added auditor may have kept the auditee informed about initial observations throughout the audit. The auditee shouldn't be surprised at this stage.

The post audit conference is a sit-down meeting. Atmosphere and attitude are important in the success of the post audit conference. These meetings can result in disputes and conflicts and it may be difficult to maintain an atmosphere of continual improvement in those circumstances.

The value added auditor may consider the following in conducting the exit meeting:

- Develop meeting agenda.
- Thank the auditee for assistance.
- Provide assurance of audit confidentiality.
- Be sensitive of auditee's needs.
- State the facts and nothing but the facts.
- Know where the auditor/auditee/customer lines of authority and responsibility are.

DEVELOP MEETING AGENDA

The agenda for the post audit conference may cover the following:

- List the people to attend.
- Show appreciation of the auditee's organization.
- Discuss the scope of the conference.
- Review and discuss the implementation of the audit.
- Review and discuss the findings.

- Review and discuss the opinions.

In a formal audit exit interview, the value added auditor asks the area managers and supervisors to attend. If the area being audited has hourly and exempt people, then they may be included in the meeting. If the size of the group is unwieldy, the meeting can be limited to team leaders and supervisory personnel. The audit team leader or area manager, or both, cochair the meeting. If only the audit team leader chairs the meeting, it may present the impression that this is an occasion where the auditee is expected to sit passively as the auditor reveals control deficiencies or nonconformances.

THANK THE AUDITEE FOR ASSISTANCE

The value added auditor should thank the auditee for his or her cooperation. Without the auditee's cooperation, the audit may not have been possible or at least would have been more difficult, have taken more time, or have required more resources.

PROVIDE ASSURANCE OF AUDIT CONFIDENTIALITY

Audit information may be confidential and may have limited distribution to specific stakeholders. The information may be privileged, proprietary, or report illegal acts. Such information should be handled carefully. The chief audit executive may have to review and approve the distribution of the audit report. The chief audit executive may report the information through some other means to senior management or the board of directors.

Audit confidentiality was not traditionally critical in many types of auditing. Today's audit challenge is that many audit reports are being used for compliance, due diligence, and legal proceedings. Records retention has become a criminal issue in many industries. Companies are required to develop and follow a records retention policy. It's critical that all audit working papers are filed and controlled as any other intellectual property. They should be accessible only to authorized personnel. If the reports are used for process improvement and are not deemed confidential then the organization should have a procedure detailing access, distribution, and destruction of working papers.

More audit stakeholders are requesting access to audit records for a number of reasons. The audit records can be used for civil litigation, criminal prosecution, tax audits, regulatory compliance reviews, or contract compliance verification. If audit reports are provided to stakeholders outside the organization, then management should approve their release.

BE SENSITIVE OF AUDITEE'S NEEDS

When delivering findings and other information in the exit interview, the value added auditor should be aware of and sensitive to the auditee's point of view especially if there are many deficiencies. If the auditor under the guise of full disclosure and honesty rubs the auditee the wrong way, then this will be communicated throughout the organization and subsequent audits will be more difficult. Again, the goal of an audit is to examine management, systems, processes, or products in order to improve them, not to fix blame.

The following are some tips for maintaining sensitivity in the exit interview:

- Acknowledge auditee assistance.
- Seek understanding and agreement.
- Look and accentuate commonalities.
- Don't prosecute or accuse.
- Focus on systems, not people.
- Show good manners.
- Maintain respect for everyone.
- Never be racist, homophobic, or sexist.

STATE THE FACTS AND NOTHING BUT THE FACTS

Value added audit reports serve two important purposes: (1) to convey information in a dispassionate, authoritative manner, and (2) if necessary to promote action on a system, process and/or product. The first purpose is the traditional role of value added auditors, to be independent and objective and to serve as the eyes and ears of the audit customer. In this role, the audit report answers the basic questions of journalism: what, who, when, why, and how. Management is especially interested in the high points, including the following:

- What is the problem or circumstances being investigated?
- What are the consequences or risks of the deficiency or nonconformance?
- What is the likelihood and magnitude of a risk occurrence?
- How much does the risk cost or may cost the organization?
- How much will it cost in the future?
- Will it recur?
- How will it be fixed in the short term?
- How will it be resolved so it does not recur?
- How much will it cost to fix and resolve?
- Who will do it?
- What are the consequences of not fixing it?
- What are the trade-offs?
- Who should correct it?
- How long will it take?
- Will the event or nonconformance recur?

Not all of these questions have to addressed in the exit meeting or final report. Depending on the risks of the deficiency and the needs of the customer, the auditor will use his or her judgment on what to address.

KNOW WHERE THE AUDITOR/AUDITEE/CUSTOMER LINES OF AUTHORITY AND RESPONSIBILITY ARE

From the above discussion, several critical factors come through. The value added auditor while working with the auditee should still be objective with the auditee. Depending on the audit scope, impartiality may be the most critical feature of the auditor and should not be compromised. The auditor shouldn't cross these lines.

The auditor must know where the lines of discretion, rightness, independence, closeness, and objectivity are. This is difficult in a collaborative value added audit or if the auditor serves in a consulting capacity.

EXIT MEETING/INTERVIEW AGENDA

Everything is connected to everything else.
Barry Commoner

Depending on the nature of the finding and customer needs, the value added auditor may want to address the following in the exit interview:

- Restate the business and value added audit objective (s).
- List the stated criteria, documents, standards, and drawings.
- Summarize findings.
- List the audit conditions.
- Prioritize nonconformances or risks.
- List effects and consequences of risks or failures.
- Identify root causes of the failure.
- Recommend corrective/preventive actions.
- Resolve disagreements, disputes, and factual inaccuracies.

All of the above points or only some of the above points can be discussed or summarized in the value added report.

RESTATE THE BUSINESS AND VALUE ADDED AUDIT OBJECTIVE (S)

The audit was authorized and funded by the audit customer or stakeholder. The value added auditor should restate the value added audit objective or objectives. This serves several purposes. It provides the context, rationale, and authorization for the value added audit. It confirms and reaffirms the purpose of the audit. It details who will receive the audit report. It also details what may be the outcomes of the value added audit.

LIST THE STATED CRITERIA, DOCUMENTS, STANDARDS, AND DRAWINGS

The audit finding may check for conformance or some degree of adherence to the audit objective specified in a document. The objective criterion against which the finding is compared and measured is identified; sometimes the standard, section, subsection, and line should be cited, Otherwise, the direct applicable line is quoted. Using the above example, the value added auditor might cite the following quality-related standard:

> "Per engineering procedure, 2.5, Section 3, Line 6.3: 'Each engineering change order (ECO) will identify the originator and be reviewed and approved by the chair of the engineering change board (ECB).'"

CONDUCTING STEP 7 – CONDUCT EXIT MEETING

SUMMARIZE FINDINGS

Findings usually include risk analysis, control effectiveness, listing of nonconformances, and opportunities for improvement. It is important to emphasize the positive when reporting unfavorable findings. The auditor should summarize value added audit findings and results in as formal or informal a manner as the audit requires. If the audit resulted from a mistake resulting in loss of life, the exit interview is formal and structured. If the audit was a routine examination of operational systems and processes and no major or critical risk/control deficiencies were discovered, then the exit interview can be an informal debriefing. The more important the audit, the more the auditor needs to adhere to a structured, formal format.

The summary briefly describes the deficiency or nonconformance. It can be one sentence:

> "Engineering change orders on Special Project SX did not go through 2 signoffs because of the requirements of rapid product development."

LIST THE AUDIT CONDITIONS

The findings are all-positive and refer to the condition, location, and specifying document. Depending on customer requirements, the audit report may also specify what type of tests were conducted, by whom, and when. Should the value added auditor compliment an auditee? There are several schools of thought on this. One says that the value added auditor should be totally independent and objective. Another believes that the auditee is required to have controls and pursue continual improvement, so why should this be complimented when it is expected of everyone? Another says that only exemplary or accelerated improvement or customer service are acknowledged and rewarded.

The value added auditor may identify the conditions, facts, or evidence obtained from testing the system or process that prompted the finding. The evidence could have been gathered through any of the six techniques discussed in the previous chapter. Again using this example, the auditor's report would describe the following condition:

> "Based on a random sample of engineering drawings from Project SX, 6 Engineering Change Orders (Echo's) out of 10 did not have the proper Engineering Change Board (ECB) authorization and approval."

The conditions, situations, or circumstances of the audit can be favorable or unfavorable. If conditions are favorable, nothing may be mentioned, or a simple statement may be issued on each process and system condition such as:

- "Process controls on production line x, Plant 3, are working properly."
- "Process capability is maintained at a minimum of 1.33 and satisfies quality standard Q2."
- "Corrective action is effective and efficient as prescribed by procedure 1.2.5."

- "Product delivery from suppliers X, Y, Z was within prescribed JIT windows."

PRIORITIZE NONCONFORMANCES OR RISKS

Nonconformances or risks can be prioritized in terms of critical, major, or minor. Critical usually refers to a life-threatening situation or condition. A major deficiency can affect product function. Minor deficiency is cosmetic, such as a blemish on the product.

LIST EFFECTS OR CONSEQUENCES OF RISKS OR FAILURES

The consequence and effects of failure should be identified if they are significant. State the consequences of not checking the engineering print. The value added auditor doesn't have to state the consequences of a minor deficiency. Again, using this example, the auditor's report might list the following effect of a control failure.

> "As a result of not checking one engineering print, SX 1-235B, the design flaw resulted in extra rework on a prototype part. Total cost was $12,435."

IDENTIFY THE ROOT CAUSE OF THE FAILURE

The value added auditor can usually identify the symptom easily, but finding the root cause may require extensive examination and consultation with workers, supervision, process engineers, and other departmental personnel. The symptom is the immediate problem and consequence; the root cause is the original source of the problem. In some cases, an interdisciplinary problem-solving team may be formed to identify the root cause.

> "In our example, symptoms are engineering change orders that are not checked and approved by the proper authorities. These resulted in a high cost. A special waiver was issued to fast track the SX project. This resulted in some (Eros) not being approved. The root cause is the Material Review Board (MRB) or the Engineering Control Board (ECB) not controlling the change-order process and a corporate environment, which required a project to be developed quickly and thereby loosening some of the MRB oversight authority."

RECOMMEND CORRECTIVE/PREVENTIVE ACTIONS

An experienced value added auditor can recommend corrective actions to eliminate symptoms and the root cause correct the deficiency. Corrective action can stabilize the system and also lead to future process improvements. As discussed, it is often better to point out the need for corrective action and allow the process owner to determine what should be done. They have more information, knowledge, and personal investment and also they are in place to monitor the corrective action to ensure the problem does not recur.

If the auditor has to recommend corrective action, the report might state the following:

> "The following are recommended: On fast-track special projects, the MRB or ECB can authorize two senior or supervising engineers to review and approve

engineering change orders. If these people believe that the changes will affect suppliers or downstream operations, the change order is approved within one day. Also, the MRB periodically samples Eros to determine approvals. This recommendation strengthens internal controls by having the MRB or ECB authorize two engineers to oversee the approval process."

In general, the audit report answers what, why, where, who, and when questions about a risk, deficiency or nonconformance. If the reader of the value added report is still wondering about the answers, then the report is not communicating information properly. If the reader understands the problem, consequences, and corrective action, then auditor has done his or her job.

RESOLVE DISAGREEMENTS, DISPUTES, AND FACTUAL INACCURACIES

The value added auditor should try to resolve disagreements, disputes, and factual inaccuracies. Disputes of one form or another can arise over any of the following:

- Disagreement over findings.
- Differences over test results, methodologies, etc.
- Differences over interpretation.
- Personality conflicts.
- Political and power conflicts.

Disputes can be seen negatively or positively. Disputes start when the value added auditor and the auditee disagree. If the disputes are genuine and not simply a power play or posturing, then resolution can lead to greater understanding by both parties. The opposite can also arise, resulting in further distance and enmity. This damages the company wide, continual improvement and risk management efforts.

If the value added auditor knows there is going to be a dispute, the value added audit should be prepared to support the audit findings and opinions. The auditor can rehearse the major elements of the audit and be prepared to discuss each point of contention. It is embarrassing to have senior management and other parties attend the post audit meeting and the auditor is unprepared for the auditee's objections. Nothing can destroy the auditor's credibility more than to ramble, speak poorly, and make mistakes.

Resolution is not always possible and there may well be irreconcilable differences. If common ground cannot be found, then the auditee and the value added auditor can agree to disagree. A higher authority, the customer or someone else, then decides on the outcome of the findings. Does this rebuttal decrease the effectiveness of the audit? If it occurs to an auditor often, it calls into question the auditor's effectiveness and proficiency. Word will get out that the auditor is unnecessarily stringent, abrasive, unprofessional, political, or ineffective.

If an audit team and the auditee reach a consensus on most of the points during the progress meetings, then there is less chance for major disagreements to arise at the end. If there are major errors in information, facts, or judgment, they are often revealed during the progress meetings. If the auditor and auditee share the vision of company wide continual improvement and customer satisfaction, there is less chance of divisiveness and irreconcilable disputes. All parties will be able to subordinate individual, departmental, or political concerns to the goal of audit customer satisfaction and organizational effectiveness.

SUMMARY

In the exit meeting, the value added auditor should thank the auditee and ensure that there are no loose ends. The auditor should review and discuss preliminary findings, opinions, and corrective actions. There should be 'no surprises' to the auditee. The exit meeting is also an opportunity for the auditor to resolve disagreements and inaccuracies. It is better to catch these during the conduct of the audit instead of after the audit report has been issued.

PART V: REPORTING VALUE ADDED AUDIT RESULTS

VALUE ADDED AUDITING LIFECYCLE(TM)

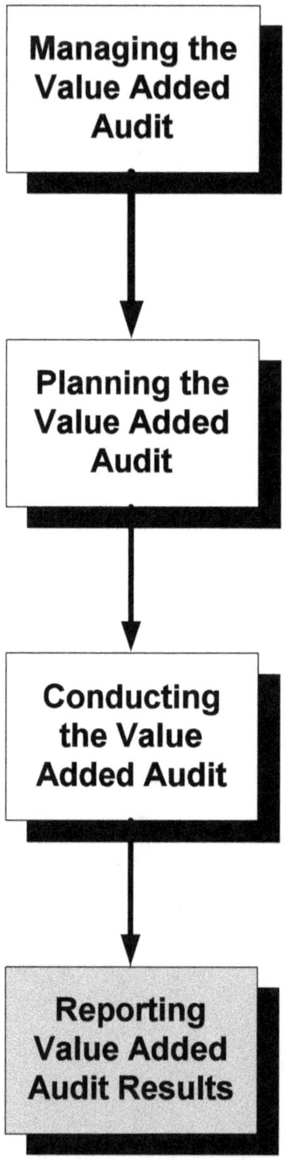

REPORTING AUDIT RESULTS

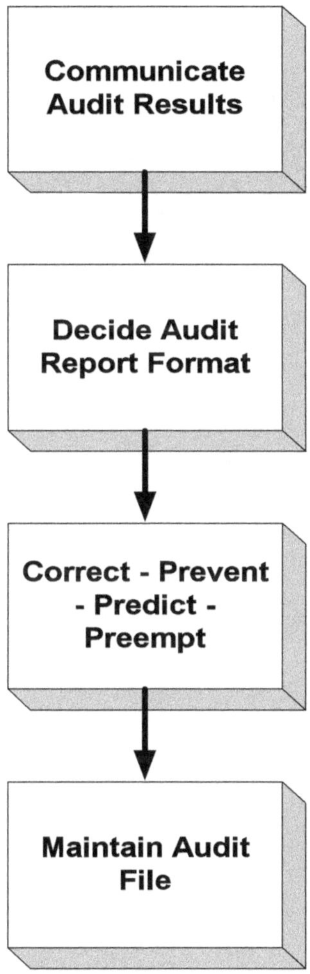

CHAPTER 19
STEP 1: COMMUNICATE AUDIT RESULTS

The value added audit report communicates the information, findings, and opinions derived from the audit. The report can communicate the acceptability of the current status of organizational, system, and process controls. Or, the report can communicate risk and nonconformances and recommend corrective/preventive actions and risk mitigations. It is imperative that the report is understandable and recommends workable actions.

CHIEF AUDIT EXECUTIVE RESPONSIBILITIES

Leadership appears to be the art of getting others to want to do something you are convinced should be done.
Vance Packard

The chief audit executive is responsible for communicating internal audit plans and reports to senior management and the board of directors. The quality, customer-supplier and other audit managers should communicate their audit plans and reports similarly to the chief audit executive. The chief audit executive can then submit a consolidated report to senior management and the board for review, approval and action. The CAE also uses these to obtain sufficient funding and resources to conduct the audits.

CAE RESPONSIBILITIES

The chief audit executive (CAE) is also responsible for managing, reporting and reconciling audit reports. This offers many benefits:

- Eliminates duplication of audits.
- Centralizes audit information.
- Communicates information and results to the required stakeholders.
- Ensures action of audit results.
- Ensures there is unified audit reporting to the board of directors and senior management.

The chief audit executive also aligns the audit's purpose and objectives to that of the organization's core mission. This ensures audits add value. As part of this alignment, the chief audit executive establishes risk based plans to target and prioritize audits to areas where they will add the most value.

AUDIT DOCUMENTATION

GAGAS 6.79 Auditors must prepare audit documentation related to planning, conducting, and reporting for each audit. Auditors should prepare audit documentation in sufficient detail to enable an experienced auditor, having no previous connection to the audit, to understand from the audit documentation the nature, timing, extent, and results of audit procedures performed, the audit evidence obtained and its source and the conclusions reached, including evidence that supports the auditors' significant judgments and conclusions. An experienced auditor means an individual (whether internal or external to the audit organization) who possesses the competencies and skills that would have enabled him or her to conduct the performance audit. These competencies and skills include an understanding of (1) the performance audit processes, (2) GAGAS and applicable legal and regulatory requirements, (3) the subject matter associated with achieving the audit objectives, and (4) issues related to the audited entity's environment.

6.80 Auditors should prepare audit documentation that contains evidence that supports the findings, conclusions, and recommendations before they issue their report.

6.81 Auditors should design the form and content of audit documentation to meet the circumstances of the particular audit. The audit documentation constitutes the principal record of the work that the auditors have performed in accordance with standards and the conclusions that the auditors have reached. The quantity, type, and content of audit documentation are a matter of the auditors' professional judgment.

6.82 Audit documentation is an essential element of audit quality. The process of preparing and reviewing audit documentation contributes to the quality of an audit. Audit documentation serves to (1) provide the principal support for the auditors' report, (2) aid auditors in conducting and supervising the audit, and (3) allow for the review of audit quality.

6.83 Auditors should document the following:
a. The objectives, scope, and methodology of the audit.
b. The work performed and evidence obtained to support significant judgments and conclusions, including descriptions of transactions and records examined (for example, by listing file numbers, case numbers, or other means of identifying specific documents examined, but copies of documents examined or detailed listings of information from those documents are not required).
c. Supervisory review, before the audit report is issued, of the evidence that supports the findings, conclusions, and recommendations contained in the audit report.

REPORTING VALUE ADDED AUDIT RESULTS TO THE BOARD

The CAE is also responsible for informing the board of directors and senior management of significant risk and problems. The CAE or the value added auditing function does not 'own' the problem. The CAE is responsible for reporting observations and even suggesting solutions to the risks or irregularities. The process owners are responsible for acting on the audit observations, findings, or recommendations. Senior management decides on the risk management and governance strategy they want to pursue.

The head of quality or ISO 9001 systems auditing usually does not have much contact with senior management including the chief financial officer or the board of directors. This will probably change as quality, systems, regulatory, and other types of become consolidated. In order to develop a unified audit report, quality and other operational auditors should communicate critical and significant issues to the CAE or senior management, including these enterprise level issues:

- Fraud.
- Illegal acts.
- Out of control processes.
- Incapable processes to meet requirements.
- Significant risks.
- Poor or no internal controls.
- Significant disagreements.
- Pending corrective and preventive actions.
- Inefficiencies.
- Waste.
- Irregularities.
- Regulatory noncompliance.
- Conflicts of weakness.
- Materiality issues.
- Ethical lapses.

UNDERSTAND CUSTOMER AUDIT REQUIREMENTS

Always do right. This will surprise some people and astonish the rest.
Mark Twain

The value added auditor should always keep in mind organizational objectives and customer requirements when conducting audits. Internal controls minimize risks that impede an organization from reaching its mission critical objectives.

The value added auditor should attempt to:

- Develop communication protocols.
- Develop consistent audit language.
- Understand audit reader requirements.
- Manage expectations.
- Communicate results.
- Use audit report to sell corrective action.
- Distribute the value added audit report.

DEVELOP COMMUNICATION PROTOCOLS

The value added auditor should develop a high standard for reporting audit results involving:

REPORTING STEP 1 – COMMUNICATE AUDIT RESULTS

REPORTING REQUIREMENTS

GAGAS 7.03 Auditors must issue audit reports communicating the results of each completed performance audit.

7.04 Auditors should use a form of the audit report that is appropriate for its intended use and is in writing or in some other retrievable form. For example, auditors may present audit reports using electronic media that are retrievable by report users and the audit organization. The users' needs will influence the form of the audit report. Different forms of audit reports include written reports, letters, briefing slides, or other presentation materials.

7.05 The purposes of audit reports are to

a. Communicate the results of audits to those charged with governance, the appropriate officials of the audited entity, and the appropriate oversight officials.
b. Make the results less susceptible to misunderstanding.
c. Make the results available to the public, unless specifically limited.
d. Facilitate follow-up to determine whether appropriate corrective actions have been taken.

7.06 If an audit is terminated before it is completed and an audit report is not issued, auditors should follow the guidance in paragraph 6.50.

7.07 If, after the report is issued, the auditors discover that they did not have sufficient, appropriate evidence to support the reported findings or conclusions, they should communicate in the same manner as that used to originally distribute the report to those charged with governance, the appropriate officials of the audited entity, the appropriate officials of the organizations requiring or arranging for the audits, and other known users, so that they do not continue to rely on the findings or conclusions that were not supported. If the report was previously posted to the auditors' publicly accessible website, the auditors should remove the report and post a public notification that the report was removed. The auditors should then determine whether to conduct additional audit work necessary to reissue the report, including any revised findings or conclusions or repost the original report if the additional audit work does not result in a change in findings or conclusions.

- **Objective communications.** The audit report is factual, unbiased, and free from distortion. The auditor should include observations, conclusions, and recommendations without prejudice.
- **Clear communications.** The audit report is easily understood and logical. The auditor should avoid unnecessary technical language and should provide sufficient supportive information to improve clarity.
- **Concise communications.** The audit report is to the point and avoids unnecessary details.

OBJECTIVES, SCOPE AND METHODOLOGY

GAGAS 7.09 Auditors should include in the report a description of the audit objectives and the scope and methodology used for addressing the audit objectives. Report users need this information to understand the purpose of the audit, the nature and extent of the audit work performed, the context and perspective regarding what is reported, and any significant limitations in audit objectives, scope, or methodology.

7.10 Audit objectives for performance audits may vary widely. Auditors should communicate audit objectives in the audit report in a clear, specific, neutral, and unbiased manner that includes relevant assumptions. When audit objectives are limited but broader objectives could be inferred by users, auditors should state in the audit report that certain issues were outside the scope of the audit in order to avoid potential misunderstanding.

7.11 Auditors should describe the scope of the work performed and any limitations, including issues that would be relevant to likely users, so that they could reasonably interpret the findings, conclusions, and recommendations in the report without being misled. Auditors should also report any significant constraints imposed on the audit approach by information limitations or scope impairments, including denials or excessive delays of access to certain records or individuals.

7.12 In describing the work conducted to address the audit objectives and support the reported findings and conclusions, auditors should, as applicable, explain the relationship between the population and the items tested; identify organizations, geographic locations, and the period covered; report the kinds and sources of evidence; and explain any significant limitations or uncertainties based on the auditors' overall assessment of the sufficiency and appropriateness of the evidence in the aggregate.

7.13 In reporting audit methodology, auditors should explain how the completed audit work supports the audit objectives, including the evidence gathering and analysis techniques, in sufficient detail to allow knowledgeable users of their reports to understand how the auditors addressed the audit objectives. Auditors may include a description of the procedures performed as part of their assessment of the sufficiency and appropriateness of information used as audit evidence. Auditors should identify significant assumptions made in conducting the audit; describe comparative techniques applied; describe the criteria used; and, when sampling significantly supports the auditors' findings, conclusions, or recommendations, describe the sample design and state why the design was chosen, including whether the results can be projected to the intended population.

- **Constructive communications.** The audit report provides the proper context, content and tone. The report also helps the engagement client and the organization with sufficient and accurate information for intelligent decision-making. The report leads to improvements where needed.

- **Timely communications.** The report is issued without undue delay in order to enable prompt and effective management decision-making.[1]

REPORTING STEP 1 – COMMUNICATE AUDIT RESULTS

DEVELOP CONSISTENT AUDIT LANGUAGE

Value added audit reports can be reported in several ways. One way is to frame the report in terms of negative assurance, specifically that there are noncompliances or observations. Or, the auditor may report that the audit did not uncover any significant weakness in the control system or significant weaknesses in risk management.

Also, the audit report should not be confusing and open to interpretation. Value added auditors should report specific risks, deficiencies, or weaknesses. The auditors should detail, which are significant and pervasive. Auditors should also prioritize audit noncompliances, nonconformances, or findings. Auditors may also issue a qualified or adverse opinion, reporting on the deficiencies. Or, auditors may also report opportunities for improvement (OFIs).

UNDERSTAND AUDIT READER REQUIREMENTS

It is critical to understand the needs and requirements of the reader of the audit report, so it can be tailored to the reader's requirements. For example, operational management has different information needs than senior management or the board of directors who may want a different level of assurance.

So, audit reports have to be targeted, framed and edited to be understandable and informative to the specific reader depending on reader requirements. The audit report may include:

- Noncompliances or nonconformances.
- Controls issues.
- Significant trends.
- Recommendations for improvement.
- Information on risks and priority.
- Patterns of abuse, risk, or nonconformances.
- State of corrective action and preventive actions.
- Patterns of control risk.

MANAGE EXPECTATIONS

Audit stakeholders have different requirements and expectations. Senior management is fundamentally responsible for authorizing and overseeing risk management and control. Operating management is responsible for assigning and controlling risks they own. Auditors are responsible for analyzing risks and assuring the effectiveness of the controls. Value added auditors should manage these requirements and expectations.

A common condition is there may be a 'gap' between the expectations and requirements of the audit report reader and the auditor's allotted time to conduct the audit. Also, the audit reader may have expected a higher level of control, while the auditor may have been time constrained to generate sufficient evidence to support the audit judgments the customer expected.[2]

REPORTING FINDINGS

GAGAS 7.14 In the audit report, auditors should present sufficient, appropriate evidence to support the findings and conclusions in relation to the audit objectives. Clearly developed findings assist management and oversight officials of the audited entity in understanding the need for taking corrective action. If auditors are able to sufficiently develop the elements of a finding, they should provide recommendations for corrective action if they are significant within the context of the audit objectives. However, the extent to which the elements for a finding are developed depends on the audit objectives. Thus, a finding or set of findings is complete to the extent that the auditors address the audit objectives.

7.15 Auditors should describe in their report limitations or uncertainties with the reliability or validity of evidence if (1) the evidence is significant to the findings and conclusions within the context of the audit objectives and (2) such disclosure is necessary to avoid misleading the report users about the findings and conclusions. As discussed in paragraphs 6.69 through 6.72, even though the auditors may have some uncertainty about the sufficiency or appropriateness of some of the evidence, they may nonetheless determine that in total there is sufficient, appropriate evidence given the findings and conclusions. Auditors should describe the limitations or uncertainties regarding evidence in conjunction with the findings and conclusions, in addition to describing those limitations or uncertainties as part of the objectives, scope, and methodology. Additionally, this description provides report users with a clear understanding regarding how much responsibility the auditors are taking for the information.

7.16 Auditors should place their findings in perspective by describing the nature and extent of the issues being reported and the extent of the work performed that resulted in the finding. To give the reader a basis for judging the prevalence and consequences of these findings, auditors should, as appropriate, relate the instances identified to the population or the number of cases examined and quantify the results in terms of dollar value, or other measures. If the results cannot be projected, auditors should limit their conclusions appropriately.

7.17 Auditors may provide background information to establish the context for the overall message and to help the reader understand the findings and significance of the issues discussed. Appropriate background information may include information on how programs and operations work; the significance of programs and operations (e.g., dollars, impact, purposes, and past audit work, if relevant); a description of the audited entity's responsibilities; and explanation of terms, organizational structure, and the statutory basis for the program and operations. When reporting on the results of their work, auditors should disclose significant facts relevant to the objectives of their work and known to them which, if not disclosed, could mislead knowledgeable users, misrepresent the results, or conceal significant improper or illegal practices.

7.18 Auditors should also report deficiencies in internal control, instances of fraud, noncompliance with provisions of laws, regulations, contracts, or grant agreements, or abuse that have occurred or are likely to have occurred and are significant within the context of the audit objectives

REPORTING STEP 1 – COMMUNICATE AUDIT RESULTS

COMMUNICATE RESULTS

Audit results should be distilled in the value added audit report. The value added auditor should use the audit report as an opportunity to communicate results and to influence the customer and auditee to improve organizational systems and controls. Unfortunately, many value added auditors miss an excellent opportunity to influence the organization because:

- Poor audit report packaging.
- Poor organization of the audit report.
- Poor writing.
- Mistakes, typos.
- Extraneous information.
- Auditee disagreement.
- Unsupportable allegations and opinions.
- Poor logic.
- Poor background knowledge.
- Impractical corrective action.

USE AUDIT REPORT TO SELL CORRECTIVE ACTION

The value added audit report may also advocate corrective or prevention action. The audit may highlight information on internal and external customer satisfaction, continual improvement, effectiveness, and cost reduction. The need for action should be evident as well as the consequences for not acting. This type of audit in some ways is a sales tool for value added auditors to promote their organizational value. Not only does it communicate the facts of an operation, but it also encourages company wide risk management and continual process improvement.

The advantages of value added auditing then become evident. The auditor and auditee relationship is collaborative instead of adversarial. The auditee sees the value added auditor as an ally in improving the auditee's work environment and is encouraged to work closely with the value added auditor to convince executive management to invest resources and time to eliminate systemic and chronic problems. The value added auditor may use the audit report to present the point of view of the auditee - for example, to obtain additional resources for equipment or training to eliminate persistent problems. This lessens the auditee's work by eliminating recurring problems, easing the flow of material, or eliminating redundancies. The result is that everyone's quality of work life is improved, risks are mitigated, and organizational competitiveness is enhanced.

The value added auditor should keep in mind that only management can change the underlying organizational system. Sometimes identifying the root cause is not obvious. The value added auditor may not discover the underlying cause except by understanding the pattern of failure. Often, the root cause is under supervisory or managerial control. As we've mentioned, 15% of the activity of the system is within the operator's control and the rest is in management's hands.[3] Management may empower employees and teams, but only senior management can:

- Develop risk plans, registers, goals, and objectives.
- Develop accountabilities, and standards.
- Authorize and allocate resources, including people, information, materials, or equipment.
- Establish risk – control standards or benchmarks.
- Train and develop employees.
- Provide straight messages and rewards.
- Establish internal process management, assurance, and controls.
- Establish an approval system for documentation.
- Insist on continual improvement and enterprise risk management.
- Establish a risk measurement system.
- Establish documentation and approval system.

DISTRIBUTE THE VALUE ADDED AUDIT REPORT

The circulation of the audit report depends on the type of audit. If a value added audit report deals with sensitive or proprietary information, distribution is usually to those with the 'need to know' and required security clearances. On the commercial side, if the value added audit reports contain proprietary material, including product specifications that could give a competitor an advantage, these reports are distributed to specified and approved parties.

As a rule, the value added auditor should circulate value added audit reports to a predetermined list of people, including the customer, the supervisor, and the auditee. Each recipient of an audit report should know the distribution policy. Distribution for different types of value added reports is established in a policy statement. Another type of value added audit report is the continual process improvement report, which should be distributed widely. It focuses more on customer satisfaction and continual improvement than on deficiencies. There is nothing secret or hidden about these reports. The important point is that their tone is constructive and not blaming.

REPORTING VALUE ADDED AUDITS FINDINGS

Measure a thousand times and cut once.
Proverb

There are few hard reporting requirements for value added auditors. Common sense, risk levels, and assurance requirements dictate the level and type of reporting. A signed, written report is issued after the audit is completed. Interim reports may be written or oral and may be transmitted formally or informally.

The following discusses various reporting challenges:

- Reporting improvement.
- Reporting opportunities for improvement.
- Reporting findings.

- Reporting nonconformances.
- Reporting corrective action.
- Reporting fraud and negligence.
- Reporting out-of-control conditions.

REPORTING IMPROVEMENTS

The language of a value added audit is a discussion or conversation with the auditee. The auditor may first discuss what is being done right and improvements from the last audit.

"... Critical process improvements were made in software development using the concurrent engineering methodology. Software defects declined 25% and time to market was decreased 10%".

"... Process capability increased in the screw machining area from 1.00 to 1.33."

"... Customer satisfaction at the help desk increased from 85% to 89% in the last quarter."

REPORTING OPPORTUNITIES FOR IMPROVEMENT

The auditor in the value added audit might make 'observations' and might recommend opportunities for improvement such as:

".... The accounts payable (AP) process has undergone certain improvements. The AP process should be flowcharted again to indicate the recent changes."

"... There is an increase in machinist turnover in the welding area. New welders need to be trained on all the machines."

" ... New robotic wire welders have been installed on the manufacturing floor. Welders need to be trained on amperage control and feed rate. "

REPORTING FINDINGS

The findings section of the value added audit report distills observations and opinions derived during the conduct of the audit. The findings reflect the value added auditor's professional assessment of what was reviewed. The value added auditor may state that systems and processes were operating satisfactorily or unsatisfactorily.

A finding can report a situation or a condition that indicates risk, does not conform or shows some variance from a business objective, metric, standard, drawing, procedure, policy, purchase order, instruction, contract, or other performance standard. A finding usually relates to a critical or major problem, which needs to be corrected so it does not recur, therefore, the focus is on prime findings or root causes, not symptoms or secondary causes.

REPORTING STEP 1 – COMMUNICATE AUDIT RESULTS

REPORTING CONCLUSIONS

GAGAS 7.27 Auditors should report conclusions based on the audit objectives and the audit findings. Report conclusions are logical inferences about the program based on the auditors' findings, not merely a summary of the findings. The strength of the auditors' conclusions depends on the sufficiency and appropriateness of the evidence supporting the findings and the soundness of the logic used to formulate the conclusions. Conclusions are more compelling if they lead to the auditors' recommendations and convince the knowledgeable user of the report that action is necessary.

A useful starting point in any value added audit is to summarize past deficiencies, areas of concern, and results of corrective action requests. The value added auditor may inquire and review the effectiveness of past corrective and preventive actions. These items may indicate problems waiting to occur or high risk areas where problems should never occur.

As much as possible, the value added auditor should prioritize audit findings. If there are many critical or major deficiencies, then the auditor can use a Pareto analysis to prioritize them. The Pareto principle basically asserts that 20 percent of the items cause 80 percent of the problems. This focuses the reader's attention. Deficiencies may be prioritized in terms of critical, major, and minor risk factors. Critical and major deficiencies deal with health, safety, and customer satisfaction, and these require immediate attention and resolution. A time requirement is often stated. Chronic, systemic, and systematic problems are also confronted and dealt with quickly. They can also be prioritized in terms of opportunities of improvement.

Are general observations or opportunities for improvement addressed in the value added audit? This is handled differently depending on the industry and the way audits have evolved. Observations describe the general nature of the area or organization being audited. In some industries, observations are a separate section after the introduction.

Some value added auditors conduct audits by the book. They follow a philosophy of 'just the facts.' These audits state whether risk - controls are in place and working properly, and both minor and major deficiencies are identified and corrective action is requested. There is little personal opinion or judgment beyond this. In other audits, the value added auditor may be required to write a mini-essay on each major deficiency, referring to the tests that were conducted, standards used, who conducted the test, and recommended corrective actions.

REPORTING NONCONFORMANCES

After discussing the issue with process owners, then there are ways to report a nonconformance, such as:

REPORTING RECOMMENDATIONS

GAGAS 7.28 Auditors should recommend actions to correct deficiencies and other findings identified during the audit and to improve programs and operations when the potential for improvement in programs, operations, and performance is substantiated by the reported findings and conclusions. Auditors should make recommendations that flow logically from the findings and conclusions, are directed at resolving the cause of identified deficiencies and findings, and clearly state the actions recommended.

7.29 Effective recommendations encourage improvements in the conduct of government programs and operations. Recommendations are effective when they are addressed to parties that have the authority to act and when the recommended actions are specific, practical, cost effective, and measurable.

"... We discussed the issues with process management and they agree with these conclusions. In our opinion, internal controls in accounts receivable should be improved with more checks and balances."

"... In our opinion, more resources should be dedicated to the help desk during peak hours. This will decrease the queue and will increase customer satisfaction."

Sometimes, the value added auditor can't smooth over the situation and must state the obvious in an adverse assessment. In such situations, the value added auditor states the requirements or objectives and states how activities don't conform to requirements such as:

"... New welders for ASME pressure vessels are not certified per code. The code states: '....'"

"... Systemic and chronic safety conditions exist at the construction site. OSHA and WISHA require the following: '....'"

"... Effective internal controls are not in place in Contracts Administration. These compromise the effective control of monies and result in a high risk of fraud."

"... Segregation of duties does not exist and signature authority is not defined in accounts receivable. Corporate procedures state: '...'"

REPORTING CORRECTIVE ACTION

Corrective action recommendations and opinions flow from findings. Findings are the facts that result from the auditor's investigations. They are the end product of fieldwork. Findings confirm conformance or detail deficiencies. The former requires little or no explanation. Often, a favorable condition is a 'yes' response on the questionnaire. A negative finding requires further investigation and corrective/prevention action.

FRAUD, NONCOMPLIANCE WITH PROVISIONS OF LAWS, REGULATIONS, CONTRACTS, AND GRANT AGREEMENTS AND ABUSE

GAGAS 7.21 When auditors conclude, based on sufficient, appropriate evidence, that fraud noncompliance with provisions of laws, regulations, contracts or grant agreements, or abuse either has occurred or is likely to have occurred which is significant within the context of the audit objectives, they should report the matter as a finding. Whether a particular act is, in fact, fraud or noncompliance with provisions of laws, regulations, contracts or grant agreements may have to await final determination by a court of law or other adjudicative body.

7.22 When auditors detect instances of fraud, noncompliance with provisions of laws, regulations, contracts, or grant agreements, or abuse that are not significant within the context of the audit objectives but warrant the attention of those charged with governance, they should communicate those findings in writing to audited entity officials. When auditors detect any instances of fraud, noncompliance with provisions of laws, regulations, contracts, or grant agreements, or abuse that do not warrant the attention of those charged with governance, the auditors' determination of whether and how to communicate such instances to audited entity officials is a matter of professional judgment.

7.23 When fraud, noncompliance with provisions of laws, regulations, contracts, or grant agreements, or abuse either have occurred or are likely to have occurred, auditors may consult with authorities or legal counsel about whether publicly reporting such information would compromise investigative or legal proceedings. Auditors may limit their public reporting to matters that would not compromise those proceedings and, for example, report only on information that is already a part of the public record.

The audit customer usually wants to know the level and type of risks as well as the effectiveness of compliance, performance, and internal control. The customer also wants to know why a deficiency occurred. This is not always easy. As projects, products, processes, and organizations become complex, there is seldom only one causal factor. The prime cause can be hidden by many minor causes that have to be understood and removed before the major cause or causes are uncovered. Determining the factors can require a multidisciplinary team to determine the root cause.

This leads to an important question: 'How does the auditor know if the issue is a prime or secondary cause?' Sometimes, it is obvious. For example, if a pump fails, the obvious cause may be no lubricant. Further investigation indicates the pump sealant decomposes when the mixture, a slurry is pumped. A second similar failure indicates that the maintenance operation does not adjust the packing and monitor the sealant. The prime cause is therefore poor or no preventive maintenance.

REPORTING FRAUD AND NEGLIGENCE

Fraud or gross negligence unfortunately exists in the imperfect world of auditing. There are many types of fraud, including:

- Fraudulent certification reports.
- Edited statistical process control charts.
- Lack of full disclosure.

REPORTING DEFICIENCIES IN INTERNAL CONTROL

GAGAS 7.19 Auditors should include in the audit report (1) the scope of their work on internal control and (2) any deficiencies in internal control that are significant within the context of the audit objectives and based upon the audit work performed. When auditors detect deficiencies in internal control that are not significant to the objectives of the audit but warrant the attention of those charged with governance, they should include those deficiencies either in the report or communicate those deficiencies in writing to audited entity officials. Auditors should refer to that written communication in the audit report if the written communication is separate from the audit report. When auditors detect deficiencies that do not warrant the attention of those charged with governance, the determination of whether and how to communicate such deficiencies to audited entity officials is a matter of professional judgment.

7.20 In a performance audit, auditors may conclude that identified deficiencies in internal control that are significant within the context of the audit objectives are the cause of deficient performance of the program or operations being audited. In reporting this type of finding, the internal control deficiency would be described as the cause.

- Incorrect financial reporting.
- Covered-up deficiencies or risks.
- Fake tests.
- Unreasonable responses to self-audit questions.
- Overbilling.
- Feigned annoyance to reasonable inquiries and searches.
- Rewriting and predating report.
- Failure to record information.

These result in precarious and risky situations, especially when trying to improve systems and finding intentional deficiencies. In external audits, these are easier to handle because the value added auditor has less risk of betraying confidences or ending up in a political quandary.

If the auditor suspects fraud or gross negligence, the value added auditor should be especially cautious. The auditor shouldn't finger point and immediately accuse someone of transgressions. The auditor should retrace the audit and check for mistakes or oversights. If the results are the same, the auditor should consult with the chief audit executive, supervisor or a high-level manager. This person should be able to advise the value added auditor of the next actions. Nothing can affect the auditor's credibility, as well as the credibility of the value added audit, than to be wrong in this circumstance.

REPORTING OUT-OF-CONTROL CONDITIONS OR DEFICIENCIES IN INTERNAL CONTROL

There are many possible causes of out of control, risky, or unexpected variation in materials, operators, environment, machines, or methods process variables. For example, the specific causes may be due to:

- Lack of plans, goals, objectives, and, accountabilities.

- Lack of internal controls.
- Lack of resources, including people, information, materials, equipment.
- No standards or benchmarks.
- Poor or no training and development.
- Supervision providing mixed messages and no rewards.
- No or inadequate internal process management, assurance, and controls.
- No or inadequate audit or corrective/preventive action processes.
- No or inadequate approval or escalation system.
- No or inadequate continual improvement.
- No or inadequate measurement system.
- No or inadequate documentation systems.

These causes are basically organizational. As W. Edwards Deming said, the operator or person on the line has little control over these factors. The person on the line only has the resources at his or her disposal. Management must be involved if the fundamental system or process must be changed.

POST AUDIT REVIEW

Trust, but verify.
President Ronald Reagan

Value added audit reports can go through several drafts. Especially audits dealing with environmental, safety and health issues may go through extensive reviews until wording and different views have been distilled so they reflect consensus and a balanced point of view. It makes sense to have someone with special knowledge review the audit.

WHO SHOULD REVIEW THE VALUE ADDED AUDIT REPORT

The following may review the entire audit or parts of the value added audit report:

- Auditee.
- Interviewees.
- Managers and supervisors.
- Process owners.
- Upstream departments.
- Internal customers.
- External stakeholders.
- Inspection, testing, and laboratory personnel.
- Stakeholders involved with corrective/preventive action.

ASKING THE AUDITEE TO REVIEW THE REPORT

The courtesy the auditor shows and the assistance the auditor gives the auditee are important for maintaining the culture of continual improvement and risk management. If the value added auditor conducted the audit cooperatively, fairly, and objectively, there should be no rancor, excuses, or denials. So, the value added auditor may ask the auditee to review the audit report. The purpose of asking the auditee to review the audit

REPORTING VIEW OF RESPONSIBLE OFFICIALS

GAGAS 7.32 Auditors should obtain and report the views of responsible officials of the audited entity concerning the findings, conclusions, and recommendations included in the audit report, as well as any planned corrective actions.

7.33 Providing a draft report with findings for review and comment by responsible officials of the audited entity and others helps the auditors develop a report that is fair, complete, and objective. Including the views of responsible officials results in a report that presents not only the auditors' findings, conclusions, and recommendations, but also the perspectives of the responsible officials of the audited entity and the corrective actions they plan to take. Obtaining the comments in writing is preferred, but oral comments are acceptable.

7.34 When auditors receive written comments from the responsible officials, they should include in their report a copy of the officials' written comments, or a summary of the comments received. When the responsible officials provide oral comments only, auditors should prepare a summary of the oral comments and provide a copy of the summary to the responsible officials to verify that the comments are accurately stated.

7.35 Auditors should also include in the report an evaluation of the comments, as appropriate. In cases in which the audited entity provides technical comments in addition to its written or oral comments on the report, auditors may disclose in the report that such comments were received.

7.36 Obtaining oral comments may be appropriate when, for example, there is a reporting date critical to meeting a user's needs; auditors have worked closely with the responsible officials throughout the work and the parties are familiar with the findings and issues addressed in the draft report; or the auditors do not expect major disagreements with the findings, conclusions, and recommendations in the draft, or major controversies with regard to the issues discussed in the draft report.

7.37 When the audited entity's comments are inconsistent or in conflict with the findings, conclusions, or recommendations in the draft report, or when planned corrective actions do not adequately address the auditors' recommendations, the auditors should evaluate the validity of the audited entity's comments. If the auditors disagree with the comments, they should explain in the report their reasons for disagreement. Conversely, the auditors should modify their report as necessary if they find the comments valid and supported with sufficient, appropriate evidence.

7.38 If the audited entity refuses to provide comments or is unable to provide comments within a reasonable period of time, the auditors may issue the report without receiving comments from the audited entity. In such cases, the auditors should indicate in the report that the audited entity did not provide comments.

report is more than simple courtesy, it also is another form of internal control to assure that the report is accurate, fair, and complete. The value added auditor may have:

- Made a mistake.
- Misinterpreted a situation.

REPORTNG RISK CONTROL DEFICIENCIES

GAGAS A.06 The following are examples of control deficiencies:

a. Insufficient control consciousness within the organization. For example, the tone at the top and the control environment. Control deficiencies in other components of internal control could lead the auditor to conclude that weaknesses exist in the control environment.
b. Ineffective oversight by those charged with governance of the entity's financial reporting, performance reporting, or internal control, or an ineffective overall governance structure.
c. Control systems that did not prevent, or detect and correct material misstatements so that it was necessary to restate previously issued financial statements or operational results. Control systems that did not prevent or detect material misstatements in performance or operational results so that it was later necessary to make significant corrections to those results.
d. Control systems that did not prevent, or detect and correct material misstatements identified by the auditor. This includes misstatements involving estimation and judgment for which the auditor identifies potential material adjustments and corrections of the recorded amounts.
e. An ineffective internal audit function or risk assessment function at an entity for which such functions are important to the monitoring or risk assessment component of internal control, such as for a large or complex entity.
f. Identification of fraud of any magnitude on the part of senior management.
g. Failure by management or those charged with governance to assess the effect of a significant deficiency previously communicated to them and either to correct it or to conclude that it does not need to be corrected.
h. Inadequate controls for the safeguarding of assets.
i. Evidence of intentional override of internal control by those in authority to the detriment of the overall objectives of the system.
j. Deficiencies in the design or operation of internal control that could fail to prevent, or detect and correct, fraud, noncompliance with provisions of laws, regulations, contracts or grant agreements, or abuse having a material effect on the financial statements or the audit objective.
k. Inadequate design of information systems general, application, and user controls that prevent the information system from providing complete and accurate information consistent with financial, compliance, or performance reporting objectives or other current needs.
l. Failure of an application control caused by a deficiency in the design or operation of an information systems general control.
m. Employees or management who lack the qualifications and training to fulfill their assigned functions.

- Not received sufficient information.
- Made a calculation error.
- Sampled an unrepresentative population.
- Not understood nuances especially dealing with people.
- Not understood the political forces within the organization.
- Not identified or understood risks and internal controls.

BENEFITS OF AUDIT REVIEW

If the value added auditor only evaluates an area periodically, then the auditor may not understand or have all the information of what occurs between audits. In general, the

REPORTING CONFIDENTIAL AND SENSITIVE INFORMATION

GAGAS 7.39 If certain pertinent information is prohibited from public disclosure or is excluded from a report due to the confidential or sensitive nature of the information, auditors should disclose in the report that certain information has been omitted and the reason or other circumstances that make the omission necessary.

7.40 Certain information may be classified or may be otherwise prohibited from general disclosure by federal, state, or local laws or regulations. In such circumstances, auditors may issue a separate, classified or limited use report containing such information and distribute the report only to persons authorized by law or regulation to receive it.

7.41 Additional circumstances associated with public safety, privacy, or security concerns could also justify the exclusion of certain information from a publicly available or widely distributed report. For example, detailed information related to computer security for a particular program may be excluded from publicly available reports because of the potential damage that could be caused by the misuse of this information. In such circumstances, auditors may issue a limited use report containing such information and distribute the report only to those parties responsible for acting on the auditors' recommendations. In some instances, it may be appropriate to issue both a publicly available report with the sensitive information excluded and a limited use report. The auditors may consult with legal counsel regarding any requirements or other circumstances that may necessitate the omission of certain information.

7.42 Considering the broad public interest in the program or activity under audit assists auditors when deciding whether to exclude certain information from publicly available reports. When circumstances call for omission of certain information, auditors should evaluate whether this omission could distort the audit results or conceal improper or illegal practices.

7.43 When audit organizations are subject to public records laws, auditors should determine whether public records laws could impact the availability of classified or limited use reports and determine whether other means of communicating with management and those charged with governance would be more appropriate. For example, the auditors may communicate general information in a written report and communicate detailed information orally. The auditor may consult with legal counsel regarding applicable public records laws.

auditee's review of the final report before its distribution allows the value added auditor to:

- Obtain more information.
- Understand the auditee's position.
- Resolve differences.
- Agree on the facts.
- Review report findings and opinions.
- Add rebuttal comments to the report.
- Initiate corrective action.

Preliminary findings can be discussed with the auditee's area supervision and also at the working level - the manufacturing operator, administrative assistant, sales representative or a professional person. The reason for discussing issues at the working level is that responsibility and authority for work is being pushed to the lowest organizational level that can handle problems. It is commonly thought that these process owners have first-hand knowledge of the problem and they are responsible for eliminating its recurrence.

General issues or questions that can come up are:

- Is the risk, deficiency or nonconformance critical or major?
- Why didn't internal risk - controls detect the problem?
- Who is responsible for the nonconformance?
- Is the nonconformance systemic or isolated?
- Is it statistically irrelevant?
- Is the control system operating effectively?
- How can the nonconformance be corrected so it does not recur?

RESOLVING DIFFERENCES

If after reviewing the audit report, the auditee offers explanations and perspectives that differ from the auditor and these cannot be resolved, these can be incorporated into the report. The auditee's response can be a virulent attack on the auditor's competence or can be a defensive denial of the situation described by the auditor. If these occur often, a message is being communicated. One interpretation is that the auditee's organization is not ready for value added auditing and does not believe the risk management or continual improvement ethic. Another interpretation may reflect on the ability or character of the auditor. In the prior case, nothing can be done because it is up to the board and senior management to establish and reinforce a consistent quality, risk management, and internal control ethic. In the latter case, the auditor's performance should be evaluated and if deficient should be addressed.

When the auditee has received the formal report, the auditee is expected to respond to each of the findings, recommendations, and conclusions. If the auditee does not agree with the audit's conclusions or recommendations then the auditee can present countervailing evidence. If this happens often, then the purpose and use of the value added audit is being subverted and it becomes useless.

SUMMARY

The value added report should communicate information that is aligned with the organization's mission critical objectives and addresses the specific objectives of the audit that were developed in the audit scope or brief statement.

The value added auditor may be assessing systems or processes that are new to him or her. Since more value added audits are participatory, the auditor may want to share the audit results and conclusions with the auditee or other stakeholders for review. The value added auditor may incorporate the stakeholder's comments in the final report or address them in the final report to the audit customers.

REPORTING STEP 1 – COMMUNICATE AUDIT RESULTS

The following are general suggestions for reporting and communicating value added audit results:

- Value added auditor should discuss conclusions and recommendations at appropriate levels of management before issuing final written reports.
- Reports are objective, clear, concise, constructive, and timely.
- Reports present the purpose, scope, and results of the audit.
- Reports support and may contain an expression of the auditor's opinion.
- Reports may include recommendations for potential improvement, acknowledge satisfactory performance and report corrective/preventive action.
- Auditee's views about audit conclusions or recommendations may be included in the audit report.
- Chief audit executive or designee should review and approve the final report.
- Chief audit executive should decide on audit report distribution.[4]

CHAPTER 20
STEP 2: DECIDE AUDIT REPORT FORMAT

An audit report can be a 1-page memo or a 50-page report with attached test results. The format and length of the report depend on customer requirements. In this section, various commonly used value added audit formats are discussed as well as the Malcolm Baldrige National Quality Award (MBNQA) report format.

TIPS ON WRITING THE AUDIT REPORT

Just the facts, ma'am.
Dragnet TV show

The only fixed rule for writing a value added audit report is that the report must add value to the audit customer. However, there are some common sense guidelines the value added auditor may want to consider, such as:

- Establish the proper tone of the final report.
- Make the audit report understandable.
- Provide information that people can act upon.
- Use graphics.
- Present independent and objective information.
- Benchmark the company against the best.
- Package audit report to communicate professionalism and authority.
- Never rush to judgment.
- Deliver the report on time.

ESTABLISH THE PROPER TONE OF FINAL REPORT

The value added auditor should establish the proper tone in the audit report. In the value added audit, the customer and the auditee are partners in the risk management and continual process improvement effort, so the tone of the report can be friendly and informal. If the audit is of an external supplier, the tone can be more formal. If the audit investigates a health or safety issue where there is a risk of loss of life, then the tone of the report is more formal. The auditor should use common sense as a guide.

The tone of the overall assessment depends on the audit's objective. A regulatory or statutory compliance audit has a different 'look and feel' than a value added audit. The tone of the compliance audit uses enforcement language such as 'findings,' and 'non-conformance reports.'

Another important factor is to focus on issues that are important to the organization. Each organization has its hot issues. Issues important to an organization are risk management, product quality, customer satisfaction, productivity, and cycle time reduction.

MAKE THE AUDIT REPORT UNDERSTANDABLE

If the subject of the audit is technical and very complex, the value added auditor should simplify it so that it is understandable to the reader, the customer. Management often suffers from information overload from data that needs to be distilled.

Audit clarity is achieved by organizing information logically and writing clearly. Organization can be learned by following the formats recommended in this book. Clear writing means stating results simply and clearly and is a natural result of understanding the auditee's systems and processes. If the auditor doesn't understand the auditee's quality processes or systems and doesn't have a grasp of auditing fundamentals, the value added auditor won't be able to communicate information clearly to the customer.

PROVIDE INFORMATION THAT PEOPLE CAN ACT UPON

Management wants information that can be acted upon. Sometimes, this is called 'do-able' information. There is nothing more frustrating than to read or listen to a debriefing of an excellently planned and conducted audit just to find that corrective action is impossible or too expensive. This is commonly caused by the auditor not consulting with the auditee so as to jointly develop practical corrective action and recommendations.

USE GRAPHICS

Pictures and figures illustrate concepts and resource flows much better than words. Desktop publishing (DTP) makes professional looking reports and illustrations easy. DTP is accessible to many small companies and departments. Even photographs can be digitized and used in the audit report.

PRESENT INDEPENDENT AND OBJECTIVE INFORMATION

The customer is relying on the auditor for accurate, complete, and current information. By providing this, the value added auditing group gains a reputation for service, quality, and reliability. The customer can rely on the report because of documentation and logic. The auditor bases opinions on observed facts, interviews, or tests. Statements as much as possible are corroborated by others or by test results. The value added auditor wants to know that statements are factual.

BENCHMARK THE COMPANY AGAINST THE BEST

The value added auditor provides an important service to the organization. So, the value added auditor should practice what he or she preaches. The auditor should strive to continually improve, and if possible, benchmark against the best auditing practices in other industries.

REPORTING STEP 2 – DECIDE AUDIT REPORT FORMAT

PACKAGE AUDIT REPORT TO COMMUNICATE PROFESSIONALISM AND AUTHORITY

A well-packaged value added audit report communicates professionalism and authority. A poorly packaged report can decrease its image, even if the information in the report is excellent. If the report is long, then it should be presented professionally. If the report is longer than 5 pages, it can be hard or paper cover, comb or spiral bound. Again, the purpose is to make the value added auditing report inviting to read and have it exude authority.

NEVER RUSH TO JUDGMENT

Value added auditing requires profound knowledge of the variables in the process, including people, products, suppliers, materials, and many other variables. This profound knowledge develops over time and comes from experience, education, and persistence. What seems to be the situation or the cause of a nonconformance on the surface may be different the more the auditor understands the variables of the process.

Also the auditor should fight the tendency of 'obviousness' such as reporting: 'It's obvious that this is the situation and this is my recommendation.' There may be many independent and dependent process variables at work. What is the main cause of variation and risk may require more study.

DELIVER THE REPORT ON TIME

The customer or the organization requested an audit because there is a need for specific information. Time-to-delivery is an important quality characteristic of all service providers. There is now more need for this. If the customer is to be satisfied, reports should be well thought out and delivered quickly. The laptop computer and Internet facilitate communicating audits results quickly.

A value added audit may require immediate action if it provides the customer with important information. If there is excessive delay in sending the report to the customer, then its value depreciates. How long should it be from the end of the audit until it is distributed? Some companies insist that it is within 2 weeks or less. More often, there is a drive for just-in-time audit reports. Reports are developed on a laptop and sent to headquarters via email. Assistant auditors or administrative assistants tidy up the report on a word processor and send the report to the customer within 1 day. This assumes that the report does not need major alterations.

TYPES OF AUDIT REPORTS

Less is more. God is in the details.
Mies van der Rohe

The format of value added audits depends on the type of audit and the special requirements of the customer. A regulatory customer may require that the value added auditor

MILITARY PROCUREMENT AUDIT REPORT[1]

To: Director of Procurement
From: Director of Quality Assurance
Subject: On-Site Assessment of XYZ Manufacturing Co.

1. An on-site assessment was conducted at XYZ Manufacturing Company from AAA to BBB. XYZ produces the M-20 Machine Gun and repair parts under contract EEE-A-1234. The assessment was conducted in response to the contractor's request for certification approval under provisions of quality standard ZI and contract EEE-A-1234. See enclosure 1 for list of people interviewed. See enclosure 2 for list of assessment team members.

2. Overall Assessment Summary and Ratings:

a. **Summary.** The assessment identified several deficiencies that preclude awarding certified status to XYZ Manufacturing Company at this time; however, XYZ is conditionally accepted on a 6-month basis contingent of significant progress being made to achieve the objectives of the Continual Process Improvement Program. The contractor should request a follow-up audit within 6 months. Of five rating quality systems, two did not receive the minimum score that warrants contractor certification. Engineering processes and manufacturing processes were below the minimum rating criteria. Rating sheets are enclosed. Results from the ratings are summarized below:

b. **Engineering processes.** This category received a score of 70% - 80% is required for certification. The primary problem is that Engineering does not fully use an Engineering Control Board (ECB) so engineering changes and distribution of the changes is late and not monitored. Also, design and calculation reviews are weak.

c. **Manufacturing processes.** This category received a score of 75% - 80% is required for certification. Critical and major product quality characteristics are capable to Cpk = 1.33. Approximately 50% of the processes are automated. Tooling and fixtures are well designed. Of the total processes 50% are capable to Cpk = 1.33. Quality standard specifies that for certification, 60% must be capable to this level.

d. **Control of processes.** This category received a score of 85%. Gages and inspection equipment are available, stored securely, and calibrated. Most critical and major processes are in control. 65% of the overall manufacturing processes are in control.

e. **Engineering/manufacturing support.** This category received a score of 85%. Procedures, documentation, and quality manual were extensive, accurate, and current. Quality assurance supports all of the organization's quality efforts. Quality metrics are extensive. However, automation is not fully utilized in capturing SPC data. Material Requirement Planning is weak. Several deliveries did not meet the master schedule.

f. **Supplier control.** This category received a score of 85%. Approximately 65% of the manufacturing dollar is outsourced. Contractor initiated quality effort internally and simultaneously with key suppliers. Supplier quality program has existed for 4 years. Suppliers of products with critical and major quality characteristics are controlling processes and are capable to a Cpk = 1.0

3. Four representative parts were selected, and contractor's process sheets were used to track the parts through the engineering design and manufacturing process. The four parts are based on the Weapon Part Name Part Number

- M20MG Receiver 1234.
- M20MG Bolt 1245.
- M20MG Barrel 1235.
- M20MG Pin 2345.

REPORTING STEP 2 – DECIDE AUDIT REPORT FORMAT

STANDARD AUDIT REPORT

To: Customer Name
From: Lead Auditor
Subject: Audit number, title, subject, auditee's name
Date: Audit report date
Audit implementation dates:
References: List of documentation referenced in report

Summary
Summary of significant findings, observations, recommendations. If necessary, corrective actions are also listed.
Introduction
The introduction describes the scope and purpose of the audit. Specifically, the following is addressed:
Who is the customer?
When was the audit conducted?
Who conducted the audit?
Where was it conducted?
Why was it conducted?
What are the audit objectives?
Findings
Deficiencies and risks are listed and prioritized. For each deficiency and/or risk, the finding refers to a specific requirement or objective not being satisfied. Control weaknesses are also listed.
Corrective Action
Corrective action is requested or recommended to remove symptoms, eliminate root causes, or improve the system within a set time. The responsible manager or team is identified in each finding.
Background Information
Background information includes:
Statement of the audit purpose, scope, and objectives.
Location and times of audit.
Areas investigated.
Methodology and tests used.
People interviewed.
Supplementary Information
This section includes the following:
Auditee's rebuttal.
Additional information (test, inspection, etc.).
Results of preliminary corrective or preventive action.

Signature:
Lead Auditor
Enclosures:

follow a prescribed audit outline or an internal engineering group may want a report with test results. This section discusses the following types of audit reports:

- Standard audit report.
- Prescribed (regulatory) format.
- Abbreviated audit.
- Qualitative risk management audit.
- Progress audit report.
- Exception report.
- Corrective action request (CAR).
- Audit assessment rating plan.
- Post audit survey.

STANDARD AUDIT REPORT

For more extensive or intensive value added audits, there are certain elements that are common to all audits, specifically:

- Summary.
- Introduction.
- Findings.
- Corrective action.
- Background information.
- Supplementary information.

A value added audit report starts with a formal executive summary, which highlights important findings, developments, and opinions. Not all elements of the audit report have to be addressed. However the audit customer or stakeholders should have sufficient, accurate and reliable information about the audit objectives to make intelligent decisions. Some value added auditors use this format as a letter or memo of transmittal communicating the completion of the value added audit and reporting significant findings. Depending on interest, the customer may then request further information. The 'Standard Audit Format' figure illustrates the key questions to be addressed at each part of the value added audit.

PRESCRIBED (REGULATORY) FORMAT

The 'Military Procurement Audit Report' sidebar on the previous page shows the results of auditing a supplier that has just applied for certification. Military contractors as well as many commercial suppliers are expected or required to participate in a certification process. In this process, suppliers start as candidates and move on to approved, preferred, certified, and partner levels. Sometimes these levels are called by different names.

REPORTING STEP 2 – DECIDE AUDIT REPORT FORMAT

ABBREVIATED REPORT

To: Director of Engineering
From: Lead Auditor
Subject: Project X
Date: X-X-XX
Auditee's name: Parts Plant I
Audit number: 123
Audit conducted on: X-X-XX
Audit objective: Observation of tensile tests on five specimens by independent, Smith Testing Laboratory performed on 1-1-22.
Audit plan: Not Required (N/A)
Process/product name: Raw material from Supplier X, identification code ab-bc-de.
Contact person: Lab supervisor, Tom Smith (123) 456- 7890.
Audit Findings: Per engineering specification 10-AA-B, Product Tensile, Smith Testing Laboratory received, reviewed, and conducted tensile tests on five sample specimens per the above engineering specification. The specimens met requirements. Charts are appended to this report.
Is corrective action request required: No
If yes: N/A
Prepared by: Jane Seymore
Lead Auditor

Enclosures:
1. Tensile test results of specimens 23, 24, 25, 26, 27

Regulatory authorities or certain customers can specify audit format, length, appended material, and other factors. The 'Military Procurement Report' sidebar on page 376 shows the recommended report format for a government contractor. This report has elements of many audit reports. In this report, five processes are evaluated:

- Engineering processes.
- Manufacturing processes.
- Control of processes.
- Engineering/manufacturing support.
- Supplier control.

These rating plans, like the Malcolm Baldrige National Quality Award, are common in value added auditing and are discussed later in the chapter. The goal of these certification efforts is company wide continual improvement.

Improvement is based on scores from a questionnaire. Each category has a maximum number of points. The auditee, usually a supplier, is expected to achieve a specific number of points in each category. In the certification process, a supplier moves from candidate to certified to supply-partner, the supplier achieves a higher score at each hurdle. The following is a typical list of ratings, based on a total of l00 percent:

- Excellent requires 90 percent or more.

- Very good requires 80 percent or more.
- Good requires 70 percent or more.
- Fair requires 50 percent or more.
- Poor is 49 percent or less.

The 'Military Procurement Audit' sidebar illustrates the format of a typical prescribed audit format. If the value added auditor practices in the public sector then the auditor should seek information on recommended audit formats which may be described in the customer's handbook, policy, or standard.

ABBREVIATED AUDIT

An abbreviated audit is a limited, focused assessment of specific information requested by the customer and is used to report on:

- Observing a test.
- Checking a process capability index.
- Interviewing people.
- Following up on corrective action.
- Tracking nonconforming material.
- Checking a measurement equipment's calibration.

This type of audit report can be a verbal report with a memo sent to file or it can be a brief written memo to the customer as shown in the 'Abbreviated Report' sidebar on the previous page.

QUALITATIVE RISK MANAGEMENT AUDIT

A qualitative risk management audit is shown on the next page. An auditor or analyst evaluates the 'likelihood' and 'consequence' of a risk event. The analyst evaluates each on a 5 point scale. Based on the evaluation, the risk event is mapped and is 'high,' 'medium,' or 'low.'

Qualitative enterprise risk management audits verify and validate enterprise risks, programmatic/process/project risks, and product/event/transactional risks. These were discussed at length in Chapter 4 of this book. Again, the value added auditor does not determine risk appetite or develop risk-controls. The auditor verifies and validates the appropriate controls are in place controlling risks within the risk tolerance of the organization or process.

The qualitative risk management report can be issued using the simple qualitative Risk Management Audit form shown on the next page. However, if more granularity or specificity is required, then additional data to support the risk management audit form can be gathered. The following pages provide forms and decision criteria that the auditor or process owner can use to evaluate risks and controls.

QUALITATIVE RISK MANAGEMENT AUDIT

Process/Program/Project Title:

Date:

Process/Program/Project Owner:

Analyst:

Risk Description:

Symptom and Root Cause: ______________________________

Likelihood: ______________________________

Consequence: ______________________________

Risk Analysis:

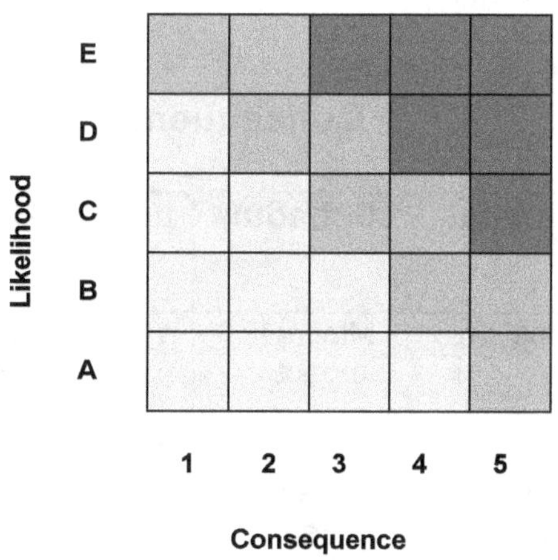

High Risk – Points ______

Medium Risk – Points ______

Low Risk – Points ______

Risk Analysis

____

____

Risk Management Strategies:

- ❑ Avoid risk
- ❑ Mitigate risk
- ❑ Accept risk
- ❑ Share risk
- ❑ Diversify risk
- ❑ Control risk
- ❑ Increase risk

Risk Resolution Date:

Description of Implementation:

Assessment of Implementation Effectiveness:

Post Implementation Risks

Likelihood: ______________________________

Consequence: ______________________________

REPORTING STEP 2 – DECIDE AUDIT REPORT FORMAT

Likelihood

Level		Approach and Process
E	Near Certainty	Cannot mitigate this type of risk; no known processes; or workarounds are available.
D	Highly Likely	Cannot mitigate this risk, but a different approach might.
C	Likely	May mitigate this risk, but workarounds will be required.
B	Low Likelihood	Have usually mitigated this type of risk with minimal oversight in similar cases.
A	Not Likely	Will effectively avoid or mitigate this risk based on standard practices.

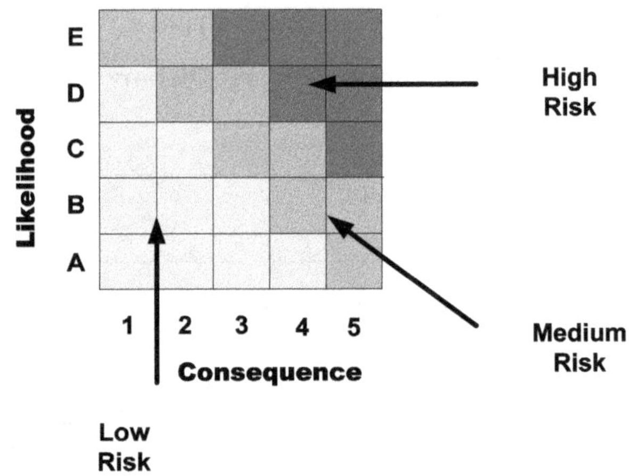

Consequence

Technical	Schedule	Level	Cost
Minimal impact	Minimal impact	**1**	Minimal impact
Minor performance shortfall, same approach retained	Additional tasks required, able to meet key dates	**2**	Development or acquisition cost increase < 1%
Moderate performance shortfall, workarounds available	Minor schedule slip, will miss date without workaround	**3**	Development or acquisition cost increase > 1% & < 5%
Unacceptable performance but workarounds available	Program critical path impact but workaround available	**4**	Development or acquisition cost increase > 5% & < 10%
Unacceptable performance, no alternatives exist	No known way to achieve program milestones	**5**	Development or acquisition cost increase > 10%

Notes:

- Consider always interfaces and interdependences
- Remember risk is at the deltas or variances
- Err on the side of being conservative
- Consult with non advocates to . . .

PROGRAM/PROJECT//PROCESS TECHNICAL CONSEQUENCE DEFINITIONS
Given the risk becomes real, what would the magnitude of the impact on system performance?
1. **Low:** Given that the risk is realized, there would be minimal impact. A successful outcome is not dependent on this issue: the technical performance goals will be met. There would be no impact on the success of the program.
2. **Minor:** Given that the risk is realized, there would be a minor performance shortfall but the same approach could retained. The resulting technical performance would be below the goal but within acceptable limits. There would be no need to change the basic design, process or approach. There would be no impact on the success of the program.
3. **Moderate:** Given that the risk is realized, there would be a moderate performance shortfall but workarounds would be available. The result technical performance would be below the goal. The basic design, process, or approach could be retained with only minor changes, and the overall system performance would still be acceptable as a result of workarounds such as the reallocation of functions or performance goals. There would be only a limited impact on the success of the program.
4. **Significant:** Given that the risk is realized, the performance would be unacceptable but workarounds would be available. The resulting technical performance would be unacceptably below the goal. The design, process, or approach would require a significant change to achieve an acceptable performance level. Additional workarounds such as the reallocation of function or performance goals could also be required. The success of the program could be jeopardized.
5. **High:** Given that the risk is realized, the performance would be unacceptable with no known workarounds. The resulting technical performance would be unacceptably below the goal. There are no known alternatives or solutions. The success of the program would be in doubt.

Given the risk becomes real, what would be the magnitude of the impact on the schedule?
1. Low: Given that the risk is realized, there would be minimal impact. The program/project schedule is not dependent on this issue. There would be no impact on the success of the program/project.
2. Minor: Given that the risk is realized, additional activities would be required to meet key dates. One or more key dates in the program schedule, but no critical path events, would be jeopardized; there are identified schedule workarounds that would be sufficient to mitigate the schedule impact. There would be no impact on the success of the program.
3. Moderate: Given that the risk is realized, there would be a minor schedule slip and one or more need dates would be missed. One or more key need dates in the program schedule, but not critical path events, would be at least one month late; there are identified schedule workarounds that would be sufficient to keep the program critical path from being affected. There would be only a limited impact on the success of the program.
4. Significant: Given that the risk is realized, the program critical path would be affected. One or more events on the program/project critical path would be at least one month late. There are identified schedule workarounds that would be sufficient to meet major program milestones. The success of the program could be jeopardized.
5. High: Given that the risk is realized, a key program milestone cannot be achieved. Completion of a key program milestone would be late and the success of the program would be in doubt. The slip requires a re-baseline of the program.

PROGRAM/PROJECT/ PROCESS COST CONSEQUENCE DEFINITIONS
Given the risk becomes real, what would be the magnitude of the impact on cost?
1. Low: Given that the risk is realized, there would be minimal cost impact. Program cost is not dependent on this issue. There would be no impact on the success of the program.
2. Minor: Given that the risk is realized, the total cost, operating cost or unit production cost would increase by 1%. The program costs and/or the production unit cost would increase by 1%. There would be no impact on the success of the program.
3. Moderate: Given that the risk realized, there would be a minor increase in financial need. The program cost, operating cost or unit production cost could increase from 1% up to 5%. The program costs and/or the production unit cost would increase from 1% up to 5%. There would be only a limited impact on the success of the program.
4. Significant: Given that the risk is realized, the total costs, operating, cost or unit production cost would increase from 5% up to 10%. The program cost and/or the production unit cost would increase from 5% to up to 10%. The success of the program could be jeopardized.
5. High: Given that the risk is realized, the total cost, operating cost, or unit production cost would increase by greater than 10%. The program costs and/or the production unit cost would increase by greater than 10%. The success of the program would be in doubt.

PROGRAM/PROJECT RISK/PROCESS LIKELIHOOD DEFINITIONS
What is the likelihood the risk will happen?
A. Low: Approach and processes will effectively avoid or mitigate the risk based on standard practices and stable processes. The chance of a negative outcome based on existing plans is not likely. This likelihood level assessment should be based on evidence or previous experience and not on subjective confidence. This assessment level requires the approach and processes to be well understood and documented. Little or no management oversight will be required.
B. Minor: Approach and process have usually mitigated this type of risk with minimal oversight in similar cases. There is a low likelihood but reasonable probability that a negative outcome is possible. Present plans include adequate margins (technical, schedule, or cost) to handle a typical problem. This assessment level requires that approach and processes to be well understood and documented. Limited management oversight will be required.
C. Moderate: Approach and processes may mitigate this risk, but workarounds will be required. A negative outcome is likely, or the current approach and processes are only partially documented. Alternative plans or methods exist to achieve an acceptable outcome even if the risk is realized. Present plans include adequate margins (technical, schedule, or cost) to implement the workarounds or alternatives to overcome typical problems. Significant management oversight will be required.
D. Significant: Approach and processes cannot mitigate this risk, but a different approach might. A negative outcome is highly likely to occur, or the current approach and process are not documented. While alternative plans or methods are believed to exist to achieve an acceptable outcome, there are not adequate margins (technical, schedule, or cost) to implement the workarounds without impacting the performance, schedule, or cost reserves. Significant management involvement is required.
E. High: Approach and processes cannot mitigate this type of risk; no known processes or workarounds are available. A negative outcome is going to occur with near certainty. No alternative plans or methods have been documented. Alternatively, the risk item has yet to be evaluated adequately to be well understood, so there is a high level of uncertainty about program success. Urgent management involvement is required.

REPORTING STEP 2 – DECIDE AUDIT REPORT FORMAT

COMPLETION OF THE QUALITATIVE RISK MANAGEMENT AUDIT FORM

The Qualitative Risk Management Audit Form consists of the following sidebars:

- **Qualitative Risk Management Audit Form.** This is the final form for reporting and treating programmatic, project, or process risks.
- **Likelihood and Consequence Form.** This is the form for evaluating the likelihood and consequence data that is inputted to the Risk Profile Matrix.
- **Program/Project/Process Likelihood Definitions.** This table provides granularity for determining various likelihood levels.
- **Program/Project/Process Cost Consequence Definitions.** This table provides granularity for determining various cost consequence levels.
- **Program/Project/Process Technical Consequence Definitions.** This table provides granularity for determining various technical consequence levels.
- **Program/Project/Process Schedule Consequences Definitions.** This table provides granularity for determining various schedule consequence levels.[1]
- **Process/Program/Project Title and Date:** Self explanatory.
- **Process/Program/Project Owners:** Name of process owner or project manager.
- **Analyst:** Name of the auditor.
- **Risk Description:** Description of the hazard or risk.
- **Symptom/Root Cause:** Symptomatic cause is what is seen on the surface and the root cause is deeper in the project or system.
- **Risk Profile Matrix:** Matrix consists of likelihood and consequence of the event. Each is evaluated in terms of a 5x5 matrix. This is then mapped to the Risk Profile Matrix. This includes a cursory 'Risk Analysis', which includes analysis of risk tolerance, priority, etc.
- **Risk Management Strategies.** Process owner or project manager determines appropriate treatment to mitigate risk.
- **Risk Resolution Date:** Date when risks will be treated and managed at the symptomatic and root cause levels. This is determined by the process owner or project manager.
- **Description of Implementation:** Process owner or project manager determine risk treatment.
- **Assessment of Implementation:** Post audit evaluation of risk management and treatment effectiveness.
- **Post Implementation Risks:** Evaluation of residual risks compared to enterprise risk appetite.

PROGRESS AUDIT REPORTS

Monthly or quarterly reports summarize what the value added audit organization accomplished during a time period. Summary reports can be in a memo format or in the abbreviated audit format shown in the above 'Progress Audit Report' sidebar on the next page. Summary reports can detail progress in terms of:

PROGRESS AUDIT REPORT

To: Internal Audit Supervision
From: Resident Auditor
Subject: Project X Budget and Corrective Action Report
Date: X-X-XX
Budget to date. $500,000
Planned budget: $550,000
Variance: $ 50,000
Project 92% complete. CPM chart attached.
Expected completion date: X-X-XX
Initial findings: Project X within budget and on time.
Level and type of nonconformances: N/ A
Required corrective actions: None
Results of corrective actions: N/A
Potential project problems: None foreseen
Signature: Resident Auditor

Enclosures:
CPM project chart

- Number of organizational areas covered.
  - Number of audits conducted.
  - Budget variance (risk).
  - Benchmarks achieved.
  - Improvement savings.
  - Regulatory compliance.
  - Internal customer satisfaction.
  - Auditee utilization.

Value added audits of large installations, multidivisional companies, and offshore suppliers require time. Sometimes a resident auditor is stationed on site for the duration of the construction, installation, or production run. The resident auditor prepares daily, weekly, and monthly progress reports. Again, report frequency depends on customer requirements, project size, and project importance. Daily reports can be onerous, so weekly or monthly reports may be preferred. The interim audit reports can follow the progress audit report format illustrated in the 'Progress Audit Report' sidebar.

If any elements of the project are suspect or a source of potential problems, then more space can be dedicated to explain the problem, to explain the actions required to correct it, or to describe how to prevent the problem from occurring or recurring.

These reports can also detail audit effectiveness. Two of the more effective audit measures are auditee utilization and internal customer satisfaction. Auditee's utilization refers

EXCEPTION AUDIT REPORT **To:** Chair, Engineering Control Board **From:** Lead Auditor **Subject:** Need for Additional Resources - Special Audit of Project SX **Date:** X-X-XX **Results:** The special audit of project SX revealed nonconformances that are beyond the initial scope of the audit. Nonconformances include: • Calculation errors. • Unapproved drawings. Preliminary findings indicate that nonconformances resulted from section X engineers not communicating requirements on the recent print release to internal manufacturing and external suppliers. Since the audit team is on site and consequences may be major, authorization of $10,000 is requested to pursue investigation. **Signature:** Lead Auditor

to whether the auditee used the information in the audit to correct problems, improve systems, or follow the report's recommendations. Internal customer satisfaction, similar to auditee utilization, refers to how well the organization perceived the benefit of value added audits.

EXCEPTION REPORT

Sometimes a report is only necessary to inform the customer or organization of major changes, problems, or corrective actions. This is called 'reporting by exception'. It saves time, energy and paperwork; however, the customer should approve reporting by exception. Exception reports work best when processes are capable, in control, and risks are managed. If the exception occurs sufficiently early, the auditee can eliminate the exception and ensure controls manage risks so that no nonconforming parts are produced or risks are realized. Also, management needs to be informed of problems sufficiently early so risk identification, monitoring, management, and contingency plans can be made. If problems occur, they require immediate follow-up and resolution.

The challenge with reporting by exception is that supervision does not know how the project is proceeding until there is an unusual occurrence or risk. What constitutes an unusual occurrence or risk? This can be worked out between the auditor and auditee. If the auditor only reports major events, the customer may not be in a position to resolve the problem or mitigate consequences. Exception reports can be used to report the following circumstances:

- Immediate need for resources or higher-level approval requirements.

REPORTING STEP 2 – DECIDE AUDIT REPORT FORMAT

CORRECTIVE ACTION REQUEST (CAR)

To: Auditee Name
From: Customer
Subject: Corrective Action Request (CAR)
Audit number, title, subject, and auditee's name
Date: Audit report date
Audit implementation dates
References: List of related documentation, which is referenced in report.
CAR # 10
Customer: Manufacturing
Auditee: Plant X maintenance section
Part number/s: Pumps 23, 45, and 48
Process description/procedure #: Slurry process pumps
Apparent cause: Failure of packing in all three pumps
Root cause: Lack of preventive maintenance
Suggested corrective action: Periodic and preventive monitoring of pumps
Action required by: ASAP
Comments of auditee:
Signature of auditor: Jane Doe
Audit Date: X-X-XX
Findings:
Corrective action:
Background information:

- Immediate corrective action.
- Additional problems.
- High risk events.
- Waivers and deviations.
- Minor problems becoming bigger problems.
- Chronic and systemic problem recurrence.
- Product development schedule slippage.
- Unanticipated interruptions.
- Mistakes.
- Over-budget projects.
- Corrective action failure.

An exception audit report is often a one-page memo, directly communicating the need for prompt attention to an urgent matter. A sample exception report is illustrated in 'Exception Audit Report' sidebar on page 387.

CORRECTIVE ACTION REQUEST

Corrective action is the recommended action to fix, remove, or correct a problem. It can be minor, such as additional employee training, or it can be major, such as the purchase of a new machine. The latter requires further analysis to justify the machine purchase.

REPORTING STEP 2 – DECIDE AUDIT REPORT FORMAT

Sometimes the value added auditor lists the findings or deficiencies and leaves it up to the customer or the auditee to determine the appropriate follow-up. Or, the value added audit report lists findings and the auditee is responsible for correcting the problems or controlling the risks reported in the findings. The value added auditor may be empowered to recommend but not to enforce the corrective action. There are several reasons for this. The value added auditor may not have the knowledge to recommend action; the people responsible for the problem have the most knowledge and the highest interest to see that it is eliminated, and they are on site to ensure that it does not recur. A Corrective Action Request (CAR) memo is shown in the 'Corrective Action Request' sidebar on page 388.

AUDIT ASSESSMENT RATING PLANS

Value added audits by their nature are sometimes subjective and must rely on the professional proficiency and due care. The purpose of the rating plan is to reduce subjectivity and variability by having value added auditors use similar, measurable questionnaires. It is an attempt to be numerically objective. In this way if two auditors use the same audit format, their evaluation process and scores should be consistent.

Rating plans are based on a similar structure. An organization is segmented or chunked into separate processes or systems. A numerical rating is developed for each process or system. For example, incoming material control, first article inspection, and traceability are different types of processes or systems with their own process management controls. There would be a questionnaire for each of these with a numerical rating. Two types of rating plans are generally used:

- Series of yes/no types of questionnaires relating to a specific quality system, functional area, process, product, and organizational structure, such as was common with ISO 9001 – 2000 revision.
- Single open-ended questions or statements that are rated in terms of points.

Questionnaires and surveys have already been explained and illustrated. Rating systems are used with questionnaires to develop a numerical rating for an area or system. This system rates questions in terms of 0 - 10 or 0 - 5. A score of 10 or 5, the highest score for a question, indicates the auditee has extensive and intensive process management and controls in place that have been working successfully over a period of time and is considered one of the best in its industry. For example in terms of a Quality Management System, a score of 0 may indicate one or more of the following:

- No process understanding, high risks, no controls.
- Processes are very limited or do not exist.
- Quality management system is not defined or specified.
- Quality management system has no supporting documentation
- Little or no evidence of quality exists.
- Little or no standardization exists.
- Significant business or quality elements are missing.

0-10 QUALITY ASSESSMENT RATING PLAN

Score

One (I) Point

A. Customer satisfaction is not measured.
B. COQ is not used.
C. Benchmarks have not been identified.
D. Major deficiencies exist in many areas.
E. Quality requirements are defined and documented in 10 percent of the organization.
F. Quality requirements are understood and implemented in 10 percent of the organization.
G. Inspection is sometimes used.
H. There is 10 percent conformance to significant quality requirements.
I. There is no evidence of quality planning.
J. Resources are limited.
K. Quality effort is very limited.
L. Quality is not discussed with suppliers.
M. Improvement is not pursued.
N. Company is not innovative.
0. Only implemented suggestions are recognized.

Zero (0) Points

A. – 0. Quality systems or processes don't exist.

Value Scoring

10 All critical processes are in control and capable.
9 At least 75 percent are in control and capable.
8 At least 50 percent are in control and capable.
7 At least 25 percent are in control and capable.
6 At least 10 percent are in control and capable.
5 Quantitative measures are used to identify all critical processes. There is quantifiable evidence as to which processes are critical.
4 At least 40 percent of critical processes are assessed for control and capability.
3 At least 20 percent of critical processes are assessed for control and capability.
2 Critical processes are subjectively identified.
1 Concepts of control and capability are understood. The concepts required to assess control and capability are understood by several people at the facility.
0 None.[2]

- Quality management system follows no systematic approach.

Scoring between 0 and 10 is graded in terms of organizational, performance, risk management, effectiveness, and efficiency. Gradations between 0 and 10 for the following quality objectives are illustrated in the above '0 - 10 Quality Assessment Rating Plan:

- Customer satisfaction.
- Cost of Quality (COQ).
- Benchmarking.

REPORTING STEP 2 – DECIDE AUDIT REPORT FORMAT

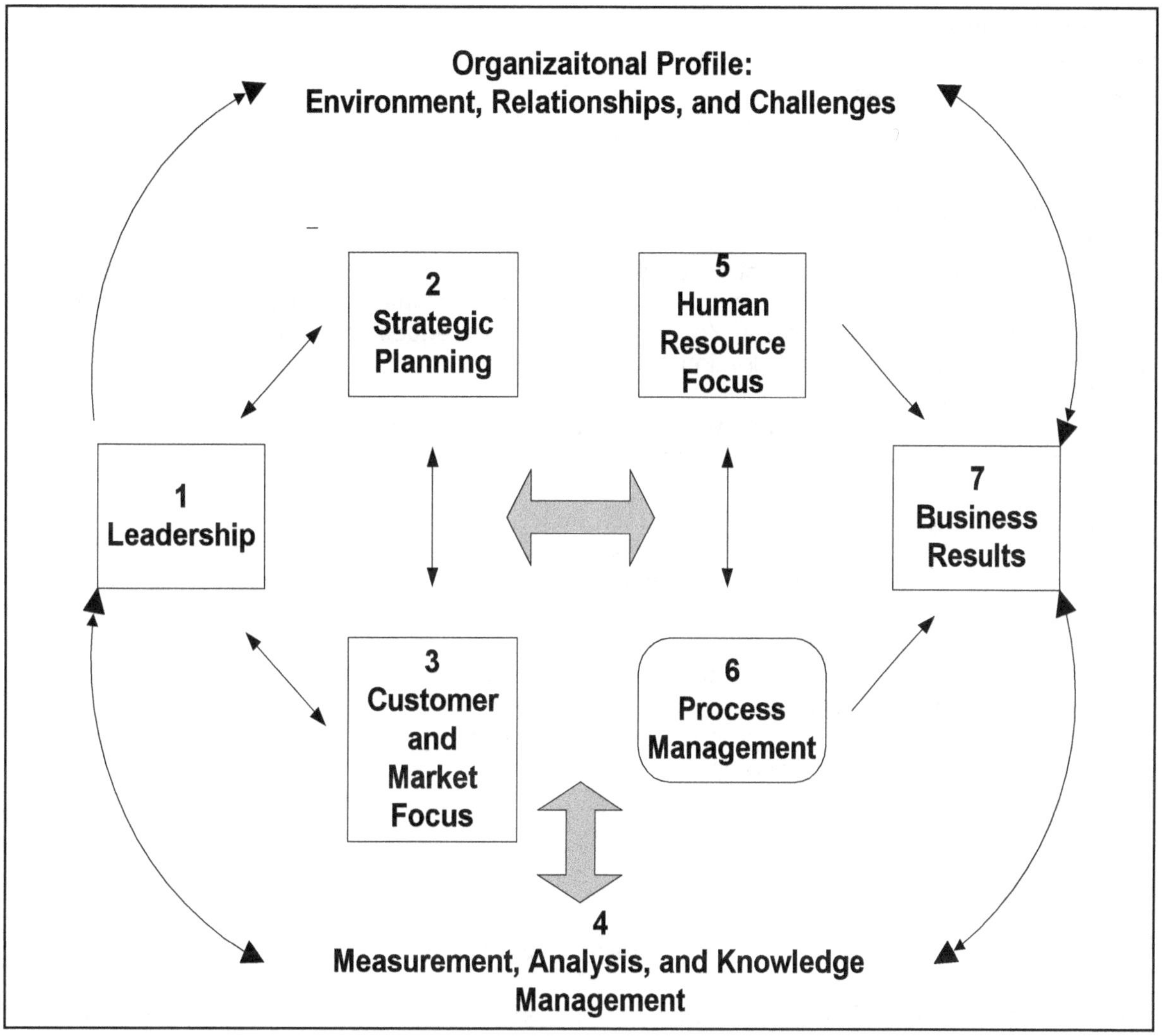

- Quality systems.
- Quality requirements.
- Quality implementation.
- Statistical driven prevention.
- Conformance to quality requirements.
- Quality planning.
- Resource utilization.
- Company wide implementation.
- Supply wide implementation.
- Quality improvement.
- Innovation.
- Quality achievement recognition.

REPORTING STEP 2 – DECIDE AUDIT REPORT FORMAT

Audit assignment #
Audit sponsor:
Business unit or function:
Location:
Audit areas covered:
Purpose:
Objective audit standards:
Audit team:(Lead/members):

Pease rate on a scale of 1 (lowest) to 5 (highest) the following statements. There is space at the end of the survey for your comments. Your reviews are for internal improvement purposes and will be held in strictest confidence. Your approval will be requested before any information is released. Thank you for your participation in this survey.

Audit planning

You were notified sufficiently in advance of the audit and its purpose?
All your questions were answered regarding the purpose, scope, timing, and benefits of the audit?

Audit Implementation

The audit team explained the audit objectives, plan, and implementation to you?
The audit team reviewed all the areas that are critical to process management owners and stakeholders?
The audit was conducted in a timely, cooperative, and efficient manner?
The audit team addressed business and process risks and reviewed risk controls?

Audit team performance

The audit team understood your processes and concerns?
The audit team was technically competent?
The audit team was sensitive to your concerns?
The audit team focused on:

a. Enterprise/Business issues?
b. Process issues?
c. Activity/product issues?

The audit team focused on improvement, specifically process effective, efficiency and economy?

Audit results

The closing meeting sufficiently debriefed you on results and findings. The audit findings were:

a. Discussed with you?
b. Accurate, relevant, and objective?
c. Focused on value and improvement?
d. Factual and significant?
e. Focused on risk management, assurance, and control?
f. Clearly expressed and written?

Was there anything that you especially thought was valuable to you, your processes, and team? Please elaborate:
Was there anything that you disliked and didn't add value to you, your process and team? Please elaborate:
Do you want the auditors to consult with you regarding how to improve and add value to your processes?
How would you rate from 1 – 10 the overall value of the audit?

TOTAL SCORE _____

Name: ______________________________
Title: ______________________________
Date: ______________________________

POST AUDIT SURVEY

Once the value added audit is conducted, how do auditors know that value was added and clients are satisfied? There are some proxy measures for evaluating audit performance. Are auditors called back for additional guidance? Are audit stakeholders satisfied? What does the organization say about the audit?

Another way to determine audit customer satisfaction is by surveying audit stakeholders after the audit. This information can be used to improve audit performance. This is illustrated in the 'Post Audit Survey' sidebar.

BALDRIGE 'WORLD CLASS' SYSTEM ASSESSMENT

You can't stand still.
Proverb

The value added auditor may have to develop and use a scoring methodology to evaluate and score the processes in the valued added audit. The Malcolm Baldrige National Quality Award (MBNQA) is an excellent example on how this is done.

WHAT IS THE MALCOLM BALDRIGE NATIONAL QUALITY AWARD?

The Malcolm Baldrige National Quality Award (MBNQA) is a value added audit that uses a numerical score. The MBNQA is also a Capability Maturity Model (CMM) and is a numerical assessment of an organization's business and quality processes. Value is achieved by a gap analysis between the auditee's processes and MBNQA criteria.

The value added auditor can develop a tailored scoring system for the internal value added audit using the Baldrige criteria. The reporting and the scoring of the Baldrige criteria have been based on three evaluation dimensions: 1. Approach; 2. Deployment; and 3. Results, which will be discussed in this section.

The MBNQA criteria are results-oriented requirements. However, the criteria don't prescribe specific tools, techniques, technologies, systems, measures, or starting points. The criteria are not prescriptive because the focus of the award is on effective processes and results regardless of the structure of the organization. Organizations being assessed are responsible for demonstrating breakthrough improvement regardless of the tools used.

BALDRIGE SCORING SYSTEM

The criteria list below are older award criteria but we've found they are useful for value added audits. MBNQA categories 1 - 6, from 'Leadership' to 'Process Management' are evaluated based on 'approach/deployment' rating criteria shown in the 'MBNQA Approach/Deployment Criteria' figure on page 397. The MBNQA 'approach' refers to how the value added auditor can address the Baldrige requirements and the 'methods' used. The factors used to evaluate the 'approach' include:

0-10 QUALITY ASSESSMENT RATING PLAN

Score

Ten (10) Points
A. External and internal customers are very highly satisfied.
B. Cost of Quality (COQ) is used company wide.
C. World class benchmarks have been surpassed.
D. Quality systems are functioning excellently.
E. Quality requirements are extremely well defined and documented.
F. Quality requirements are extremely well understood and implemented.
G. Statistical driven prevention is pursued company- and supplier wide.
H. There is l00 percent conformance to quality requirements.
I. Quality is extremely well planned.
J. Resources are used very effectively and very efficiently.
K. Quality effort is company wide.
L. Quality effort is supplier wide.
M. Improvement is measurable, rapid, and continual.
N. Company is extremely innovative.
0. Quality achievements are always recognized.

Nine (9) Points
A. External and internal customers are highly satisfied.
B. COQ is used company wide.
C. National benchmarks have been surpassed.
D. Quality systems are functioning excellently.
E. Quality requirements are well defined and documented.
F. Quality requirements are well understood and implemented.
G. Statistical driven prevention is pursued company wide.
H. There is 100 percent conformance to quality requirements.
I. Quality is well planned.
J. Resources are used effectively and efficiently.
K. Quality effort is company wide.
L. Quality effort is supplier wide.
M. Improvement is measurable, rapid, and continual.
N. Company is innovative.
0. Quality achievements are always recognized.

Eight (8) Points
A. External and internal customers are satisfied.
B. COQ is used in most areas.
C. National benchmarks have been met.
D. Quality systems are functioning excellently.
E. Quality requirements are defined and documented.
F. Quality requirements are well understood and implemented.
H. There is 90 percent conformance to quality requirements.
I. Quality is well planned.
J. Resources are used effectively and efficiently.
K. Quality effort is company wide.
L. Quality effort focuses on first-tier suppliers.
M. Improvement is measurable and continual.
N. Company is innovative.
0. Quality achievements are always recognized.

REPORTING STEP 2 – DECIDE AUDIT REPORT FORMAT

0-10 QUALITY ASSESSMENT RATING PLAN

Score

Six to Seven (6 - 7) Points

A. External and internal customers are usually satisfied.
B. COQ is used in selected areas.
C. Industry benchmarks have been met and surpassed.
D. Quality systems are functioning well.
E. Quality requirements are defined and documented in 80 percent of the organization.
F. Quality requirements are understood and implemented.
G. Statistical driven prevention is pursued in 60 percent of the organization.
H. There is 80 percent conformance to significant quality requirements.
I. Quality is planned in 80 percent of the organization.
J. Resources are used effectively and efficiently.
K. Quality effort is focused in major areas.
L. Quality effort focuses on significant quality characteristics with first-tier suppliers.
M. Some improvement is measurable and continual.
N. Company is innovative.
0. Quality achievements are recognized.

Four to Five (4 - 5) Points

A. External and internal customer satisfaction is being measured.
B. COQ effort is just starting.
C. Industry benchmarks have been met.
D. Major deficiencies exist in one area.
E. Quality requirements are defined and documented in 60 percent of the organization.
F. Quality requirements are understood and implemented in 50 percent of the organization.
G. Statistical driven prevention is pursued in 40 percent of the organization.
H. There is 50 percent conformance to significant quality requirements.
I. Quality is planned in 50 percent of the organization.
J. Resources are usually used effectively and efficiently.
K. Quality effort is focused in selected areas.
L. Quality effort focuses on some significant quality characteristics with first-tier suppliers.
M. Some improvement is measurable.
N. Company is innovative.
0. Major team achievements are recognized.

Two to Three (2 -3) Points

A. Customer satisfaction is not measured.
B. COQ is not used.
C. Benchmarks have not been identified.
D. Major deficiencies exist in eight areas.
E. Quality requirements are defined and documented in 20 percent of the organization.
F. Quality requirements are understood and implemented in 30 percent of the organization.
G. Limited statistical analysis and inspection is widely used.
H. There is 50 percent conformance to significant quality requirements.
I. There is no evidence of quality planning.
J. Resources are available for special or pilot projects.
K. Quality effort is very limited.
L. Quality effort is at pilot stage with first-tier suppliers.
M. Some improvement is measurable.
N. Company is not innovative.
0. Implemented suggestions are recognized.

0-10 QUALITY ASSESSMENT RATING PLAN

Score

One (I) Point

A. Customer satisfaction is not measured.
B. COQ is not used.
C. Benchmarks have not been identified.
D. Major deficiencies exist in many areas.
E. Quality requirements are defined and documented in 10 percent of the organization.
F. Quality requirements are understood and implemented in 10 percent of the organization.
G. Inspection is sometimes used.
H. There is 10 percent conformance to significant quality requirements.
I. There is no evidence of quality planning.
J. Resources are limited.
K. Quality effort is very limited.
L. Quality is not discussed with suppliers.
M. Improvement is not pursued.
N. Company is not innovative.
0. Only implemented suggestions are recognized.

Zero (0) Points

A. – 0. Quality systems or processes don't exist.

Value Scoring

10 All critical processes are in control and capable.
9 At least 75 percent are in control and capable.
8 At least 50 percent are in control and capable.
7 At least 25 percent are in control and capable.
6 At least 10 percent are in control and capable.
5 Quantitative measures are used to identify all critical processes. There is quantifiable evidence as to which processes are critical.
4 At least 40 percent of critical processes are assessed for control and capability.
3 At least 20 percent of critical processes are assessed for control and capability.
2 Critical processes are subjectively identified.
1 Concepts of control and capability are understood. The concepts required to assess control and capability are understood by several people at the facility.
0 None.[3]

- Appropriateness of the methods to the requirements.
- Effectiveness of the use of the methods.
- Degree to which the approach is repeatable, integrated, and consistently applied.
    - Embodies evaluation/improvement/learning cycles.
    - Based on reliable information and data.
- Alignment with organizational needs.
- Evidence of innovation.

The MBNQA 'deployment' refers to the extent to which the 'approach' is applied to all requirements of the Baldrige criteria. The factors used to evaluate 'deployment' include:

REPORTING STEP 2 – DECIDE AUDIT REPORT FORMAT

MALCOLM BALDRIGE NATIONAL QUALITY AWARD CRITERIA			
2000 Categories/Items		**Point Values**	
Leadership Approach/Deployment		**120**	
1.1	Organizational Leadership	80	
1.2	Social Responsibility	40	
Strategic Planning			**85**
2.1	Strategy Development	40	
2.2	Strategy Deployment	45	
Customer and Market Focus			**85**
3.1	Customer and Market Knowledge	40	
3.2	Customer Relationships and Satisfaction	45	
Measurement, Analysis, and Knowledge Management		**90**	
4.1	Measurement and Analysis of Organizational Performance	45	
4.2	Information and Knowledge Management	45	
Human Resource Focus			**85**
5.1	Work Systems	35	
5.2	Employee Learning and Development	25	
5.3	Employee Well-Being and Satisfaction	25	
Process Management Approach/Deployment			**85**
6.1	Value Creation Processes	50	
6.2	Support Processes	35	
Business Results			**450**
7.1	Customer Focused Results	75	
7.2	Product and Service Results	75	
7.3	Financial and Market Results	75	
7.4	Human Resource Results	75	
7.5	Organizational Effectiveness Results	75	
7.6	Governance and Social Responsibility Results	75	
	TOTAL POINTS		**1000**

REPORTING STEP 2 – DECIDE AUDIT REPORT FORMAT

MBNQA 'APPROACH/DEPLOYMENT CRITERIA	
SCORE	**APPROACH/DEPLOYMENT**
0%	• No systematic approach evident; anecdotal information.
10% to 20%	• Beginning of a systematic approach to the basic requirements of the Item is evident. • Major gaps exist in deployment that would inhibit progress in achieving the basic requirements of the Item. • Early stages of a transition from reacting to problems to a general improvement orientation are evident.
30% to 40%	• Effective, systematic approach, responsive to the basic requirements of the Item, is evident. • Approach is deployed, although some areas or work units are in early stages of deployment. • Beginning of a systematic approach to evaluation and improvement of key processes is evident.
50% to 60%	• Effective, systematic approach, responsive to the overall requirements of the Item. • Approach is well-deployed, although deployment may vary in some areas or work units. • Fact-based, systematic evaluation and improvement process is in place for improving the efficiency and effectiveness of key processes.. • Approach is aligned with basic organizational needs identified in the other Criteria Categories.
70% to 80%	• Effective, systematic approach, responsive to the multiple requirements of the Item and current and changing business needs, is evident. • Approach is well-deployed, with no significant gaps. • Fact-based, systematic evaluation and improvement process and organizational learning/sharing are key management tools; clear evidence of refinement, innovation, and improved integration as a result of organizational-level analysis and sharing. • Approach is well-integrated with organizational needs identified in the other Criteria Categories.
90% to 100%	• Effective, systematic approach, fully responsive to all the requirements of the Item and all current and changing business needs, is evident. • Approach is fully deployed without significant weaknesses or gaps in any areas or work units. • Very strong, fact-based, systematic evaluation and improvement process and extensive organizational learning/sharing are key management tools; strong refinement, innovation, and integration, backed by excellent organizational-level analysis and sharing, are evident. • Approach is fully integrated with organizational needs identified in the other Criteria Categories.

REPORTING STEP 2 – DECIDE AUDIT REPORT FORMAT

MBNQA 'RESULTS' CRITERIA	
SCORE	**RESULTS**
0%	• No results or poor results in areas reported.
10% to 20%	• Some improvements *and/or* early good performance levels in a few areas. • Results not reported for many to most areas of importance to the organization's key business requirements.
30% to 40%	• Improvements *and/or* good performance levels in many areas of importance to the organization's key business requirements. • Early stages of developing trends and obtaining comparative information are evident. • Results reported for many to most areas of importance to the organization's key business requirements.
50% to 60%	• Improvement trends *and/or* good performance levels reported for most areas of importance to the organization's key business requirements. • No pattern of adverse trends and no poor performance levels are evident in areas of importance to the organization's key business requirements. • Some trends *and/or* current performance levels — evaluated against relevant comparisons *and/or* benchmarks — show areas of strength *and/or* good to very good relative performance levels. • Business results address most key customer, market, and process requirements.
70% to 80%	• Current performance is good to excellent in areas of importance to the organization's key business requirements. • Most improvement trends *and/or* current performance levels are sustained. • Many to most trends *and/or* current performance levels — evaluated against relevant comparisons *and/or* benchmarks — show areas of leadership and very good relative performance levels. • Business results address most key customer, market, process, and action plan requirements.
90% to 100%	• Current performance is excellent in most areas of importance to the organization's key business requirements. • Excellent improvement trends *and/or* sustained excellent performance levels are reported in most areas. • Evidence of industry and benchmark leadership demonstrated in many areas. • Business results fully address key customer, market, process, and action plan requirements.

- Use of the 'approach' in addressing MBNQA item requirements relevant to the organization.
- Use of the 'approach' by all appropriate work units.

REPORTING STEP 2 – DECIDE AUDIT REPORT FORMAT

MBNQA category 7, 'results,' is evaluated based on 'results' category. The MBNQA 'results' refers to outcomes in achieving the purposes given in the MBNQA item. The factors used to evaluate results include:

- Current performance.
- Performance relative to appropriate comparisons and/or benchmarks.

'Effectiveness' is a critical element of the Baldrige approach to conducting assessments or audits. In Baldrige, 'Effectiveness' relates to producing the desired result and to the appropriateness for intended use. What's the best way to analyze and report on 'effectiveness?' To many, the best way is probably through some scoring mechanism. If the value added auditor is going to give a process or system a numerical score, the value added auditor may follow the Baldrige scoring criteria and recommendations discussed.

SCORING DIMENSIONS

The Baldrige uses a weighted average system to evaluate processes. Baldrige items are classified according to the kinds of information and/or data a company has relative to the above three evaluation dimensions. 'Approach and Deployment' are linked to emphasize that description of the 'Approach' should always indicate that 'Deployment' is consistent with the specific requirements of the Baldrige criterion being assessed.

The 'results' criteria require data showing performance levels and trends on key measures and/or indicators of organizational performance. 'Results' criteria also call for data on the breadth of performance improvements, specifically indicating how effectiveness of the improvement results.

This is directly related to the 'deployment' dimension. If improvement processes are widely deployed, there should be corresponding results. A score for a 'results' item is thus a composite based upon overall performance, taking into account the breadth of the improvements and their importance.

'IMPORTANCE' AS A SCORING FACTOR

Companies first self-assess to the Baldrige criteria or items. The three evaluation dimensions described previously are critical to the evaluation and feedback. However, evaluation and feedback also must consider the importance of their 'approach', 'deployment', and 'results' to their key success factors. Understanding of key customer requirements and strategic objectives, and how they are addressed in action plans are particularly important in the self-assessment.

Baldrige scores can be difficult to assign. How does the auditor decide if performance or process criteria are at the 40%, 50%, or 60% level? As in many value added audits, there is a subjective element. The Baldrige scoring criteria offers the following general tips:

REPORTING STEP 2 – DECIDE AUDIT REPORT FORMAT

- In assigning a score to a Baldrige item, the value added auditor first decides which scoring range (e.g., 50% to 60%) best fits the overall Baldrige item response. Overall 'best fit' does not require total agreement with each of the statements for that scoring range. Actual score within the range depends upon judgment of the closeness of the 'item' response in relation to the statements in the next higher and next lower scoring range.
- An 'approach/deployment' item score of 50% represents an approach that meets the overall objectives of the item and that is deployed to the principal activities and work units covered in the item. Higher scores reflect maturity (cycles of improvement), integration, and broader deployment.

- A 'results' item score of 50% represents a clear indication of improvement trends and/or good levels of performance in the principal results areas covered in the item. Higher scores reflect better improvement rates and/or levels of performance, and better comparative performance as well as broader coverage and integration with business requirements.

SUMMARY

Value added audits must be communicated suitably and accurately. Compliance audit reporting is different than performance effectiveness or risk management reporting. Compliance audit communications should include audit objectives, audit scope, conclusions, recommendations, and corrective/preventive actions.

There is no minimum list of information to be included in the final value added audit report. However, the audit report should provide sufficient, useful, and usable information so audit stakeholders can make informed decisions. Information in the final report may include:

- Stakeholder requirements.
- Audit standards.
- State of internal risk – control.
- Risk appetite and tolerance of organization.
- Audit scope and objectives.
- Past audit reports.
- Process flows analyzed.
- Background information.
- Previous audit results.
- Relevant information.
- Objective supporting information.
- Tests conducted.
- Observations/findings/opinion.
- Conclusions and recommendations.
- Corrective/preventive action requests.

CHAPTER 21
STEP 3: CORRECT – PREVENT – PREDICT – PREEMPT

Boards of directors, CEOs and CFOs don't want surprises in their organizations. CEOs usually are 8th level management. They want to know what's happening at the lowest levels of the organization as well as throughout the supply chain, especially if these events can impact the financial statement, reputation of the organization, or its share price. In other words, senior management wants to know of possible risk events and know there are processes in place to correct, prevent, predict, and even preempt their occurrence and recurrence. Surprises involve risk to the organizations as well as personal exposure and consequences to individual senior management.

Value added audits can be one of the most effective and extensive risk monitoring and improvement techniques used by organizations. Corrective and preventive action reports are among the valuable reporting mechanisms of a value added report.

CORRECTIVE AND/OR PREVENTIVE ACTION

Quality is commitment made real. It is not perfection. But, rather, the dedication to perfection.
Shearson Lehman Brother ad

Once the value added audit is conducted, audit findings are documented in a report that is circulated to authorized clients. The audit report may recommend intervention, risk management, corrective action, prevention action, no action, or other actions. Usually, if corrective action or preventive action is indicated, a corrective action request (CAR) or preventive action request (PAR) is written to initiate further action.

CORRECTIVE ACTION

Corrective and preventive actions are essential elements of all processes and systems. Corrective and preventive actions ensure symptoms as well as their root causes are eliminated so problems do not recur. ISO 9001 and other quality standards require the elimination of the causes of actual and potential nonconformances depending on the magnitude of the problem as well as potential risks. Once the cause of the difference, risk, or discrepancy is discovered, corrective action is undertaken to eliminate the recurrence of the risk, defect, flaw, problem, or discrepancy.

Corrective action may involve:

- Redesigning or modifying the product.
- Redesigning or modifying fixtures and dies.
- Training suppliers.
- Updating specifications or changing tolerances.
- Increasing incoming material inspection.
- Additional risk – controls or other risk mitigations.

When to Implement Corrective Action

Corrective action can be initiated while the auditor is still conducting the audit or when management has approved the report and the customer authorizes it. The former is encouraged because it indicates a proactive approach to problem solving, risk management, and continual improvement. The latter indicates a static approach to management and improvement where corrective action is only initiated upon a formal request by the customer or as the result of an audit.

Corrective action is not always easy or simple. It may require a change of operations. People may have to learn new techniques. New suppliers may have to be chosen. Machines may have to be purchased. Certain deficiencies or risks may require suspension of operations and immediate resolution, since they may involve regulatory, safety, health, and other significant matters.

Sometimes, correction is required of the audit customer because he or she made a mistake or wants to retain the auditee's services or products, so the customer may apply the following risk mitigations:

- Visit auditee's plant.
- Improve the design.
- Redesign or modify the product.
- Redesign or modify the process.
- Train suppliers.
- Update specifications.
- Increase incoming inspection.
- Tighten incoming inspection.
- Conduct additional auditing.

Corrective action is considered closed when the problem or risk event does not recur. Many problems are intractable, whereby the obvious, simple solution did not fix the basic problem. It may take more time and resources to eliminate the root cause, or a simple fix may eliminate one problem but creates another problem. Normally, the auditor doesn't have to monitor corrective action. However for some critical risks or deficiencies, a formal post audit may be conducted to ensure the problem was corrected.

REPORTING STEP 3 – CORRECT – PREVENT – PREDICT – PREEMPT

PREVENTIVE ACTION

Corrective and preventive action are sometimes confused. Corrective action is the process of eliminating a nonconformance and ensuring that it doesn't recur. To do so, process owners must understand the nature of the problem and be able to suggest a solution such as purchasing a new fixture, developing a new procedure, or training employees. The need for corrective action may be indicated by recurring customer complaints, nonconformances, unstable processes, or product rework.

On the other hand, preventive action is anticipating a potential risk or problem and identifying steps to eliminate its possible occurrence. How can potential problems be anticipated is the key question? One method is to identify the areas of highest risk and to ensure there are compensating controls to minimize or eliminate the risk. Another method is to use internal resources to see if there are any trends, which may indicate a potential problem. Typical examples of trends that may indicate potential problems include high rework, customer service reports, high product waivers, inspection problems, and high waste.

PROCESS IMPROVEMENT

Process improvement or breakthrough transformation is a continual process. 'Event' focused continual improvement is an oxymoron. W. Edwards Deming called this the 'instant pudding' syndrome. Many process problems are chronic, systemic, or intractable. Quick fixes often don't work. The problem resurfaces again exactly or in some other form. People may believe the problem was fixed or can't be fixed. Management and employees become cynical about the entire process.

Feedback loops ensure that the responsible parties were notified of the problem and that if a correction was implemented, the problem did not recur. Feedback loops may involve the continual reporting the status of:

- Product development.
- Nonconforming material disposition.
- Effective corrective action.
- Occurrence or recurrence of deficiencies.
- Resource utilization, including people, budgets, and progress.
- Project progress.

In any of the above instances, a decision must be made. For example there may be dissatisfied audit stakeholders, lack of responsiveness, poorly designed products, or unsatisfactory services. These must be rerouted to an earlier process step for corrective action, disposition, or some other decision. This backward looping is costly and wasteful. The further a product or service moves from the source of the mistake, the more value has been added to the product and the more it costs to rectify the problem or mitigate the risk.

REPORTING STEP 3 – CORRECT – PREVENT – PREDICT – PREEMPT

If a problem occurs frequently, this chronic problem signals an opportunity for improvement. When a problem, error, or other deficiency is detected, the root cause should be determined and root-cause corrected. There should be closed-action feedback loops where the deficiency is discovered and eliminated so the deficiency does not recur. This often involves changing people's hearts and minds.

PREDICTION - PREEMPTION

The future of value added auditing is 24 – 7 continuous process auditing. Companies want the ability to predict risk events and to develop possible preemptions. Companies and key stakeholders want this level of real time information and assurance. In this situation, the role of the value added auditor is to analyze enterprise risks and ensure controls manage organizational risks effectively.

Continual process auditing is critical in all organizational activities to ensure a project is progressing smoothly; processes are controlled and capable; product development considers all critical and major customer requirements; and customers are satisfied. For example in improvement projects, monitoring may involve weekly progress meetings with the team leader or team members; periodic review of project controls, such as budgets, project management charts; and periodic auditing of the process improvement team.

AUDIT ESCALATION AND FOLLOW-UP

Progress means taking risks, for you can't steal home and keep your foot on third base.
Herbert Prochnow

In a value added audit, if the auditor sees critical nonconformances, high risks, or lack on controls, then it's incumbent on the auditor to 'say it like it is' and escalate the issue to senior management. This is very difficult for any auditor. This can create an adversarial relationship and even antagonism within the organization.

UNDERSTANDING AUDIT OBJECTIVES

The auditor should be sure he or she really understands the enterprise and specific purpose of the value added audit. If there are some doubts as to the audit's purpose or objective, the value added auditor should seek clarification from audit stakeholders.

If there are questions about audit evidence or facts, the value added auditor should not jump to conclusions. The value added auditor should discuss the issue with process owners and process management. If there are disagreements, the auditor should iron these out. Process owners may have 'soft' or other controls that the auditor may miss. The auditor also shouldn't fall into the 'know it all trap,' and quickly escalate an audit issue or finding to management for action or resolution. This can make a collaborative audit an adversarial one very quickly. And if the auditor is wrong, he or she is toast. The auditor's credibility and integrity are major personal values and will tank if the auditor is wrong.

Also, the auditor shouldn't leap to conclusions or escalate issues without authorization. The chief audit executive is ultimately responsible for reporting to senior management and the board of directors significant engagement observations and recommendations. What issues should be escalated to the board level? Significant corrective actions, preventive actions, and conditions that can put an organization at risk or can adversely affect the organization should be reported. This is sometimes a judgment call. However, these top-level issues always should be brought to senior management or the board of director's attention.

AUDIT CONCLUSIONS AND RECOMMENDATIONS

Audit conclusions are the value added auditor's observations, recommendations, opinions or evaluations of the auditee's activities and processes reviewed. The auditor's conclusions may encompass the entire scope of the audit or specific elements of the audit. Audit conclusions may indicate:

- Policies, procedures and work instructions are in compliance.
- Business objectives are met.
- Activities under review are working as intended.
- Process is in control.
- Process is capable of meeting requirements or mitigating risks.
- Risks are managed.
- Controls are working.
- Waste is minimized.

Audit recommendations may include that the following are implemented:

- Corrective action requests.
- Preventive action requests.
- Indication of satisfactory or improving performance.
- Risks identification.
- Control effectiveness analysis.
- Suggestions for process improvement.
- Suggestions for improving business performance.

AUDIT FOLLOW-UP

The chief audit executive should establish a process to assure the process owners have followed upon the recommended actions. Several factors should be considered in determining follow up:

- Significance of observation or recommendation.
- Risk of not implementing actions.
- Stakeholder requirements.
- Complexity of the corrective/preventive action.

REPORTING STEP 3 – CORRECT – PREVENT – PREDICT – PREEMPT

FOLLOW UP DECISION MATRIX						
	Redo Audit	Confirm Action Locally	Confirm Promissory Action	Reaudit Specific Attribute	Audit Manager Resolve	Resolve With Auditee
Supervision Did Not Know the Rules						
Procedure Change Required to Fix Condition						
Improvements Implemented in Critical Areas						
Reply Indicates Noncurrence with Finding						
Reply Indicates Corrective Action Not Required						
Trade Training Briefing Given to Corrective Problem						
System Process Correct is Weak – Problem Not Localized						
Audit Reply or Division Management Requests a Follow Up Audit[1]						

- Level of cost or effort to correct/prevent the deficiency.
- Time priorities.
- Regulatory or other requirements.
- Level and probability of risks.

How does the chief audit executive know that audit recommendations have been implemented? The audit organization may keep a log of outstanding corrective and preventive actions. The log may:

REPORTING STEP 3 – CORRECT – PREVENT – PREDICT – PREEMPT

- Establish time frame for process owners to respond to the auditor's recommendations.
- Evaluate process owner's response to recommendations.
- Verify suitability, effectiveness, and efficiency of the response.
- Follow up to the audit if appropriate.
- Reassess if the nonconformance is chronic and systemic.
- Escalate if risks increase or actions are inappropriate.

Senior management is ultimately responsible for determining the appropriate actions to be taken in response to audit recommendations. The chief audit executive may have to work with senior management to ensure that the organization can accept the risks of not taking action. Or expressed another way, the chief audit executive may discuss with senior management if they can accept the current level of risk.

Internal auditing generally evaluates efficiency and effectiveness of a business's risk management and controls. Organizational stakeholders want to know about the risks, hazards, and other conditions that may impact the organization.

DIFFERENT TYPES OF FOLLOW-UP

When the auditee receives the value added audit, the value added auditor or the customer may want to conduct some type of follow up, such as:

- Confirming local action.
- Confirming promises for corrective action.
- Reauditing a specific process, system, or attribute.
- Resolving differences with the audit manager.
- Resolving differences with the auditee.

Follow up can be seen in the 'Follow Decision Matrix' figure on page 408.

SUMMARY

Corrective and preventive actions may result from a value added auditing report. These are basically the intervention and improvement instruments resulting from a value added audit. Corrective and preventive actions may result in:

- Minimizing risks.
- Improving controls.
- Minimizing wastes.
- Decreasing unwanted process variability.
- Redesign of a product to increase robustness.

REPORTING STEP 4 – MAINTAIN AUDIT FILE

CHAPTER 22
STEP 4: MAINTAIN AUDIT FILE

Value added auditors document their work in working papers. These working papers may consist of written reports, checklists, product analyses, sampling, testing, tracing, or other analyses that support the basis of observations, recommendations, and corrective/preventive actions.

WORKING PAPERS

The trouble with facts is that there are so many of them.
Samuel Crothers, author

Working papers document the value added audit. They are the records of preliminary planning, questionnaires, and testing results. These papers become part of the audit permanent file. This does not imply that every scrap of paper is considered a working paper. Working papers include primarily critical pieces of information.

VALUE ADDED AUDIT DOCUMENTS

Working papers are records that the auditor keeps as to what happened during the audit addressing where, why, how, and when issues. Working papers include all the information that the auditor thinks is necessary to substantiate the audit report. Examples may include:

- Test results.
- Interview notes.
- Meeting minutes.
- Agendas for interviews.
- Personal thoughts.
- Phone conversations.
- Analytical tests.
- Cost studies.
- Sampling procedures and results.
- Internal control evaluations.
- Confirmations.
- Audit plan.
- Preliminary audit results.
- Internal control questionnaire.
- Project scope.
- Project work breakdown structure.
- Product test results.

- Organizational charts.
- Process analyses.
- Checklists.
- Flow charts.
- Standards, specifications, policies and other compliance documents.
- Results of audit.
- Corrective/Preventive action requests.
- Final audit report.

The working papers form an audit trail. The organization of the working papers is the mark of a professional auditor. So if another auditor takes over, he or she should be able to review the work, follow the mental processes and reach similar conclusions as the original auditor.

WORKING PAPER PURPOSE AND VALUE

Working papers serve a number of purposes, specifically:

- Provide legal support if necessary.
- Document deficiencies and risks.
- Record audit results.
- Remember and reconstruct events.
- Form an audit trail of the auditor's thinking, planning, implementing, and reporting.
- Identify and report deficiencies and nonconformances.
- Support audit findings and opinions if they are challenged.
- Provide documentation for management review of progress and methodology.
- Provide background and reference data for subsequent value added audits.
- Provide a roadmap for revising the audit.
- Help communicate the purpose and direction of the audit.
- Assist others in understanding controls, risks, and opportunities.
- Assist in third party reviews.
- Provide due diligence audit trail.
- Can be used in lawsuits.
- Demonstrate compliance with regulations.
- Provide evidence of tests.

MAINTAIN PERMANENT AUDIT FILE

The first piece of intelligent tinkering is to save all the parts.
Paul Ehrlich

The permanent file is the repository of all pertinent data from previous audits. Every company should have a data retention policy and procedure. Much data are accumulated in value added audits, so the files have to be purged, reduced photographically, or digitized. They are purged because information becomes irrelevant or incorrect - divisions and suppliers may have merged, been sold off, or acquired.

REPORTING STEP 4 – MAINTAIN AUDIT FILE

There are no set or hard rules of what to retain. Decisions of what to keep and what to purge are made carefully. The files will become cluttered with SPC charts of processes that no longer supply products, drawings of outdated products, and old standards. The following discussion may reveal what to retain:

- Previous value added audit reports.
- Multidivisional reports.
- Follow-up audit reports.
- Working papers.
- Interview notes.
- Background information.

PREVIOUS VALUE ADDED AUDIT REPORTS

Previous audit reports and working papers distill much information. These audits provide a historical perspective of risks, flaws, discoveries, recommendations, conclusions, corrective actions, and recurring problems. They provide a benchmark of improvement as well as the rates of improvement. Sufficient improvement indicates that less audit testing is required. Insufficient or no improvement indicates increased testing. Chronic or systemic problems also indicate increased audit attention.

MULTIDIVISIONAL REPORTS

More often, companies are inducing entrepreneurial thinking by making business units, internal departments, and project groups independent profit centers. These groups are responsible for product development, sourcing, marketing, distribution, and service. Each is almost a separate company with its own systems and controls. Information about each unit should be retained because it provides specific useful information. Assumptions made about the entire organization won't apply to a specific unit. Individual information can be used to compare and contrast systems within the entire organization as well as those in a similar industry. Process improvement or deficiencies in one operation can lead to insights in other operations.

FOLLOW UP AUDIT REPORTS

Sometimes the value added auditor may recommend a follow-up audit. This audit can be a separate investigation or can be part of the next scheduled audit. Often, the latter is preferred to optimize resources. These post audit reviews serve important purposes. They reveal whether audit results were taken seriously and recommended corrective actions were implemented. They also demonstrate whether the corrective action was effective in terms of eliminating the symptom and the root cause of the problem.

WORKING PAPERS

Different auditors may have performed previous audits. Their perspectives and observations provide guidelines for conducting the present value added audit. These comments aren't incorporated into the report, but can be put in the margins of the working papers.

In many cases, these comments provide insights and a point of view that illuminates the status of the auditee.

INTERVIEW NOTES

Interview notes can reveal much strategic and tactical process and product information. Senior management interviews reveal critical information and perspectives on process improvement strategies.

BACKGROUND INFORMATION

Background information is all the information, data, documentation, test results, or other information that the auditor derived during the tests to arrive at the findings and opinions in the audit report. Background information can be summarized in the audit report, referred to only, or placed in an appendix. Normally the audit customer is not concerned about background information, and only wants to know that professional proficiency and due care were exercised. Unless the information deals with health, safety, or critical items, most senior executives only read the executive summary and the findings sections of the audit report.

SUMMARY

Working papers should be consistent. Some organizations insist that each paper be dated, signed, and numbered while others follow a less disciplined approach where papers are just dated. If important tests or interviews are conducted, they should at least be dated. As much as possible, the papers should follow a logical, systematic format.

Neatness is a mark of a professional auditor. This is a form of internal discipline, that the value added auditor develops through years of practice.

Value added auditors should also be technically literate. Value added auditors often use laptop computers to develop their audit reports and to take notes. The computers have files of past audits and other information in memory. More extensive information or previous reports can be reviewed electronically from the audit site. Laptop computers also can store the entire contents of the value added audit, including flowcharts, standards, engineering prints, specifications, purchase orders, and test results.

An audit report can be a long document. Many people don't like to write, and there is a tendency to wait until the end of the audit to start writing. A tip for developing timely audits is to draft reports on a laptop as the audit progresses. If writing is a problem, the value added auditor can dictate the audit into a recorder and have an administrative person transcribe it. It is a good idea to discuss results with the auditee at each step of the audit, so the auditee's comments or concerns can also addressed as the audit progresses.

INDEX

ENDNOTES

ENDNOTES

ENDNOTES

ENDNOTES

PREFACE
[1] PwC trademarked COSO approach is Objective – Risk – Control – Alignment. In order to develop a 'how to' book on value added auditing, we have modified this model into Objective – Risk – Control – Analysis/Assurance.
[2] "*About Internal Auditing*," IIA Web Site, 2000.
[1] Robert, Michel, Strategy - Pure and Simple, New York: McGraw Hill, 1993.

GOVERNANCE AND AUDITING
[1] "Sarbanes/Oxley Act," Section 302, 2002.
[2] NIST, "Baldrige National Quality Program: 2000 Criteria for Performance Excellence," p. 40.
[3] Lawrence B. Sawyer, Sawyer's Internal Auditing (Altamonte Springs: The Institute of Internal Auditors, 1988),7
[4] Seid, Gerald and Harwood, John, "What Could Bring 1930s Style Reform of US Businesses?", *Wall Street Journal*, July 24, 2002, p. A 8.
[5] "*About Internal Auditing*," IIA Web Site, 2000.
[6] IIA, "What is Internal Auditing," IIA Web Site, 2000.
[7] Roth, James, **Best Practices**, IIA, 2000.
[8] This has previously been published in ASQ's *Quality Progress*, March, 2001.
[9] IBM's 1998 Annual Report.
[1] IIA Professional Practices Framework, "Glossary," 2000.
[2] "Management Discovers the Human Side of Automation," Business Week, September 29, 1986, p. 71.
[3] Chimerine, Lawrence, "The New Economic Realities in Business," *Management Review*, January, 1997, p. 13.
[4] Yellow Book, GAGAS 2011.
[5] COSO Web site, www.coso.org, "Internal Control – Integrated Framework, 2000.
[6] "Executive Summary of the Integrated Framework," www.COSO.com, p. 1, 2003.
[7] "Executive Summary of the Integrated Framework," www.COSO.com, p. 2, 2003.
[8] "Executive Summary of the Integrated Framework," www.COSO.com, p. 3, 2003.

ENTERPRISE RISK MANAGEMENT 101
[1] Australian/New Zealand Standard, AS/NZA 4360, Risk Management, 1999.
[2] FAA PROGRAMMATIC RISK MANAGEMENT, 2002.
[3] COSO, ENTERPRISE RISK MANAGEMENT FRAMEWORK, COSO WEBPAGE, 2003.
[4] Trembley, Ara, "Raising the Level of Risk Assessment," *National Underwriter*, May 1, 2000.
[5] IIA, "Standards for the Professional Practice of Internal Auditing," Glossary, November 14, 2000.
[6] Australian/New Zealand Standard, AS/NZA 4360, Risk Management, 1999.
[7] COSO, ENTERPRISE RISK MANAGEMENT FRAMEWORK, COSO WEBPAGE, 2003.
[8] COSO, Enterprise Risk Management Framework, COSO 2003, p. 3.
[9] IIA, "Standards for the Professional Practice of Internal Auditing," Glossary, November 14, 2000.
[10] Roth, James, Best Practices, COSO definition, IIA, 2000, p. 55.
[11] Dictionary web site, www.dictionary.com, 2006.
[12] COSO, Enterprise Risk Management Framework, COSO 2003.
[13] FAA Order 1100.161; Air Traffic Safety Oversight; p. 2.

ENDNOTES

PROCESS MANAGEMENT 101

[1] Nortel Networks, *Process Assessment Methodology Training*, 2001.

MANAGING THE VALUE ADDED AUDIT

[1] IIA, "In Partnership with Management, IIA Web Site, 2000.
[2] IIA, Red Book, 230.02.
[3] IIA, Red Book, 550.01.8.
[4]Chambers, G. Selim, and G. Vinten, **Internal Auditing,** Altamonte Springs, IL: Commerce Clearing House,1988,
[5] IIA Practice Advisory, 2051-1, 2002.
[6] IIA, Redbook 300.05.

PLANNING STEP 1 – UNDERSTAND AUDIT AND BUSINESS OBJECTIVES

1.G. Dessler, Management Fundamentals, Reston, VA: Reston Publishing Co., 1985, pp. 30-31.
[2] IIA, Practice Advisory 2130.A1.
[3] IIA, Practice Advisory 2210.A1.
[4] IIA, Redbook 300.02.06.
[5] IIA, Redbook 300.02.06.
[6.]"Why Enrollment is Up at Quality College," *Fortune*, April 29, 1985, p. 170.
[7.]J. Hickey, "What Price Quality," *I&CS*, August, 1987.
[8.]"Why Enrollment is Up at Quality College," *Fortune*, April 29, 1985, p. 170.
[9] D. Garvin, **Managing Quality**, NY: Free Press, 1988, pp. 49-69.
[10] Johnson, Sharon, "Tour the Software Chamber of Horrors," *Technology Review*, June, 1994, p. 6.
[11.]D. Garvin, "What Does Product Quality Really Mean?", *Sloan Management Review*, Fall, 1984, p.32.

PLANNING STEP 2 – NOTIFY/VISIT AUDITEE

[1] Source; Adapted from L. B. Sawyer and G. E. Summers, **Sawyer's Internal Auditing** (Altamont Springs, FL: Institute of Internal Auditors, 1988), 141.
[2] c. Chambers, G. M. Selim, and G. Vinten, **Internal Auditing** (Chicago: Commerce Clearing House, 1987).

PLANNING STEP 3 – UNDERSTAND AUDITEE'S SYSTEM, PROCESS, AND PRODUCT DOCUMENTATION

[1] ISO CD2/ISO 19011, ASQ Quality Press, 2000.

PLANNING STEP 4 – DEVELOP AUDIT PLAN

[1] IIA Red Book 520.04.
[2] IIA, RedBook 300.02.

PLANNING STEP 5 – DEVELOP AUDIT SURVEY

[1] IIA, Practice Advisory 2120,A1.
[1] Carnegie Mellon University, The Capability Maturity Model: Guidelines for Improving the Software Process, Addison Wesley, 1994, p.3.
[2] Carnegie Mellon University, The Capability Maturity Model: Guidelines for Improving the Software Process, Addison Wesley, 1994, p. 15-17.

ENDNOTES

[3] Nortel Networks, *Process Assessment Methodology Training*, 2001.
[4] Nortel Networks, *Process Assessment Methodology Training*, 2001.

CONDUCTING STEP 2 – ASSESS PROCESS CAPABILITIES
[1] IIA, Redbook, 300.
[2] Source: Popoff, Frank; Brache, Alan P., "The Seven Deadly Sins of Process Improvement," Chief Executive, June, 1994, p: 22-26.
[3] NIST, "Baldrige National Quality Program: 2000 Criteria for Performance Excellence," p. 39.

CONDUCTING STEP 3 – ASSESS SYSTEM AND PROCESS RISKS
[1] Bernstein, Peter, **Against the Odds: The Remarkable Story of Risk**, John Wiley, 1996.
[2] IIA, Redbook 300.

CONDUCTING STEP 4 – EVALUATE CONTROL EFFECTIVENESS
[1] ANSI/ISO/ASQ 9000 – 2000, p. 15.
[2] COSO, "Key Concepts: COSO Definition of Internal Control," COSO Home Page.
[3] IIA, Redbook 300.06.
[4] COSO, "Key Concepts: COSO Definition of Internal Control," COSO Home Page.
[5] A. Arens and J. Loebbecke, **Auditing: An Integrated Approach**, 4th ed. (Englewood Cliffs, NJ: Prentice-Hall, .Inc.), 141-146..
[6] IIA, Redbook 300.03.2.
[7] IIA, Redbook 300.02.06.
[8] IIA, Redbook 300.02.06.
[9] (Practice advisory 2110-1: assessing the adequacy of risk management processes)
[1] These six techniques are largely derived from L. Sawyer, **Sawyer's Internal Auditing** (Altamonte Springs, FL: Institute of Internal Auditors, 1988),277-281. He calls the six evaluation techniques observing, questioning, analyzing, verifying, investigating, and evaluating dangerous situations.
[2] 6D. R. Carrnichael and J. J. Willingham, **Auditing Concepts and Methods** (New York: McGraw-Hill, 1989), 134.
[3] IIA Red Book, 430.04.07.
[4] IIA, Practice Advisory 2310.
[5] IIA, Practice Advisory 2120.A1.
[1] "Practical Considerations Regarding Internal Auditing Expressing an Opinion on Internal Control," Institute of Internal Auditors, 2005.
[2] "Practical Considerations Regarding Internal Auditing Expressing an Opinion on Internal Control," Institute of Internal Auditors, 2005, p. 8.
[3] Berinato, Scott, "The Future of Security," *CIO Magazine*, December 15, 2003.
[4] Hutchins, Greg, "Higher Levels of Due Care," *Internal Auditor Magazine*, 2005.

REPORTING STEP 1 – COMMUNICATE AUDIT RESULTS
[1] IIA Red Book, 400.03.1 through 400.03.5
[2] IIA, Practice Advisory 2120.A1.
[3] W. Edwards Deming, Quality, **Productivity and Competitive Position**, MIT, Cambridge, 1982.
[4] "Standards for the Professional Practice of Internal Auditing" (Altamonte Springs, FL: Institute Internal Auditors, 1983), paragraphs 430.01-430.07.

REPORTING STEP 2 – DECIDE AUDIT REPORT FORMAT

ENDNOTES

[1] Adapted from the recommended audit format in "Program for Continual Process Improvement." U.S. Army Material Command Handbook-XXX. March 12. 1991. D2-D-6.
[1] Material from FAA Risk Management publications in public domain.
[2] Source: Derived from Alcoa, Supplier Quality Audit (1990).

REPORTING STEP 3 – VERIFY CORRECTIVE/PREVENTIVE ACTIONS
[3] Source: Derived from Alcoa, Supplier Quality Audit (1990).

REPORTING STEP 3 – VERIFY CORRECTIVE/PREVENTIVE ACTIONS
[1] ASQ Quality Audit Technical Committee, "How to Plan an Audit," *Quality Press*, 1987, p. 24.

www.ingramcontent.com/pod-product-compliance
Lightning Source LLC
LaVergne TN
LVHW060019210926
848481LV00013B/3747
* 9 7 8 0 9 6 5 4 6 6 5 8 5 *